Springer Series in Advanced Manufacturing

Series Editor

Professor D. T. Pham
Intelligent Systems Laboratory
WDA Centre of Enterprise in Manufacturing Engineering
University of Wales Cardiff
PO Box 688
Newport Road
Cardiff
CF2 3ET
UK

Other titles in this series

Assembly Line Design
B. Rekiek and A. Delchambre

Advances in Design
H.A. ElMaraghy and W.H. ElMaraghy (Eds.)

Effective Resource Management in Manufacturing Systems: Optimization Algorithms in Production Planning
M. Caramia and P. Dell'Olmo

Condition Monitoring and Control for Intelligent Manufacturing
L. Wang and R.X. Gao (Eds.)

Optimal Production Planning for PCB Assembly
W. Ho and P. Ji

Trends in Supply Chain Design and Management: Technologies and Methodologies
Hosang Jung, F. Frank Chen and Bongju Jeong (Eds.)

Process Planning and Scheduling for Distributed Manufacturing
Lihui Wang and Weiming Shen (Eds.)

W.D. Li, S.K. Ong, Andrew Y.C. Nee and
Chris McMahon (Eds.)

Collaborative Product Design and Manufacturing Methodologies and Applications

Springer

W.D. Li, PhD, Senior Lecturer
Chris McMahon, Professor
Department of Mechanical Engineering
University of Bath
Bath BA2 7AY, UK

S.K. Ong, PhD, Associate Professor
Andrew Y.C. Nee, PhD, DEng, Professor
Department of Mechanical Engineering
National University of Singapore
9 Engineering Drive 1, 117576 Singapore

British Library Cataloguing in Publication Data
Collaborative product design and manufacturing
methodologies and : applications. - (Springer series in
advanced manufacturing)
1. New products - Management 2. Teams in the workplace
I. Li, W. D.
658.5'75
ISBN-13: 9781846288012

Library of Congress Control Number: 2007923195

Springer Series in Advanced Manufacturing ISSN 1860-5168
ISBN 978-1-84628-801-2 e-ISBN 978-1-84628-802-9 Printed on acid-free paper

9 8 7 6 5 4 3 2 1

Springer Science+Business Media
springer.com

Preface

During the past few decades, there have been major innovation and paradigm shifts in product development methodologies and strategies. The current R&D trend is towards the development of collaborative design and manufacturing systems. The research theme is in line with the growing demand for global cooperative design and outsourcing in product development to gain better competitive advantage. Using the collaborative systems, designers and manufacturers can participate in global design chains and collaborate with partners locally and overseas to pursue competitive advantages. Furthermore, collaborative systems allow designers to work closely with suppliers, manufacturing partners and customers across enterprises' firewalls to obtain valuable inputs for their design and manufacturing activities.

From the early 1990s, some major R&D works have been reported, including the CyberCut system by the University of California at Berkeley; the FIPER (Federated Intelligent Product EnviRonment) system (FIPER Project, www.fiperproject.com/fiperindex.htm) funded by NIST; the Web-DPR system by the Georgia Institute of Technology), *etc*. Commercial systems include SolidWorks eDrawing™, Autodesk Streamline™, Impactxoft IX Design™, Onespace™, SmarTeam™, PTC ProjectLink™ and Windchill™, UGS TeamCentre™, *etc*. However, the developed strategies, methodologies and solutions still fall short of the expectation of the practical needs. They have not been generally accepted due to the weaknesses and limitations in collaboration management, interactive capabilities, security of data, real-time and ease of collaboration, *etc*. Different culture, educational background, or design habit of people also make it difficult to organize optimal collaborative design and outsourcing activities. To address the issues and make collaborative engineering more realistic and applicable, more efforts are being made.

The aim of this book is to update the relevant and recent research and development in this field. In this book, thirteen original and innovative chapters have been included to address the major challenges of developing collaborative design and manufacturing systems and techniques, with scientific and rigorous foundations as well as application values. The covered topics include: collaborative methodologies and strategies between humans, and between systems and humans to facilitate collaborative design and manufacture; cooperation across domains for multi-disciplinary design and manufacture; distributed system and service architectures for collaborative design and manufacture; interoperability of collaborative systems; new feature- and assembly-based methodologies for facilitating collaborative design and manufacture; workflow and conflict resolution/management in collaborative design and manufacture; design process and design change management in collaborative development, *etc.*

This book can be used as reference for mechanical/manufacturing/computer engineering graduate students and researchers in the fields of concurrent engineering and collaborative engineering for the efficient utilization, deployment and development of collaborative product design and manufacturing.

During the development of this book, we have received invaluable input and support from the chapter authors. We are also grateful to the editors of Springer-Verlag for their patience and professionalism during the editing process.

W.D. Li *(Cranfield University)*
S.K. Ong *(National University of Singapore)*
A.Y.C. Nee *(National University of Singapore)*
C.A. McMahon *(Bath University)*

January 2007

Contents

1

An Adaptable Service-based Framework for Distributed Product Realization

Jitesh H. Panchal, Hae-Jin Choi, Janet K. Allen, David Rosen and Farrokh Mistree

Systems Realization Laboratory
G.W. Woodruff School of Mechanical Engineering
Georgia Institute of Technology, USA

In this chapter, we propose a service-based engineering framework to support distributed product realization. Adaptability is the key strength of this framework, which arises from an appropriate balance between the ease of use of the framework and the flexibility for reconfiguration. *Standardization of the interfaces* between services permits communication between diverse software agents and relieves users from having to handle routine operations, resulting in the ease of use of the framework. *Flexibility of the framework's configuration* allows users to rapidly reconfigure the framework to changing design processes, and reduces the burden of customization. The capabilities for this adaptable distributed product realization framework are developed based on the Open Engineering Systems paradigm. Various existing distributed frameworks are evaluated against the requirements and missing features are identified. Our efforts towards the development of such a framework – the eXtensible Distributed Product Realization (X-DPR) environment are discussed. X-DPR is flexible and applicable to general industrial product realization processes. It is used to integrate distributed, collaborative product realization activities over the Internet. We trace the development of the framework based on design requirements. Features of X-DPR are implemented to satisfy the requirements. X-DPR is compared to existing engineering frameworks based on the required features. The key words and phrases used in this chapter are defined below.

Agent – Software component that can be invoked remotely to perform tasks in a product realization process.

Client – A software component that requests services from remote agents.

Framework – A computational backbone that facilitates deployment and utilization of agents.

Open Engineering Systems – Systems of industrial products, services, and/or processes that are readily adaptable to changes in their environment which enable producers to remain competitive in a global marketplace through continuous improvement and indefinite growth of an existing technological base.

Service – An activity that an agent can perform based upon a client's request.

1.1 Introduction

1.1.1 Need for an Adaptable Framework

Competition, globalization, a decreasing half-life of information, and greater product complexity necessitate the effective utilization of distributed resources and the management of the derived information. A distributed product realization process consists of a philosophy, a systematic approach and implementation methods to organizing product development activities. This process must be able to incorporate information from all parts of the product lifecycle. It is intended to support collaborative, concurrent decision making by geographically dispersed engineers who have different goals, knowledge, experiences, tools and resources. Software frameworks that facilitate globally distributed design and manufacturing activities are becoming more and more important, and many universities and industries have developed specific frameworks to complete specific tasks. However, in these frameworks, there is often a trade-off between agility, flexibility and implementation/customization effort.

If an engineering framework is implemented as *middleware*, it may be flexible enough to be useful for various product realization processes, but requires a significant effort to particularize it for a specific process. Middleware tools free users from having to write their own routines to handle reliable data transfer between applications or from having to worry about complexities when multiple systems are integrated. However, users still must write codes to integrate application functionalities. Examples of middleware toolkits include OMG's CORBA (Common Object Request Broker Architecture) [1] and Microsoft's DCOM (Distributed Component Object Model) [2]. On the other hand, if an engineering framework is developed as *end-user software*, the user must only put forth minimal effort but, in general, these frameworks are inflexible and cannot be modified easily when new situations arise, such as, when the company's design processes change. In other words, middleware tools provide standardization of communication protocols and leave a lot of integration work to the users whereas engineering frameworks (end-user software) provide easier integration capabilities but are not flexible. Hence, *choosing between the flexibility and ease-of-use of engineering frameworks is one of the primary challenges.*

Using a simple example, we demonstrate why an adaptable engineering framework is necessary. Imagine an engineering designer developing a simulation program and wanting to deploy it to a network so that it is available remotely for other engineers. To do this, a designer needs to do the following:

1. Implement a message and data construct to convey specifications (input and output) and data to and from the simulation program,

2. On the server side, a designer needs to develop a separate wrapper (a small main procedure to be called from other software) in addition to his or her simulation program because, in most cases, it is not possible for a framework to access directly the simulation program,
3. Develop a service description file (or documents) containing input and output specifications and related information about the deployment of the simulation program to let other engineers know how to use the service,
4. Notify the framework system that a new simulation service is available by registering the service on a registry server, and
5. On the client side, a designer needs to develop a user interface to get input parameters from a client and show output.

In this case, we assume that a pre-existing framework is configured and the deployment of the program is simple; however, the type of work to be done is not so simple. The effort for deploying these software applications is enormous and is discouraging if we think about the number of applications (*e.g.*, analysis, simulation, optimization, decision support, *etc.*) required in a general engineering design scenario. The problem is further complicated if *a)* the applications need to be changed and updated frequently, or b) the design processes in which these applications are used are changed. For example, in the case of changing design processes, the interfaces between different applications should be changed, which requires significant effort. These are some of the main obstacles preventing distributed engineering frameworks from being useful for distributed collaborative product realization in global industry. Therefore, we propose *an adaptable engineering framework for distributed product realization, which has both flexibility and usability in application for industrial product realization.*

Scope and Focus: We realize that there is a plethora of challenges related to the collaborative engineering frameworks including the development of standards for information representation, communication between heterogeneous resources, seamless flow of information between humans and computers, methods for efficient collaboration between designers, strategies for conflict resolution, engineering repositories, coordination and transaction management [3]. However, due to the extensive scope of this topic, and to limit the scope of this chapter, *we focus on addressing the challenge of balancing flexibility and ease-of-use through the appropriate standardization in engineering frameworks, thereby making the framework more adaptable.* The focus is on a framework where multiple software applications are deployed as agents, providing services to human designers. The design processes discussed in this chapter are limited to simulation-based design processes. The processes involving communication between humans will be considered in future publications.

1.1.2 An Open Engineering Systems Approach

Our approach to developing an adaptable distributed computing framework is based on the Open Engineering Systems (OESs) paradigm. We base our discussion on the following definition of OESs provided by Simpson, *et al.* [4]: "*OESs are systems of industrial products, services, and/or processes that are readily adaptable to changes in their environment which enable producers to remain competitive in a global marketplace through continuous improvement and indefinite growth of an existing technological base*" [4]. The basic OES premise is that a quality product should be brought to market as quickly as possible and then that a product line is continuously developed in an effort to remain competitive. Thus, OESs must be adaptable to changes in the market, technology, the supply/resource chain, the system environment, and government/legislation changes. As only some of these changes can be predicted, as much flexibility as possible must be maintained as long as possible to ensure product adaptability. Flexibility is achieved by incorporating the following characteristics:

1. Modularity: the relationship between the functional and physical structures of products, so that there is a one-to-one correspondence between physical structures and a minimization of unintended interactions [4].
2. Mutability: the capability of a system to be contorted or reshaped in response to changing requirements or environmental conditions [4].
3. Robustness: the capability of a system to function properly despite small environmental changes or noise [5].

A distributed computing framework also must satisfy the OESs paradigm. From a software framework perspective, *modularity* refers to the modularity of various components of the framework so that changes in any component do not require major changes in other components. *Mutability* refers to the capability of the framework to be reconfigured easily when there is a change in the requirements. *Robustness* refers to the capability of the framework to function properly despite the noise factors like network failures, unexpected usage, *etc.* In the design of an adaptable framework, each of these three characteristics provides requirements that influence the framework's form and function. It is important to realize that the word "Open" as used in this chapter is different from the "open source" software applications where the source code is freely available. In this chapter, *openness* refers to the ability of a system to be readily adaptable to changes either inside or outside it.

These requirements and desirable features of an adaptable framework are discussed in Section 1.2. A literature review of distributed computing frameworks is presented in Section 1.3 and these frameworks are evaluated based on OESs requirements. In Section 1.4, we provide a motivating design scenario of Linear Cellular Alloys design. In Section 1.5, we discuss the development of an open, adaptable framework, the eXtensible Distributed Product Realization (X-DPR) environment, X-DPR. Finally, in Section 1.6, we close the chapter with suggestions for future developments and a summary of our achievements.

1.2 Requirements and Features of an Adaptable Framework

From the OESs perspective, in this section, we discuss the requirements for an adaptable framework. There are many additional requirements if the framework is also to be distributed, but in this chapter, we emphasize only the requirements for an adaptable framework in Table 1.1. These requirements are discussed more completely following Table 1.1.

1. Adaptability to network architecture changes or malfunction (framework modularity)

To reduce the impact of network environment changes or malfunction, it is essential to reduce interdependence of communication components. The server-client communication protocol is dependent on the server. Therefore, server changes or malfunction can cause serious communication problems. Thus, the communication protocol must support *highly independent communication.*

Table 1.1. Requirements and associated features for an adaptable framework

	Requirements of an adaptable framework to support distributed product realization	**Necessary features of the framework which will satisfy the requirements**
1.	Adaptability to network architecture changes or malfunction *(modularity)*	Mutually independent communication protocols
2.	Usability on heterogeneous platforms with heterogeneous operating systems *(robustness)*	Computing platform independence
3.	Adaptability in the face of heterogeneous programming languages for different agents *(robustness)*	Interoperability interface independent of the programming language
4.	Capability to transmit message and data changes *(robustness)*	Generalized construct of message and data
5.	Rapid reconfiguration of the product realization environment *(mutability)*	Process editing capability Ease of re-assigning a task in a process to an agent service Mapping of information between tasks Maintaining consistency between the agent services' description and the client's user interface Having a standard for the description of engineering services Managements agents' services
6.	Minimizing the impact of agent service changes *(modularity)*	Process task decomposition capability
7.	Readiness for future expansion *(robustness)*	Compatibility with standard Web service frameworks
8.	Readiness for discrepancy of process information *(robustness)*	Sharing common process workspace Real-time management of process information

2. Usability on heterogeneous platforms with heterogeneous operating systems (framework robustness)

Software agents (either service providers or clients) reside on different kinds of machines, *e.g.*, desktops, mainframes, laptops or PDAs. They can be located anywhere on the globe, run with various operating systems, and can be connected through the Internet. In a general design-manufacture environment, examples of agents might include analysis codes, CAD (Computer-Aided Design) modelers, optimization routines, *etc*. These agents can also be manufacturing equipments or even human engineers providing some kinds of services. A service provider (or a client) on any platform must be able to deploy (or access) other services using framework components (server- or client-side applications) without having to implement a version of framework components, which is compatible with his or her platform. An adaptable framework must be able to support the engineering activity independent of the computing platform.

3. ***Adaptability in the face of heterogeneous programming languages for different agents (framework robustness)***

Most engineering frameworks use individual wrappers for each agent's service. These wrappers must be deployed because of incompatibility between third party software programming languages and the framework's programming languages. The implementation of separate wrappers is not only tedious but also limits accessibility to some details about the service. Therefore, *a programming language independent interoperability interface* is important for a framework that is to be adaptable.

4. ***Capability to transmit message and data changes (framework robustness)***

One of the most important issues in developing an adaptable framework is formulating standard message and data streams so that they are product and process independent. For example, a design specification of a gear needs three variables: number of teeth, gear module, and tooth thickness. When an engineer also wants to include manufacturing information in the specification, it is desirable if the framework administrator does not have to form a different message construct to convey the new gear design variables. This is impossible without a generalized message and data construct. This problem occurs whenever design specifications change and it becomes even worse when the product or process changes. The types of information transmitted are general information (such as message headers), parameters (input and output), and engineering data (such as CAD files). Ideally either this information should be encapsulated separately and attached together with the message, or, the message itself should be flexible enough to capture information about different kinds of inputs and outputs. A *generalized construct for transmitting message and data,* valid for any engineering task, is necessary.

5. ***Rapid reconfiguration of a product realization environment (framework mutability)***

Reconfiguration of a product realization environment includes remodeling the product realization process, reassigning a task to another agent, and modifying (adding, removing, or changing) agents. An adaptable framework should support users rapidly reconfiguring their environment without modifying code or recompiling the framework.

When a new product realization project begins, it is necessary to model the process rapidly and efficiently. Even in the middle of a product realization process, it may be necessary to change the process. Therefore, a convenient *product realization process* modeling *capability* is necessary. An adaptable framework should be able to easily *assign a task to an agent service* using a process representation User Interface (UI).

In a generic framework where different applications may provide vastly different functionalities, it is very likely that the outputs of one service are not identical to the required inputs of another agent. In other words, it is reasonable to assume that the interfaces between agents are not standardized; therefore, it is important to develop a *specification mapping capability* to connect the output of one task to the input of another task.

Various tasks in a process can be assigned to agents and can be executed by the agents. If an agent does not require any input from the user, it can be executed directly in the predefined process without interacting with the user. However, if the agent needs user input, a graphical user interface must be developed. As the interactions of the service with the user are different from case to case, different graphical user interfaces are required for different agents. If large numbers of agent services are incorporated within a product realization process, it becomes nearly impossible to create the required number of client-side user interfaces, also, agents for tasks change or are upgraded from time to time. Therefore, the *capability of maintaining consistency between agent service description and client's user interface* is very important.

To support rapid reconfiguration, a user of the framework should be able to search for, collect, index and archive information about available agent services inside a framework. This is not a direct requirement for an adaptable framework, but it is an essential feature, which supports the mutability of a framework. The first step in searching for appropriate available agents anywhere in the world is the definition, characterization, and standardized description of engineering services. A Web services definition language (Web Service Description Language –WSDL [6]) for e-commerce in our domain of application has already been developed. Definitions and standards of services in WSDL, however, are quite different from those required in the engineering domain and are therefore inappropriate for describing engineering Web services. Consequently, there is a need for developing *engineering service description standards* to make remote parsing of available agents possible. Contingent upon the development of an engineering service description language, further research might focus on *archiving, searching and selecting engineering agents' services*.

6. Minimizing the impact of agent service changes (framework modularity)

An adaptable framework should be capable of minimizing the impact of agent service changes. For example, while designing an automobile, one group of designers can work on the engine and another group of designers can work on the structural strength analysis. These groups can further be divided into smaller groups who are distributed across the globe. If one of the divided structural strength analysis tasks should be replaced by a new task, the user of the framework should have the capability of decomposing the tasks into small processes so only a small change in the divided task is needed. However, if the framework doesn't have a task decomposition function, the large upper level task must be modified and deployed again. Therefore, the framework should have a *task decomposition capability* to minimize the impact of service changes.

7. ***Readiness for future expansion (framework robustness)***

The system should be compatible *with other standard web services* frameworks such as MS .NET, and Sun ONE because a product realization process is not a stand-alone engineering process but is intimately related to other business frameworks such as applications for resource management, supply chain, management *etc.* Hence, an effort should be made to make the framework to conform to industry standards.

8. ***Readiness for discrepancy of process information (framework robustness)***

Even if there are no environmental changes, the distributed framework changes due to participants' input as a product realization process proceeds. Discrepancy about process information among users in distributed product realization occurs because of the ever-changing framework status during a process. A designer might want to know how other engineers' work is progressing and when that work will be done. Unlike a business framework, an engineering framework may have relatively long transaction times between service providers and clients; therefore, it is essential to share information about agent service availability. To facilitate these needs, an adaptable framework should be able to share the *common process workspace displaying real time process and agent service status*. This need leads to the requirement for *real-time management of process information* because this information is produced globally, and needs to be collected and managed systematically.

Desirable requirements and the appropriate features of an adaptable framework are discussed in this section. Although the requirements in this section are presented in the context of engineering frameworks, they are valid for any general distributed computer framework. These requirements and features are revisited in Sections 1.3 and 1.4 to review the capabilities of frameworks presented in the literature and to compare them with the X-DPR framework. Now, we move on to Section 1.3, where we review the existing frameworks with a mindset of OESs-based requirements presented in this section.

1.3 Review of Capabilities Provided by Existing Frameworks

The approaches for distributed collaborative design can be broadly classified into two categories: *Web-based systems* and *agent-based systems* [7]. These categories are discussed in Sections 1.3.1and 1.3.2 respectively.

1.3.1 Web-based Systems

The Web-based systems use the client server architecture with the Internet as a backbone for communication. The Web-based architecture supports multiple teams to access the central information base and to communicate through a central Web server. The collaboration between designers is generally through tools like chat tool, white board, Web cam *etc*. Most of the currently available Web-based tools are developed using Java technologies. These Web-based systems are used for remote usage of distributed software applications through applet-servlet pairs [8] or through other means like LISP [9], PROLOG [10], ActiveX [11] *etc*. The Web-based systems can be categorized further into domain specific software integration tools and general distributed computing tools.

1. Domain specific integration tools

The domain specific integration tools are generally CAD-CAE (Computer-Aided Engineering) integration tools. Cramer, *et al.* [12] developed a collaboration architecture to allow distributed designers to work on the same CAD model concurrently. The architecture incorporates three components: a server, a controller, and multiple members. The communication between members is through the controller and server by exchanging data packets. The system is developed specifically for exchanging information between CAD and CAE applications. The system uses CORBA for sharing information between CAD tools and CAE applications. In this system, the objects and the communication between objects are clearly defined. The system can be used with other CAE applications only if the applications provide APIs for interactions. A number of collaborative CAD tools are developed using VRML files for remote viewing of CAD models. The Virtual Web Plant [13] is developed for distributed access to engineering data at a central location. The tool integrates three-dimensional models from various CAD plant design tools and to display them interactively. It uses VRML for displaying CAD information remotely. The central data repository is an object-oriented database. The system also uses Java applets as clients for accessing the central data repository. Wang, *et al.*, [14] developed a Web-based virtual environment for mobile phone customization (VMPDS) that allows users to collaborate on conceptual design of mobile phones. VMPDS is developed using VRML, Java, and JavaScript. Lin and Afjeh [15] present an XML-based framework for Web-based aircraft engine simulation. The framework allows easier data flow across different simulation components. Simpson, *et al.*, [16] present an interactive web-based product platform customization framework for enhancing customer interaction and reducing design and manufacturing lead time for custom orders. Other similar applications for integrating CAD tools over the Internet are discussed in [17], [18], [19] and [20].

2. General distributed computing applications

Rezayat [21] introduced the notion of an e-Web portal to illustrate how Web-based standards and distributed object technologies can be integrated to provide controlled access to any type of information and resource within the extended enterprise. The author has argued that out of a number of systems that provide client-server services (including CORBA, DCOM, HTTP/CGI, RPC *etc.*), only CORBA and DCOM provide the degree of sophistication needed to implement practical object-based client server system at an enterprise level. The authors also recognized a need for using standards like XML for formalizing the semantics of the information. TeleDM [22] is an e-service test bed for verifying Web-based product design developed with the aid of a prototype-manufacturing environment. A client-server infrastructure with Web-based technologies is used in this test bed. The clients are Java applets with corresponding servlets on the server side. The clients can also be CAD tools (AutoCAD, Pro/E, *etc.*) or RP (Rapid Prototyping) planner tools (ACIS viewer *etc.*). The process of 'design for X' is modeled as a design-coordinate-redesign process. Wang, *et al.*, [23] developed a Web-based generic distributed mechanical system simulation platform based on gluing algorithm approach. The platform allows the integration of distributed simulations into a system level simulation. An XML description of individual simulation models is provided, which is a key element that links together different parts of the system level simulation model. The individual simulation models are integrated using a gluing algorithm. The benefits of such an approach include independence of subsystem models, and support for collaborative design in a supply chain. Other similar distributed computing applications are: Ansys AI workbench [24], MSC.Acumen [25], EDS Teamcenter [26], PTC Windchill [27] and Alibre Design [28].

1.3.2 Agent-based Systems

Similar to the Web-based design systems, agent-based systems also provide a collaborative environment for the sharing of design information, data and knowledge among distributed design teams. However, unlike the Web-based design systems using the client/server architecture, an agent-based system is a loosely coupled network of problem solvers that work together to solve problems that are beyond their individual capabilities. The agent-based systems are generally based on direct communication between agents instead of a client-server type communication that is common to the Web-based systems. The Web-based systems are easier to develop using the available technologies. An agent-based system is desired when the system is rapidly changing and the process is too complex. Agents are suited for ill-structured and modular systems.

Agent based technologies date back to early nineties when the Web was not very popular. The agent-based systems are based on simplified architecture. The basic aim of the agent-based systems is software reusability and flexibility in using the same software programs for different scenarios. The agents have dynamic linking with each other. This dynamic linking between agents can be achieved by having common information exchange protocols, syntax and semantics for communication. Most agent-based systems (see references [19], [29], [30], [31], [32]) have used knowledge-based standards for achieving interoperability between agents. Knowledge based standards involve defining common ontologies and/or definitions that the agents agree upon. Whenever there is a communication between different agents, they use the common ontologies. However, internally, these agents may use different software level standards for processing data. Hence, this provides flexibility to agents in terms of developing agents. One such knowledge based agent framework is PACT [29], which is one of the earliest agent based system for engineering design applications. The PACT framework is developed with focus towards integrating legacy software tools using knowledge interchange languages like KQML (Knowledge Query Modeling Language), KIF

(Knowledge Interchange Format), ACL (Agent Communication Language) *etc.* The system uses wrappers based on knowledge contained in various systems. A common ontology is defined for knowledge interoperability between agents. The PACT system provides flexibility in terms of the fact that the agents can use their own data models and the tools need to be committed to a single standard for defining data models. The SHARE project [33] was concerned with developing open systems for concurrent engineering particularly for design information and data capturing and sharing. The system provided collaboration services including multi-media mail, desktop conferencing, online catalog ordering and fabrication services. Rajagopalan, *et al.*, [34] proposed an agent-based infrastructure to provide designers with access to multiple layered manufacturing services including design, process planning and manufacturing service. Madhusudan [35] presents a Web service-based framework to expose intra-organizational information sources. In this framework, processes are dynamically composed using artificial intelligence planning mechanisms. Wu, *et al.*, [36] integrated Web services and the agent technology and developed an information framework for collaborative product development. One of the key features of this framework is its flexible client side product development environment. The framework has been developed to address the need for negotiation while managing conflicts in engineering design processes.

Four of the most recent frameworks that describe the state-of-art in distributed frameworks are DOME [34], NetBuilder [11], Web-DPR [62], and FIPER [15]. These four frameworks are selected because they represent agent-based, Web-based, product-centric, and process-centric frameworks. DOME and NetBuilder represent agent-based systems. While DOME is a product centric framework where each agent models a sub-system of the artifact, NetBuilder is a process centric framework where each agent models an activity in the design process. WebDPR is selected because it represents the Web-based systems and is a foundation for the X-DPR framework presented in this chapter. FIPER represents the current state of commercial distributed design frameworks and is so far the most advanced commercial engineering framework. In this remaining part of this section, we review these frameworks in details in the context of requirements developed in Section 1.2. In Table 1.2, the necessary features of engineering frameworks, based on the requirements presented in Section 1.2 are listed and existing frameworks evaluated for these features. From this review, we can determine what necessary features are missing in these frameworks. DOME, NetBuilder, Web-DPR and FIPER are discussed in more detail in Sections 1.3.2.1, 1.3.2.2, 1.3.2.3 and 1.3.2.4, respectively.

The capabilities of NetBuilder, Web-DPR and FIPER are summarized in Table 1.2. Each table entry is marked as ■ Fully Implemented or ▲ Partially Implemented. These three frameworks have many features that an adaptable framework should have, but in each case, the information constructs, *e.g.*, service description constructs and message constructs, are based on their own protocols instead of industry standards and the information constructs do not contain enough content to describe complex engineering tasks. The DOME framework is designed as product centric and not process centric.

Table 1.2. Review of distributed engineering frameworks with respect to desirable adaptable framework features

Necessary framework features to satisfy requirements	DOME (1998) [37]	NetBuilder (1998) [38]	Web-DPR (2001) [8]	FIPER (2006) [39]
1. Mutually independent communication protocol	Yes	Yes	No (Client-Server)	Yes
2. Computing platform independence	No	No	Yes (Java)	Yes (Java)
3. Interoperability interface independent of programming language	C++ Wrapper (CORBA)	Wrapping Toolkit (CORBA)	Agent Template	Java wrapper, FIPER SDK
4. Generalized construct for message and data	No	Mapping protocols /data types	Web-DPR message construct	Not mentioned
5. Editing product realization process	No	Metaprogram NetEditor	Editing process file	Workflow desktop
6. Assigning a task in a process to an agent service	Yes	Yes	Yes	Yes
7. Specification mapping between tasks	No	Data type mapping	No	Parameter Mapping tool
8. Maintaining consistency between agent services description and client user interface	No	No	Dynamic Web-browser UI	Yes
9. Engineering service description standard	MDL source file	Metaprogramm ing model	No	XML (FIPER's own standard)
10. Management of agents services	No	Resource Catalogue	Indexing service process db	FIPER Library
11. Process task decomposition	Yes	Yes	No	Yes
12. Compatibility with other standard web services frameworks	No	No	No	Uses XML, SOAP
13. Sharing common process workspace	Not mentioned	No	Process Web browser	Yes
14. Real-time management of process information	No	Yes	Coordinator stores process logs/Process db	Yes

Our effort towards an adaptable framework called X-DPR is discussed in Section 1.1. The X-DPR framework is described with a running example of the design of Linear Cellular Alloys. This example is described next.

1.3.2.1 Distributed Object-based Modeling and Evaluation (DOME)

The Distributed Object based Modeling and Evaluation (DOME) framework [37] is intended to integrate designer specified mathematical models for multi-disciplinary and multi-objective design problems. The focus of the DOME framework is to create a modeling scheme that handles the different variable types needed in engineering design; integrate multi-objective evaluation and optimization with design models; and provide an object based methodology to facilitate the integration of design models. In this framework, a product design problem is modeled in terms of interacting objects, called modules, each representing a specific aspect of the problem. One of the key assumptions of the framework is that product design problems are decomposable into sub-problems. The decomposition reflects both the physical subdivision of the product into components or sub-assemblies and the division of analysis expertise. Each object represents a subset of an aspect of the problem and acts as a stand-alone model managing the data and services that it can provide. An integrated design model is realized by objects representing the different parts of the problem. These objects are executed simultaneously.

In summary, the DOME environment is focused on simulation-based design and breaking down the design artifact into sub-systems that can be represented mathematically and may be distributed over the network. The framework is not designed with an open system paradigm, but with a product dependent distributed objects framework, which is more intuitive from a designer's point of view. It is platform dependent and, because it uses a CORBA protocol, requires lots of effort to create wrappers and the appropriate graphical user interfaces. DOME does not have a supporting tool for the management of objects in the framework and real-time information handling.

1.3.2.2 NetBuilder

NetBuilder [38] provides a mechanism for coordinating collaborative activities in a distributed environment. There are two key aspects to the NetBuilder approach. First, NetBuilder provides a compositional framework that allows designers to combine individual tools into *meta-programs* that capture the simulation process. These meta-programs can be executed and stored for future use. Second, NetBuilder supports wrapping individual modules so that they can be invoked as part of meta-programs in a uniform way. NetBuilder leverages mechanisms of distributed computing such as CORBA to provide a seamless integration of networked resources. NetBuilder provides the capability of capturing dependencies among simulation subtasks in terms of links connecting meta-program modules. When a meta-program is running, the NetBuilder scheduler determines which modules may be executed by checking to see whether the appropriate input data is ready. Each analysis tool is wrapped which allows it to accept input and produce output in a standard format. NetBuilder also contains a module wrapping toolkit to support the encapsulation of existing tools as CORBA-compliant modules.

NetBuilder has most of the features that are needed for an adaptable framework. Real-time management of process information is a valuable feature, as is the mapping communication protocol. However, there are some features which are only partially implemented, which limits NetBuilder's usage as an adaptable framework. CORBA itself requires that separate wrappers must be developed for all modules being integrated. The framework enables interfaces between modules on heterogeneous platforms, but components of the framework (such as meta-programs) cannot run on heterogeneous platforms. The descriptions of service assets are clearly defined in the Resource Catalog; however, there is not enough information for a user to find an appropriate service asset, and the format of the Resource Catalog is not an industry standard. In summary, NetBuilder enables the rapid and dynamic assembly of systems distributed on a large scale, but has limitations in serving as an adaptable framework. However, it represents valuable progress toward an adaptable engineering framework.

1.3.2.3 Web-DPR

Web-DPR [8] has been developed based on the communication framework of PRE-RMI [40]. The major objective of Web-DPR is to make agent services accessible with a simple Java enabled Web browser. The essential components of the Web-DPR framework are a Web server, framework database, coordinator and Agent Template [41].

The Web-DPR framework database stores information about available agents, the event channels they are registered to and other information about the design process. A client sends a request to the Web server, and this request is then transferred to event channels. The event channels then forward the request to agents. Information is transferred between various agents either as messages or as engineering data. A message is a short note or a command to other engineers, which is independent of product design domains.

Engineering data includes data files, CAD models, *etc*. This engineering data is archived in a central data vault. In Web-DPR, the event is split into message and engineering data in order to ensure that an agent's functions are totally independent of the functions of other agents.

A Java based application, *Agent Template,* is used to create and deploy agents easily into the framework. With the Agent Template, users do not need to have much knowledge about the internal implementation details about the framework. Web-DPR has features including a general message construct based on Java-RMI, dynamic Web-browser UI, standardized wrapper (Agent Template), *etc*. However, it cannot support the detailed access to remote objects since it wraps distributed modules using an Agent Template, which only provides abstract access to these remote objects. The dynamic Web browser UI cannot take range values, nor select alternatives. The Web-DPR framework does not support parameter mapping between tasks or task decomposition. This framework uses a Web server to publish services to the Web so there can be a bottleneck on the Web server.

1.3.2.4 Federated Intelligent Product EnviRonment (FIPER)

FIPER (Federated Intelligent Product EnviRonment) [39] is composed of three different layers – Core Infrastructure, Core Extensions and Application Components. The Core Infrastructure provides the foundation for the environment and is comprised of a collection of services for handling process management and data communication and storage [42]. The Core Extension contains modules that can be plugged into the Core Infrastructure and allow organizations to use the existing IT infrastructure. The Application Components provide the functionality desired by the users and can be published to the environment. FIPER uses a standard Java-based wrapping mechanism to allow easy creation of components for the environment. The FIPER Standard Development Kit (SDK) is provided to help write necessary Java code and execute it. The FIPER library is a virtually centralized and physically distributed repository for publishing, searching for and retrieving components. It facilitates collaboration by sharing the services offered by the Application Components. It also allows an engineer to assemble components into a workflow model of his/her design process. Kao, *et al.*, [43] present the use of FIPER framework for aircraft engine combustor design. FIPER enables real-time business to business collaboration at GE and Parker.

In terms of the desirable features listed in Section 2, the FIPER framework is the most advanced. However, it still has some restrictions. Although the processes in FIPER can be stored as templates and reused for designing the same product with different specifications, the main restriction is the reusability of processes for designing *different* products even if the tasks and distributed applications involved remain the same. Currently, the processes in FIPER and other similar commercial frameworks (such as iSIGHT [44] and ModelCenter [45]) are inherently defined as a series of tasks with flow of product parameters between these tasks. Hence, the processes defined at a computational level in frameworks such as FIPER cannot be used to design different products, whose parameter sets are different. The reusability of processes for different products is addressed by Panchal, *et al.*, in [46]. Further, FIPER does not support product information modeling.

1.4 Motivating Example: Design of Linear Cellular Alloys (LCAs)

LCAs are honeycomb materials (see Figure 1.1) which are processed through a formation and compounding of a slurry (binder phase mixed with metal powder oxides) which are then extruded under pressure through a multi-stage die and subjected to drying and reduction into the metallic phase in a hydrogen rich environment followed by sintering to achieve nearly fully dense metal composites [47, 48]. A wide range of cell sizes and shapes can be achieved including functionally graded structures, which provides multi-functional structural and thermal performance. Cell sizes on the order of half a millimeter and up and cell wall thicknesses on the order of 50-100 micrometer can be achieved resulting in very fine as well as very coarse structures. These metallic structures can be produced with any arbitrary two-dimensional cross-sections. These materials are suitable for multi-functional applications that involve not only good structural properties but also good thermal properties [48]. One of the main advantages of these LCAs is that any desired property can be obtained by suitably designing these materials. Some of the applications of these materials include heat sink for microprocessors, combustor liners, *etc*.

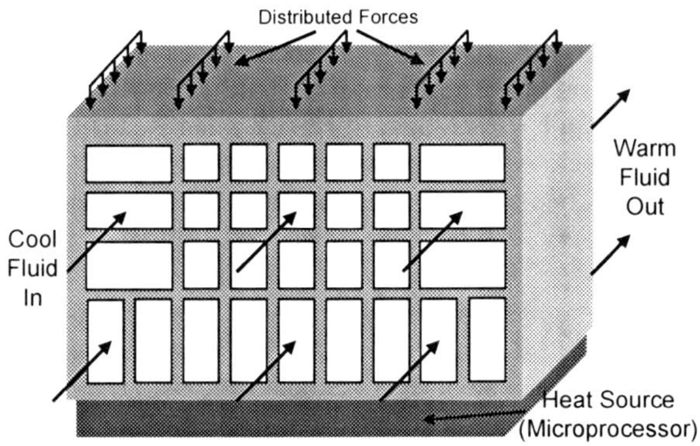

Figure 1.1. LCAs with rectangular cells

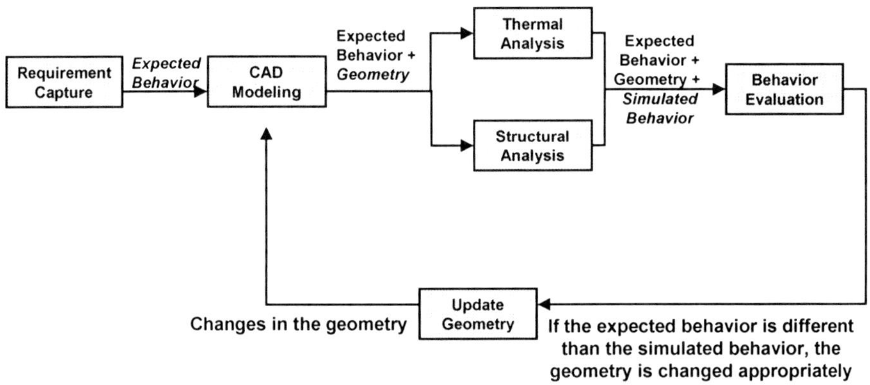

Figure 1.2. Design process for multi-functional LCAs

The process of design of LCAs is shown in Figure 1.2. It involves six steps starting from the requirements gathering phase to the final geometry of LCAs. The first step is to capture the requirements for designing these LCAs. These requirements are in terms of the expected behavior of the Linear Cellular Alloys. These requirements are used to create LCAs' geometry in a CAD modeling tool. Based on the experience, the designer starts with a cell geometry, which is modified later to match the expected behavior with a simulated behavior. The geometry and material information along with the expected behavior are used to analyze the performance of LCAs. Since this is a multi-functional application, the analysis is carried out for structural and thermal requirements. These analyses provide information about the simulated behavior for given loading (both thermal and structural). This simulated behavior is compared against the expected behavior. If these two do not match, appropriate changes are made to the geometric parameters to obtain the desired performance. Some of these steps in the design process like thermal and structural analysis are carried out using automated software applications, whereas other steps like capturing requirements require manual inputs.

In the next section, we discuss our effort towards an adaptable framework - X-DPR and show the features and capabilities of X-DPR through distributed design of LCAs following the design process shown in Figure 1.2. The features of the X-DPR framework are evaluated against the requirements discussed in Section 1.2.

1.5 X-DPR (eXtensible Distributed Product Realization) Environment

In this section, we provide an overview of the framework (Section 1.5.1), discuss the main features of the framework (Section 1.5.2), use the framework for LCAs design (Section 1.5.3), and finally, show how the framework can be characterized as an adaptable framework (Section 1.5.4).

The capabilities of X-DPR framework with respect to requirements discussed in Section 0 are summarized in Table 1.3. The framework is developed based on industry standards. The models for capturing and passing information are also based on various standards.

1.5.1 Overview of X-DPR

The X-DPR framework is designed with peer-to-peer communication between agents, where each agent is an independent entity communicating with other agents. A peer-to-peer communication framework enables independent communication between different agents (see #1 in Table 1.3). X-DPR is an open system in which different modules can be easily integrated into the system for enhancing the functionality of the overall system. Engineers can integrate their own applications residing on their machines with X-DPR, which will help to create a global library of engineering tools over the Internet. This library can then be integrated with tools from other areas such as marketing, sales or other business services to realize a global enterprise. The X-DPR framework uses the Decision Support Problem Technique [49, 50] to support meta-design, a process of designing systems that includes partitioning the system for function, partitioning the design process into a set of decisions and planning the sequence in which these decisions will be made.

The system is designed so that a designer can easily model his/her design process using visual tools. This capability for meta-design is unique in X-DPR. Engineers can then connect process models with services available in the global library using the Internet and execute complete design processes online. X-DPR provides flexibility at a design process level. It enables designers to design a process and replace entities of process with other entities later. The framework allows engineers to develop and execute process models collaboratively. Thus multiple designers distributed around the globe can work together as a team on product realization projects. A detailed discussion about each element of the framework is presented in Section 1.5.2.

Table 1.3. Capabilities of X-DPR with respect to adaptable framework features

No.	Necessary framework features to satisfy the requirements	X-DPR (2002)
1.	Mutually independent communication protocol	Peer-to-peer
2.	Computing platform independence	Yes (Java)
3.	Interoperability interface independent of programming language	Language Independent / SOAP
4.	Generalized construct of message and data	XML based
5.	Editing product realization processes	Client application process diagram
6.	Assigning a task in a process to an agent service	Task assigning tool bar and searching for available assets using tool bar
7.	Specification mapping between tasks	Interface Mapping Tool
8.	Maintaining consistency between agent service descriptions and a client's user interface	Dynamic UI based on WSDL
9.	Engineering service description standard	Partially implemented using WSDL
10.	Management of agents services	SOAP agent service database
11.	Process task decomposition	Process diagram construction toolbar
12.	Compatibility with other standard web services frameworks	Compatible with other SOAP servers
13.	Sharing common process workspace	Process diagram white-board and central process database
14.	Real-time management of process information	No

1.5.2 Elements of the Framework

In this section, we describe the elements of the X-DPR framework in further detail and show how these elements fulfill the requirements presented in Section 1.2. The elements of the X-DPR framework are shown in Figure 1.3 and the capabilities of the framework are shown in Table 1.3. In Figure 1.3, two agents are shown along with the client application. The exchange of information between various elements of the framework is also shown. In the X-DPR framework, an agent is defined as a software component that can be invoked remotely to perform tasks in a product realization process. Agents can be invoked by the client application or by other agents by sending XML messages. On receiving these messages, the agent processes the input information and replies back with XML messages. Agents in X-DPR are associated with *a)* an associated input message template in XML format, *b)* a processing mechanism in the form of a software, *c)* output message template in XML format, *d)* a WSDL description file that provides information about the location of service and way to invoke this agent and *e)* an XML-based UI description file (optional) that is used by the client application to generate a custom user interface. The details of these elements are discussed in Sections 1.5.2.1 through 1.5.2.7.

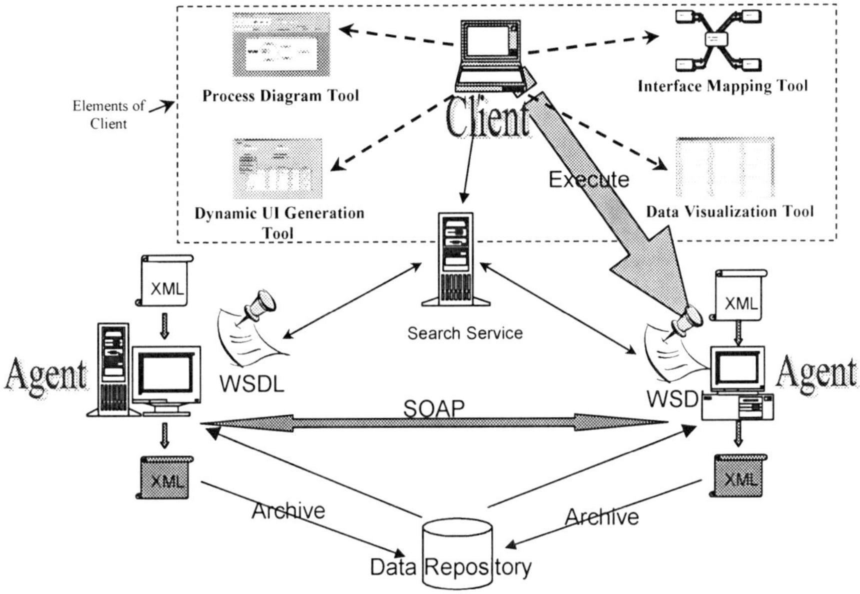

Figure 1.3. Block diagram of the X-DPR framework

The client application used to create process diagrams, manage the flow of information and access agents remotely has four elements – a process diagram tool, a dynamic UI generation tool, an XML data viewer tool and an XML mapping application. With the process diagram tool, users can create their own networks of tasks. These tasks can be assigned to agents available over the Internet. The user can search for available agents with a search service that is essentially a database containing the location and description of the agents. Once a task in the process diagram is assigned to an agent, it can be executed from the process diagram tool. The dynamic UI generation tool extracts information from the description file (WSDL) of an agent and creates a UI for the client-based on the inputs taken by the agent. The XML mapping tool maps the XML-based input-output UIs between different agents. It facilitates the smooth and seamless flow of information from one agent to another. The information generated throughout a process is archived in a data repository. One of the most important capabilities of the client application is its flexibility to execute any agent remotely by dynamically creating SOAP messages from the WSDL file and the XML-based agent input template. The

Java .class files in client application responsible for dynamic agent execution are packaged as a Java .jar file and deployed with the remote agents so that these agents can invoke other agents directly.

1.5.2.1 Data Repository

The data repository is a database of all the information processed during the design process. The data repository in the X-DPR framework is developed using STEP [51] (STandard for the Exchange of Product data) and the XML standards. STEP is an international standard for engineering information models. STEP standards have various predefined schemas that can be reused directly for application specific information models. STEP standards are used in the X-DPR framework in order to make the data repository standards based. The Express language [52], which is a part of the ISO standard (ISO 10303-11) is used for developing the information model for the product that is being designed. The highest-level schema for the LCAs product information is shown in Figure 1.4.

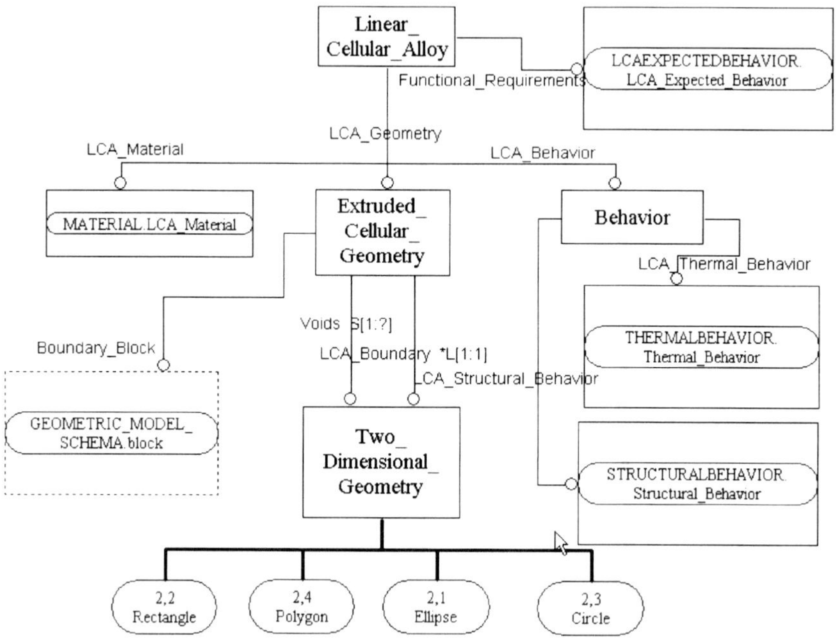

Figure 1.4. LCAs information model in Express-G form

In the LCAs design example presented in Section 1.4, the information model contains:

1. Expected Behavior (*i.e.*, the requirements) (`LCA_Expected_Behavior` entity)

2. Form, which consists of
   - Topology (`Extruded_Cellular_Geometry` entity)
   - Material used to manufacture the LCAs (`LCA_Material` entity)
3. Simulated Behavior, which consists of:
   - Thermal behavior (`Thermal_Behavior` entity)
   - Structural Behavior (`Structural_Behavior` entity)

The details of these entities are not presented in this chapter. While developing the information model supporting LCA design, the following STEP parts are used: Part 42 (for geometry and topology representation), Part 104 (for finite element analysis information), Part 45 (for materials information), Part 50 (for mathematical constructs), and part 47 (for tolerances). The schema is written in an Express file and an instance is a Part 21 file. The data access from Part 21 file is carried out using JSDAI toolkit developed by LKSoft [53] The JSDAI Express Compiler creates Java APIs from STEP Express schemas. These Java APIs are used to extract information required by different agents from Part 21 files. This extracted information is formatted as XML files and sent as inputs to agents. This method facilitates capturing the engineering information in the object-oriented STEP database and also allows information transfer through standardized, platform independent XML standard. Although the data repository is an integral component of the X-DPR framework, its functionality is similar to other agents. It accepts XML messages from agents that are stored in the repository, and sends back XML messages when requested by other agents. A user can also implement custom data repositories as agents in the X-DPR framework. However, these custom data repositories have to be explicitly instantiated as tasks in design processes using the process diagram tool discussed next in Section 1.5.2.2.

1.5.2.2 Process Diagram Tool

The process diagram tool, shown in Figure 1.5, is used to model a product realization process, and then it can be used to invoke the available agents integrated into the framework. The tool is coded in Java and hence is platform independent (see #2 in Table 1.3). This tool contains a white-board on which the process diagram can be created by simple drag and drop operations. The process diagram construction toolbar aids in this process of creating flow diagrams with blocks and connecting lines. These blocks represent various tasks in a design process and the connecting lines indicate the flow of information from task to task. The tasks can be assigned to any of the Web services available over the network (see #6 and #10 in Table 1.3). Using the process diagram tool, we can define process templates that can be edited for specific design problems (see #5 in Table 1.3).

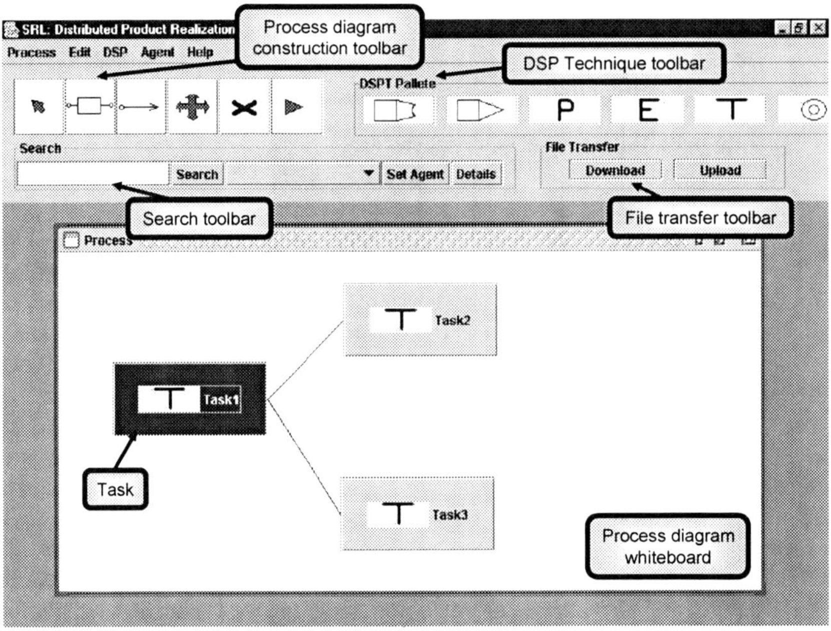

Figure 1.5. Process diagram tool in the X-DPR framework

The search toolbar is used to search for available services. The Decision Support Problem Technique (DSPT) [49] toolbar is used to model a design process in terms of phases, events and tasks and it also contains links to decision support tools for the design process. The file transfer tool is used for sending and receiving files (for example, CAD files) to various agents. The process diagram tool supports a hierarchical process development decomposing a task into sub-tasks (see #11 in Table 1.3). This means that a designer can move from a higher level in the process and then design a particular task as a network of sub-tasks. These processes are then saved in a central database such that they can be accessed by distributed team members and software agents (see #13 in Table 1.3). This process database contains information about the tasks in the process, flow of information between these tasks, agents assigned to these tasks, the tasks that are currently completed, in progress, or un-initiated. In the current implementation of X-DPR, the agents are executed automatically in a sequential fashion. All the tasks that require the outputs from finished tasks as their inputs are activated as soon as these inputs are available. Currently, complex and conditional sequencing is not available in the X-DPR framework.

Implementation of the process diagram tool – The process diagram tool is implemented using the `Swing` library in Java. The Java application extends the `JFrame` class in `javax.swing` package. The Java classes used to implement the process diagram whiteboard are: *a) frmInternal.class*, which extends the `javax.swing.JinternalFrame` class, *b) pnlInternal.class,* which extends `javax.swing.JPanel` class., and *c) block.class* which extends the `javax.swing.JButton` class. The file transfer toolbar contains two buttons to upload and download files. These files can be any file that the agent needs for execution or any file as a result of the execution of the agent. This toolbar is especially useful when the agents require processing of binary files like CAD files. The implementation of this file transfer tool is carried out using the following classes: `FileUpload.class`, `ClientPut.class`, and `ServerFileTransfer.class`. The file transfer is achieved by sending SOAP attachments between `ClientPut.class` and `ServerFileTransfer.class`.

1.5.2.3 Dynamic UI Generation

If an agent requires user input, a graphical UI must be developed for this purpose. The kind of interaction of an agent with the user varies from case to case and different graphical UIs are required for different agents. Since it is not possible to create a separate UI and code it into the client, a dynamic graphical UI is created based on the number and types of input that the agent requires.

Implementation of the dynamic UI generation - Two types of dynamic generation of UI generation are developed in X-DPR. The first type corresponds to a situation in which the inputs required from the user are very simple – for example, a few different parameters must be specified in a function. In this case, the description of the required inputs to the agent can be extracted directly from the Web service description (WSDL) file. Inputs from the user are generally taken with simple text boxes. The process of customized UI generation can be accomplished as follows: (i) the client looks for the WSDL document published by the service, (ii) from the WSDL document, the client extracts inputs and the corresponding data types, and (iii) the client generates a graphical UI for the user inputs. Based on the data entered by the user, the agent is executed.

The second type of UI generation corresponds to a situation where the inputs of an agent are complex XML tree structures. For example, the input to a design of experiments agent implemented in iSIGHT [54] is in the form of an XML file which requires complex interactions with the user. For example, the user must enter all the DOE parameters and their ranges. The user also needs to enter the type of Design of Experiment (DOE) to be performed. In this case, the complete description of the inputs and how the user inputs will be taken are not available in the Web service description (WSDL) file. In this case, a single XML file describing the user interface must to created at the agent and deployed with the agent itself. This XML file contains nodes for individual entities in the UI to be created such as text box, label, combo box, table, checkbox, radio button, *etc*. For each element, an XML tree representation is provided which contains the information required to generate the UI component. For example, for a label, the information required to generate is its location on the form, its size and the text of the label. The XML schemas for some of these elements are standardized in X-DPR. The client application accesses this description file remotely and a UI is created automatically at the client for that agent. The process is shown in Figure 1.6.

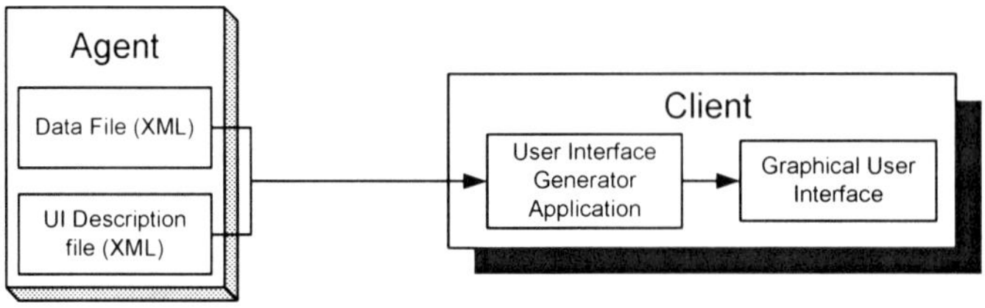

Figure 1.6. Dynamic UI generation using the UI description XML file (see Figure 1.3)

In the LCAs example scenario, the inputs to various agents are complex XML tree structures containing geometry information, boundary conditions, analysis results *etc*. The second type of the UI generation technique is used where a separate XML file describing the UI is created. This UI description XML file is used by the client application to generate the UI remotely. The dynamic UI generation tool helps in maintaining consistency between agent service descriptions and the client's UI (see #8 in Table 1.3).

1.5.2.4 Interface Mapping Tool

In a generic framework where different applications provide vastly different functionalities, it is very likely that the output of one agent will not be exactly that which is required as the input to another agent. To achieve a seamless flow of information between agents, the outputs need to be converted into a format compatible with the inputs to other agent. In general, if there are n agents, the number of conversions required will be n*(n-1). For example, in the LCAs design example, structural designer creates a response surface to investigate the effect of design variables (thickness of ribs, overall height of LCAs, *etc*.) on the overall strength. The first step in the process is to carry out a design of experiments in the design space. The design of experiments is performed using a commercial application – iSIGHT. The result of the DOE is a set of points in the design space at which the analysis is carried out. The analysis of the component at these points is carried out in an in-house code or an Finite Element Method (FEM) program such as ANSYS [55]. The output of the design of experiments from iSIGHT is in iSIGHT's own ASCII file format and the input to the ANSYS FEM solver requires the ANSYS ASCII file format.

For automatic transfer of information from iSIGHT to ANSYS, a designer must write a parser to convert one file format into another. To overcome this problem of developing separate converter applications, an interface-mapping tool is created that can be used to map information output from one agent to the inputs of another (see #7 in Table 1.3). Here, the term *interface* refers to the structure of information input and output by agents. This tool has the capability of mapping the XML structure from the output of one agent to the XML input of another agent (see Figure 1.7). The mapping tool shows the tree structures of the output XML file of one agent and the input XML file of another agent on two sides of a window. The user selects corresponding information entities (*i.e.*, XML nodes) on the input and output sides and maps them. Once the mapping is established between two agents, the mapping rules are saved in a separate XML file for future use. Hence, the users need to establish the mapping between two file formats only once. It is important to note here that the mapping tool implemented in X-DPR can only be used with XML inputs and outputs. In the case of applications such as iSIGHT and ANSYS, where the inputs/outputs are simple text files, the applications need to be wrapped (*i.e.*, text files converted into XML files) such that the agents have XML inputs and outputs. This is also illustrated in the block diagram of X-DPR (see Figure 1.3) where agents are shown with XML inputs and outputs.

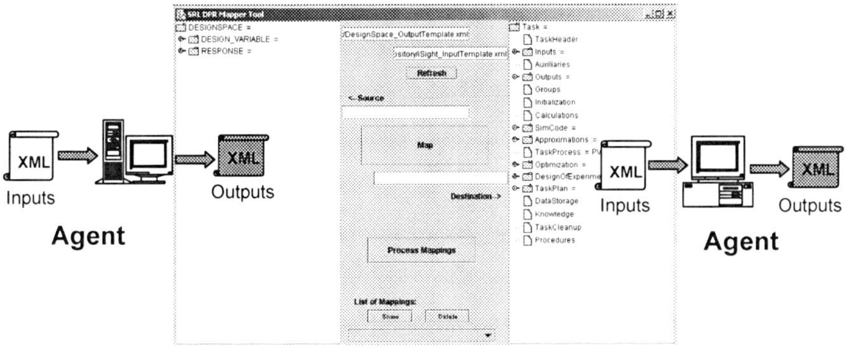

Figure 1.7. Mapping of interfaces between two agents

Implementation of interface mapping tool – The implementation of this tool is done using X-Path, which is a language for specifying the path of an element in an XML document. The tool consists of the following two components encode in Java - a) mapping definition UI (`XMLMapper.class`) and b) mapping execution class (`XMLMapperFromRelations.class`). The XML mapping definition UI is shown in Figure 1.7. The form has two tree elements corresponding to the XML output of the first agent and the XML input of the second agent. The XML structures are shown in the tree elements. The user selects an element from either side and creates the mapping between these elements using the "Map" button. Two text boxes show the source element path and the destination element path respectively. These paths correspond to the XPath of the selected elements. The information about mapping between all the elements is stored in an XML *mapping information file* that it can be later used to extract information from the outputs of the first agent and populate the XML input file of the second agent. The mapping execution class reads the *mapping information file* and extracts information from the XML output file of the first agent and fills up the XML input of the second agent.

The current implementation of the mapping tool does not support algebraic manipulation of XML nodes. For example, if we have data about variables 'a' and 'b' inside two different XML nodes, it is not possible to assign 'a+b' or 'a-b' *etc.* to a node of another XML file. This kind of a facility of algebraic manipulations is important and planned in the future versions of the framework. Further, the mapping tool currently supports information mapping only between XML files. The capabilities of the tool would immensely increase by providing support for mapping different types of schemas like mapping information from Express schema to an XML schema.

1.5.2.5 Messaging and Agent Description in X-DPR

The transfer of information between different software applications in X-DPR is through XML based standards such as the Simple Object Access Protocol (SOAP) [56]. SOAP standard for interfacing different software applications is programming language independent. XML is a platform- and language-neutral standard for representing information. The benefit of XML is that it separates data from meta-data (*i.e.*, information about the data). XML is also being adopted as a universal standard for representing information in distributed computing frameworks. SOAP is a communication mechanism based on XML. It is also a platform and language independent standard. Previous communication protocols, such as CORBA, DCOM, EJB (Enterprise Java Beans) and Java-RMI, share the common problem that they are incompatible with each other and that the applications deployed with these protocols cannot be accessed through a firewall. The SOAP protocol addresses this problem. SOAP uses a simple HTTP request/response-based communication, allowing it to pass through corporate firewalls [57]. A SOAP message typically contains an XML message along with an HTTP header.

In X-DPR, we use XML to define interfaces between different design activities (see #4 in Table 1.3), SOAP for message transfer between distributed applications over different platforms (see #3 in Table 1.3), and WSDL to describe different Web services (see #9 in Table 1.3). Since we are using standards common to all web services based frameworks, X-DPR is compatible with other similar frameworks (see #12 in Table 1.3).

1.5.2.6 Publishing a Service

The agents are published in the X-DPR framework simply by creating a description file based on the WSDL standard. The client retrieves the information from the WSDL description and creates a UI for the agent. WSDL documents can either be created manually or can be created automatically using commercially available toolkits. The Microsoft SOAP toolkit [58] can be used to create WSDL document for COM objects and Systinet Server for Java [59] can be used for creating WSDL for Java classes.

1.5.2.7 Asset Search Service

The task of searching for agents appropriate for a particular task is implemented as a Web service in itself. This Web service is called the Search Service. The Search Service maintains a database of links to WSDL files with a description of the service. Currently, the new agents in the database are populated manually and the database is created in the Microsoft Access. However, it is planned that the Search Service will perform a running search on the Web for WSDL description files.

The agent search service also maintains information about whether the service is currently in use or not. In the X-DPR framework, an agent's lifecycle is described by three states – *available, busy, and unavailable*. In the *available* state, the agent can receive requests for execution from clients or other agents. When the agent is being executed by the client or another agent, it shifts into the *busy* state. The agent shifts between the available and busy state automatically. An agent is *unavailable* when it is registered in the database but cannot execute. This may happen when the agent is physically disconnected from the network. In the current implementation, the agent developer has to manually set the state to *unavailable*. Whenever a user searches for an agent, the search service automatically gives a list of *available* and *busy* agents. Keeping the Search Service as a separate module is helpful because it can be developed independently of the framework and thus replaced with a different service at a later date. This also leaves the possibility that if commercial Web service search engines are developed in the future, they may be integrated into the framework. The agent search service can be extended in future by using the standard Universal Description, Discovery, and Integration (UDDI) protocol. UDDI describes a standard method for publishing, managing, discovering web based services [60]. UDDI is also based on other XML-based standards such as SOAP, WSDL, and XML Schema.

Implementation of Asset Search Service - The search service is implemented as a Java class (`SearchDB.class`). The Java class performs SQL queries on the database that contains the following information – *a)* Agent Name, *b)* Agent Description, *c)* Location of description file (WSDL file), *d)* Input file template (if any), *e)* Output template (if any), f) Location of the user interface description file (if any) and *g)* an entry, that specifies the current state of the agent. The interface-mapping tool described in Section 1.5.5 uses the input and output template files to map the outputs and inputs of different agents. From the process diagram tool, the search toolbar can be used to perform a keyword search on agents available for use. The process diagram tool sends a request to the search service and the search service returns a list of available agents matching the keywords. The user can select any of the agents from the list and assign it to the blocks in the process diagram.

1.5.3 Using the X-DPR Framework for LCAs Design

In the LCAs design example, seven distributed software applications are involved. These include applications for a) requirements capturing (in-house Java application), b) problem definition c) design of experiments (iSIGHT), d) thermal analysis (in-house Matlab code), e) structural analysis (in-house Matlab code), f) Response Surface Model (iSIGHT), g) updating the geometric parameters (in-house Java code). These applications are deployed as agents. In order to explain the use of the X-DPR framework in the context of LCAs design, we revisit the five tasks listed in Section 1.1 that are performed by an agent developer. The first task is to specify input and output data constructs for each application. The inputs and outputs for all the seven applications are described as individual XML files. The second step is to develop a wrapper for each of these applications. The wrapper development involves converting the input XML file into the applications native input format and converting the output from the native format into the XML format described in the previous step. The wrapper for iSIGHT has been developed in Visual Basic, and for other applications, it has been developed in Java. The third step is to develop a service description file for each application to be published as an agent. This step is performed automatically by the use of applications such as the Microsoft SOAP toolkit for Visual Basic-based wrappers and using the Systinet Server for Java-based wrappers. The fourth step is to notify the framework of the newly available service. This step is performed by adding individual entries to the agent database described in Section 1.5.8. The fifth step is to create a UI for the user interaction with agents if required. The UI for each agent is described for each agent using the UI description file (WSDL). These UIs are created at the client side only of the user wants to interact with the agent. Although all of the five steps described in Section 1.1 are required in the X-DPR framework, the prime advantage of using an adaptable framework such as X-DPR is its openness that reduces rework if there is a change in either the design process or the agents themselves. The features that provide this flexibility to the framework are discussed in Section 1.7.

Having discussed the activities performed by agent developers in deploying the applications, we now move on to the steps that the user follows for executing the LCAs design process remotely. Using the process diagram tool, the user creates a design process as a network of tasks where each task corresponds to an agent to be executed remotely. For the LCAs design process, these tasks are – capturing customer goals in terms of target heat transfer and stiffness of the LCAs, defining design variables, responses and associated ranges, which are mapped to a design of experiments task. The design of experiments task outputs a list of points in the design space where the analyses (thermal and structural) are executed. The output of the design of experiments task is mapped to the inputs of both thermal and structural analysis tasks. The outputs of analysis tasks are inputs for the response surface modeling task where an approximate surface is fit between the input variables and output variables. The output of the response surface modeling task is input to a task that updates the LCAs geometry. At this time, these tasks are not tied to any software agent. The user then performs keyword searches corresponding to each agent using the search toolbar, which results in a list of available agents. For example, when the user searches for "Design of Experiments", two agents are shown in the list – "iSIGHT" and "Minitab". Both of these agents have the similar functionality. The user then selects the suitable agents and assigns them to tasks defined in the process diagram using the search toolbar. Through this assignment, the task associated with an agent is linked with the corresponding WSDL file and also sets up a link with its input and output file template. The user then maps the outputs and inputs of agents that are linked in the process diagram using the process diagram tool. For example, the output XML template file of the design of experiments agent is mapped to the input XML template files for structural and thermal analyses. After the inputs and outputs are mapped, the process is executed from the process diagram tool. This initiates an execution chain of agents wherein the agents are executed sequentially until all the agents have executed once. Loops of execution are not supported in the current implementation of X-DPR.

1.5.4 X-DPR as an Adaptable Framework

We have discussed the elements of X-DPR, implemented as an adaptable framework. The requirements of the framework and associated capabilities of X-DPR are summarized in Table 1.3. The main features of X-DPR as a standardized yet flexible framework are:

1. Flexible mapping of information interfaces between agents using the XML mapping tool

In X-DPR, the Web is used as a backbone along with the associated technologies (Java, Web browsers, *etc.*) and standards (XML, SOAP, WSDL, *etc.*) for communication. In X-DPR, XML is used because it formalizes the semantics of the contents of information and facilitates electronic data exchange. As we have seen previously, most of the earlier frameworks focused on standardizing the structure of data models and information exchange between agents. This caused problems while integrating new tools into the framework. Any new tool to be integrated to the framework must abide by the standardized schemas in which information is communicated between agents, which limit the flexibility of agents to implement their own input/output schemas.

The interface-mapping tool helps to achieve flexibility in defining data structures for storing and passing information. Hence, the agents have flexibility in defining the structure of information flowing in and out and by using the XML standard in the X-DPR framework, and therefore, the system is more flexible, easily configurable and open (see Table 1.1). *This provides the required adaptability when the inputs/outputs of the agents change or when there is a process change resulting in changes in the agents that interface with each other* (see Section 0). The use of interface mapping tool can also be used for mapping different domain ontologies. Information schemas from various domains can be mapped to each other to accomplish an enterprise level transfer of information rather than just information transfer between software applications.

2. Standardized means of describing UIs

Another important issue while developing distributed agent-based systems is the way in which users interact with remote agents. Most of the frameworks developed until now have assumed that a fixed set of agents are deployed into the system and fixed UIs have been created for each type of agent. However, in an open engineering system in which it is not known what kinds of agents will be deployed on the framework and what kinds of interaction will be required by the system, it is difficult to create individual UIs for each agent. This problem is amplified when each agent is configured differently for different processes. In X-DPR, dynamic user interface generation at the client side is used to overcome this problem. This feature ensures that the UIs can be reused on *heterogeneous platforms* and for *different programming languages* (see Table 1.1). It provides both the adaptability to changes in UIs without changing the framework, and the ease of use because changing the UIs only requires changing the associated XML file.

3. Standardized means of representing process information (using the Decision Support Problem Technique)

In the X-DPR framework, the capability for meta-design is provided using the Decision Support Problem (DSP) Technique (see Section 1.5.3). It helps designers rapidly configure design processes and use distributed resources execute the processes. This capability fulfils the requirement of *rapid configuration of the product realization environment* (see Table 1.1).

4. XML standards-based messaging protocols

The use of the platform and language independent XML-based standards ensures that the framework is usable on *heterogeneous platforms* and *heterogeneous programming languages* (see Table 1.1).

5. Standardized description of assets using WSDL

Some of the frameworks such as Web-DPR describe some information about an agent such as the location of the agent and whether it is currently available. However, there is no information about what kinds of tasks that the agent can perform so that a remote user can determine the applicability of the agent for the task at hand. In the X-DPR framework, we use the XML-based standard WSDL to describe the capabilities of agents and ways that they can be invoked. This provides the adaptability to changes in agents and their services. This fulfils the requirement of an adaptable framework to *adapt to agent services changes* (see Table 1.1). Whenever there is a change in the services provided by an agent, these changes are automatically reflected in the corresponding WSDL files. Other features of the X-DPR framework related to the adaptability requirements are outlined the Table 1.3. In the next section, we close the chapter by providing future research directions and summary of our efforts.

1.6 Conclusions

The integration of the communication infrastructure for industry is essential. Industry and academia have tried to standardize data formats, which are platform and application independent. There has been a substantial effort to construct computational communication frameworks to integrate distributed resources (software, hardware, and human experts); however, those tend to be domain specific solutions or hard to reconfigure.

We have designed and implemented an adaptable engineering framework for distributed product realization, X-DPR, which balances the need for flexibility and standardization. The Open Engineering Systems paradigm, which includes modularity, mutability, and robustness, is the reference concept for the formulation of requirements for an adaptable framework (see Section 1.2). These are:

- Adaptability to network architecture changes or malfunction
- Usability on heterogeneous platform with heterogeneous operating systems
- Adaptability in the face of heterogeneous programming languages for different agents
- Ability to transmit message and data changes
- Rapid reconfiguration of the product realization environment
- Minimizing the impact of agent service changes
- Readiness for future expansion
- Management of process information to avoid discrepancies

These requirements help us to identify the features an adaptable framework should have. Four distributed engineering frameworks are reviewed in this chapter.

However, from the literature survey, we conclude that existing frameworks have some desirable features, but these frameworks do not fully satisfy the necessary requirements.

The X-DPR framework has been designed as an adaptable framework as discussed in Section 1.3. The features that balance the requirements discussed in Section 1.2: flexible mapping of information between agents, a standardized means for describing user interfaces, a standardized means for representing process information, a standardized message protocols based on XML and a standardized description of assets using WSDL. In X-DPR, the existing communication infrastructure is used as a backbone along with the technologies (Java, Web browsers, *etc.*) and standards (XML, SOAP, WSDL, *etc.*) for communication. These provide the flexibility and enable the future expansion of the framework. The interface-mapping tool also helps to achieve this flexibility, by easily allowing agents to reconfigure information flowing in and out. The Client application supports rapid reconfiguration of engineering task and decision-making activities and also task decomposition to formulate hierarchical product realization processes. Standardized service description files (represented in WSDL) are used for creating graphical UIs and interacting with agents. In the X-DPR framework, an emerging industry standard Remote Procedure Call (RPC) protocol, SOAP-based agent wrapper provides much more flexibility and ease of implementation than is available in the other frameworks. Workflow information is shared among distributed users by the Process Diagram Whiteboard in real-time. The most important advantage of the X-DPR framework is that it is compatible with other business frameworks. We envision the X-DPR framework as a link between an engineering framework (that manages design chains) and business framework (that manages supply chain).

X-DPR satisfies most of the features that are needed for an adaptable framework; however, there is still a need for further improvement based on the requirements addressed in Section 1.2. In the X-DPR framework, Web services are described using WSDL files. As mentioned in Section 1.3, these files are used to describe functions that can be called remotely in a software program and are used to index and search for available agents in an agent service database. These are the reference files to form the dynamic UI for remote users. However, in engineering, the amount of information currently conveyed in WSDL is inadequate if there are a number of services available, which provide the similar functionality. Description files must provide more information about the services. For example, a design of experiments agent should provide information when a specific technique should be used. A simulation program should also provide information about the range of values of input variables for which the program is valid. We are developing an Extended/Engineering Web Service Description Language (E-WSDL) to meet this need.

Co-ordination and conflict management play an important role in distributed engineering design frameworks. Brazier, *et al.*, [61] have categorized these conflicts into two broad categories: *a)* conflicts during management of information content, and *b)* coordination between activities. The first category includes conflicting design requirements, and conflicts while updating design description, whereas the latter category includes design process and agent coordination conflicts. In the X-DPR framework, coordination between agents is achieved by capturing the sequence of agent execution and maintaining their status in the database (see Section 1.5.8). The X-DPR framework takes advantage of the STEP data repository and its Java interface discussed in Section 1.5.2.1 for conflict management during design description update. The X-DPR framework does not currently implement the conflict management of requirements and the coordination issues arising when different designers share the same design variables but have conflicting objectives. Various theoretical frameworks, such as negotiation [62, 63], game theory [64-68], and Pareto optimality-based methods [69], have been developed in the literature for coordination between distributed designers and agents. There has also been a recent interest in the academia in applying different interaction protocols such as cooperative and non-cooperative game theory-based protocols for coordination between design teams in decentralized design processes, where different teams with conflicting objectives share a design space. These protocols help in the coordination of design decisions in a multi-designer scenario. So far, these protocols are studied at a theoretical level but have not been implemented in any distributed design framework. We believe that the next phase in the evolution of distributed product realization frameworks is in implementing these efficient methods for designer interactions.

1.7 Acknowledgments

We gratefully acknowledge the support of the National Science Foundation grants NSF- 0100123, and DMI-0085136.

1.8 References

[1] Schmidt, D. C., 2002, *Distributed Object Computing with CORBA Middleware,* http://www.cs.wustl.edu/~schmidt/corba.html

[2] Microsoft, 1996, *DCOM Technical Overview,* http://msdn.microsoft.com/library/default.asp?url=/library/en-us/dndcom/html/msdn_dcomtec.asp

[3] Sriram, R. D., Szykman S. and Durham, D., 2006, "Guest editorial – special issue on collaborative engineering," *Journal of Computing and Information Science in Engineering: Special Issue on Collaborative Engineering,* 6(2), pp. 93–95.

[4] Simpson, T. W., Lanutenschlager U. and Mistree, F., 1998, "Mass customization in the age of information: the case for open engineering systems," *The Information Revolution: Present and Future Consequences* (W. H. Read and A. L. Porter, Eds.), Ablex Publishing Co., Greenwich, CT, pp. 49–71.

[5] Rosen, D. W., 1998, "Progress towards a distributed product realization studio: the rapid tooling testbed," *3rd IFIP WG 5.2 Workshop, Proceedings of Workshop on Knowledge Intensive CAD (KIC-3)*, Tokyo, Japan, pp. 167–196.

[6] Christensen, E., Curbera, F., Meredith G. and Weerawarana, S., 2001, *Web Services Description Language (WSDL) 1.1,* http://www.w3.org/TR/wsdl

[7] Wang, L., Shen, W., Xie, H., Neelamkavil J. and Pardasani, A., 2001, "Collaborative conceptual design - state of the art and future trends," *Computer Aided Design,* 34(13), pp. 981–996.

[8] Xiao, A. H., Kulkarni, R., Allen, J., Rosen, D. W., Mistree F. and Feng, S. C., 2001, "A Web based distributed product realization environment," *ASME Computers in Engineering*, Pittsburgh, Pennsylvania. Paper Number: DETC2001/CIE-21766.

[9] Chalfan, K. M., 1986, "A knowledge system which integrates heterogeneous software for a design application," *Application of Artificial Intelligence to Engineering Problems*, Southhampton, England.

[10] Rodgers, P. A., Huxor, A. P. and Caldwell, N. H. M., 1999, "Design support using distributed web based AI tools," *Research in Engineering Design,* 11(1), pp. 31–44.

[11] Huang, G. Q., Lee, S. W. and Mak, K. L., 1999, "Web based product and process data modeling in concurrent 'design for X'," *Robotics and Computer Integrated Manufacturing,* 15(1), pp. 53–63.

[12] Cramer, D., Jayaram, U. and Jayaram, S., 2002, "A collaborative architecture for multiple computer aided engineering applications," *ASME 2002 Design Engineering Technical Conferences*, Montreal, Canada. Paper Number: DETC2002/CIE-34498.

[13] Ebbesmeyer, P., Gausemeier, J., Krumm, H. and Molt, T., 2001, "Virtual Web plant: an Internet-based plant engineering information system," *Journal of Computing and Information Science in Engineering,* 1(2), pp. 257–260.

[14] Wang, P., Bjarnemo, R. and Motte, D., 2005, "A Web-based interactive virtual environment for mobile phone customization," *Journal of Computing and Information Science in Engineering,* 5(1), pp. 67–70.

[15] Lin, R. and Afjeh, A. A., 2004, "Development of XML data binding integration for Web-enabled aircraft engine simulation," *Journal of Computing and Information Science in Engineering,* 4(3), pp. 186–196.

[16] Simpson, T. W., Umapathy, K., Nanda, J., Halbe, S. and Hodge, B., 2003, "Development of a framework for Web-based product platform customization," *Journal of Computing and Information Science in Engineering,* 3(2), pp. 119–129.

[17] Wang, F. and Wright, P., 1998, "Internet-based design and manufacturing on an open architecture machine center," *Japan-USA Symposium on Flexible Automation*, Otsu, Japan, pp. 221–228.

[18] Kim, C. S., Kim, N., Kim, Y., Kang, S. and O'Gardy, P., 1998, "Internet-based concurrent engineering: an interactive 3d system with markup," *Proceedings of Concurrent Engineering*, Tokyo, Japan, pp. 555–563.

[19] Gupta, S. K., Lin, E., Lo, A. J. and Xu, C., 2002, "Web-based innovation alert services to support product design evolution," *ASME 2002 Design Engineering Technical Conferences*, Montreal, Canada. Paper Number: DETC2002/CIE-34462.

[20] Shyamsundar, N., Dani, T., Sonthi, R. and Gadh, R., 1998, "Shape abstractions and representations to enable Internet-based collaborative CAD," *Japan-USA Symposium on Flexible Automation*, Otsu, Japan, pp. 229–236.
[21] Rezayat, M., 2000, "The enterprise-Web portal for life-cycle support," *Computer Aided Design,* 32(2), pp. 85–96.
[22] Jiang, P., Fukuda, S. and Raper, S. A., 2002, "TeleDM: an Internet Web service testbed for fast product design supported by prototype manufacturing," *Journal of Computing and Information Science in Engineering,* 2(2), pp. 125–131.
[23] Wang, J., Ma, Z.-D. and Hulbert, G. M., 2005, "A distributed mechanical system simulation platform based on a "gluing algorithm"," *Journal of Computing and Information Science in Engineering,* 5(1), pp. 71–76.
[24] ANSYS, 2006, *ANSYS Workbench™,* http://www.ansys.com/products/workbench.asp
[25] MSC Software, 2006, *MSC.Acumen,* http://www.mscsoftware.com.my/products/software/msc/acumen/index.htm
[26] UGS, 2006, *Teamcenter 2005,* http://www.ugs.com/products/teamcenter/
[27] PTC, 2003, *PTC Windchill,* http://www.ptc.com/products/windchill/
[28] Alibre, 2006, *Alibre Design 9.1,* http://www.alibre.com/products/
[29] Cutkosky, M., Engelmore, R., Fikes, R., Genesereth, M., Gruber, T., Mark, W., Tenenbaum, J. and Weber, J., 1993, "PACT: an experiment in integrating concurrent engineering systems," *IEEE Computer,* 26(1), pp. 28–37.
[30] Ray, S. R., 2002, "Interoperability standards in the semantic Web," *Journal of Computing and Information Science in Engineering,* 2(1), pp. 65–71.
[31] Rezayat, M., 2000, "Knowledge-based product development using XML and KC's," *Computer Aided Design,* 32(5-6), pp. 299–309.
[32] Olsen, G. R., Cutkosky, M., Tenenbaum, J. andGruber, T. R., 1995, "Collaborative engineering based on knowledge sharing agreements," *Concurrent Engineering Research and Applications,* 3(2), pp. 145–159.
[33] Toye, G., Cutkosky, M., Leifer, L., Tenenbaum, J. and Glicksman, J., 1994, "SHARE: a methodology and environment for collaborative product development," *International Journal of Intelligent and Cooperative Information Systems,* 3(2), pp. 129–153.
[34] Rajagopalan, S., Pinilla, J. M., Losleben, P., Tian, Q. and Gupta, S. K., 1998, "Integrated design and rapid manufacturing over the Internet," *ASME 1998 Design Engineering Technical Conferences*, Atlanta, G.A. Paper Number: DETC98/CIE-5519.
[35] Madhusudan, T., 2004, "An intelligent mediator-based framework for enterprise application integration," *Journal of Computing and Information Science in Engineering,* 4(4), pp. 294–304.
[36] Wu, T., Xie, N. and Blackhurst, J., 2004, "Design and implementation of a distributed information system for collaborative product development," *Journal of Computing and Information Science in Engineering,* 4(4), pp. 281–293.

[37] Pahng, F., Senin, N. and Wallace, D., 1998, "Distribution modeling and evaluation of product design problems," *Computer Aided Design,* 30(6), pp. 411–423.

[38] Dabke, P., Cox, A. and Johnson, D., 1998, "NetBuilder: an environment for integrating tools and people," *Computer Aided Design,* 30(6), pp. 465–472.

[39] Engineous Inc., 2005, *FIPER Infrastructure,* http://www.engineous.com/product_FIPERInfrastructure.htm

[40] Gerhard, F. J., Rosen, D. W., Allen, J. and Mistree, F., 2001, "A distributed product realization environment for design and manufacturing," *Journal of Computing and Information Science in Engineering,* 1(3), pp. 235–244.

[41] Choi, H.-J., 2001, *A Framework for Distributed Product Realization*, M.S. Thesis, Woodruff School of Mechanical Engineering, Georgia Institute of Technology, Atlanta.

[42] Wujek, B. A., Koch, P. N., McMillan, M. and Chiang, W.-S., 2002, "A distributed component based integration environment for multidisciplinary optimal and quality design," *9th AIAA/ISSMO Symposium on Multidisciplinary Analysis and Optimization*, Atlanta, GA. Paper Number: AIAA2002-5476.

[43] Kao, K. J., Seeley, C. E., Yin, S., Kolonay, R. M., Rus, T. and Paradis, M., 2004, "Business-to-business virtual collaboration of aircraft engine combustor design," *Journal of Computing and Information Science in Engineering,* 4(4), pp. 365–371.

[44] Engineous Inc., 2004, *iSIGHT, Version 8.0,* http://www.engineous.com/product_iSIGHT.htm

[45] Phoenix Integration Inc., 2004, *ModelCenter®, Version 5.0,* http://www.phoenix-int.com/products/ModelCenter.html

[46] Panchal, J. H., Fernández, M. G., Paredis, C. J. J. and Mistree, F., 2004, "Reusable design processes via. modular, executable, decision-centric templates," *10th AIAA/ISSMO Multidisciplinary Analysis and Optimization Conference*, Albany, New York. Paper Number: AIAA-2004-4601.

[47] Hayes, A. M., Wang, A., Dempsey, B. M. and McDowell, D. L., 2001, "Mechanics of linear cellular alloys," *Proceedings of IMECE, International Mechanical Engineering Congress and Exposition*, New York, NY.

[48] Seepersad, C. C., Dempsey, B. M., Allen, J. K., Mistree, F. and McDowell, D. L., 2002, "Design of multifunctional honeycomb materials," *9th AIAA/ISSMO Symposium on Multidisciplinary Analysis and Optimization,* Atlanta, GA. Paper Number: Paper No. AIAA-2002-5626.

[49] Bras, B. A. and Mistree, F., 1991, "Designing design process in decision-based concurrent engineering," *SAE Transactions, Journal of Materials and Manufacturing,* 100, pp. 451–458.

[50] Muster, D. and Mistree, F., 1988, "The decision support problem technique in engineering design," *International Journal of Applied Engineering Education,* 4(1), pp. 23–33.

[51] Nell, J., 2003, *STEP on a Page (ISO 10303),* http://www.nist.gov/sc5/soap/

[52] Schenck, D. A. and Wilson, P. R., 1994, *Information Modeling: The EXPRESS Way,* Oxford University Press.

[53] LKSoft, 2006, *JSDAI,* http://www.jsdai.net/

[54] Engineous Inc., 2001, *Product Overview: iSIGHT,* http://www.engineous.com/overview.html

[55] Ansys Inc., 2003, *Ansys Products,* http://www.ansys.com/ansys/index.htm

[56] Box, D., Ehnebuske, D., Kakivaya, G., Layman, A., Mendelsohn, N., Nielsen, H. F., Thatte, S. and Winer, D., 2000, *Simple Object Access Protocol (SOAP) 1.1,* http://www.w3.org/TR/SOAP/

[57] Marcato, D., 2002, *Distributed Computing With SOAP,* http://www.devx.com/upload/free/features/vcdj/2000/04apr00/dm0400/dm0400.asp

[58] Microsoft, 2003, *SOAP Toolkit 3.0,* http://msdn.microsoft.com/downloads/default.asp?URL=/downloads/sample.asp?url=/msdn-files/027/001/948/msdncompositedoc.xml

[59] Systinet Corporation, 2004, *Systinet Server for Java,* http://www.systinet.com/products/ssj/overview

[60] OASIS, 2004, *Introduction to UDDI: Important Features and Functional Concepts,* http://uddi.org/pubs/uddi-tech-wp.pdf

[61] Brazier, F. M. T., Langen, P. H. G. v. and Treur, J., 1995, "Modeling conflict management in design: an explicit approach," *Artificial Intelligence for Engineering Design, Analysis, and Manufacturing,* 9(4), pp. 353–366.

[62] Scott, M. J., 1999, *Formalizing Negotiation in Engineering Design*, PhD Dissertation, Mechanical Engineering, California Institute of Technology, Pasadena, CA, USA.

[63] Scott, M. J. and Antonsson, E. K., 1996, "Formalisms for negotiation in engineering design," *ASME Design Theory and Methodology Conference*, Irvine, CA. Paper Number: 96-DETC/DTM-1525.

[64] Badhrinath, K. and Rao, J. R. J., 1996, "Modeling for concurrent design using game theory formulations," *Concurrent Engineering: Research and Applications,* 4(4), pp. 389–399.

[65] Lewis, K. and Mistree, F., 1997, "Modeling interaction in multidisciplinary design: a game theoretic approach," *AIAA Journal,* 35(8), pp. 1387-1392.

[66] Lewis, K. and Mistree, F., 1998, "Collaborative, Sequential and Isolated Decisions in Design," *ASME Journal of Mechanical Design,* 120(4), pp. 643–652.

[67] Marston, M., 2000, *Game Based Design: A Game Theory Based Approach to Engineering Design*, PhD Dissertation, G. W. Woodruff School of Mechanical Engineering, Georgia Institute of Technology, Atlanta, GA.

[68] Rao, S. S. and Freihet, T. I., 1991, "A modified game theory approach to multiobjective optimization," *Journal of Mechanical Design,* 113(3), pp. 286–291.

[69] Petrie, C. J., Webster, T. A. and Cutkosky, M. R., 1995, "Using Pareto optimality to coordinate distributed agents," *Artificial Intelligence for Engineering Design, Analysis, and Manufacturing,* 9(4), pp. 313–323.

2

A Web-based Intelligent Collaborative System for Engineering Design

Xiaoqing (Frank) Liu and Samir Raorane

Department of Computer Science, University of Missouri-Rolla, USA

Ming C. Leu

Department of Mechanical and Aerospace Engineering
University of Missouri-Rolla, USA

Design of a modern product is often a very complicated process, which involves groups of designers, manufacturers, suppliers, and customer representatives. Conflicts are unavoidable during collaboration among multiple stakeholders, who have different objectives, requirements, and priorities. Current Web-based collaborative engineering design systems do not support collaborative conflict resolution. In this chapter, we present a Web-based intelligent system that we have developed for collaborative engineering design. It extends a collaborative solid modeling software system by adding an argumentation-based conflict resolution tool, a whiteboard, and a chat utility. We have developed an intelligent computational argumentation model to enable the management of a large scale argumentation network and resolution of conflicts based on argumentation from many participants. A Web-based collaborative engineering design system has been developed based on the above model to resolve conflicts over the Internet by enabling collaborators to select the most favored design alternative in the design argumentation from multiple perspectives. An example of collaborative design of latch mechanism for a solar car using the developed system is presented to show its effectiveness.

2.1 Introduction

Modern products are increasingly designed via collaborations that are distributed across people, organizations, and space. Because of the involvement of various disciplinary groups in decision making, numerous conflicts exist at every stage of a collaborative engineering design process [1, 2]. Decisions made by different groups may not be consistent, components may not physically fit together, and system interfaces may not be compatible. Although different tools and software support systems have been developed to facilitate collaborative engineering design [3-5], the lack of effective intelligent conflict detection and resolution capabilities still hampers effective and efficient collaborative design.

In this chapter, we present a quantitative argumentation method for collaborative engineering design. Based on the method, we have developed an intelligent Web-based collaborative engineering design system that links designers of engineering systems and facilitates effective and efficient conflict resolution among them. This system allows multiple designers to design solid models collaboratively, and facilitates resolution of conflicts effectively through argumentation.

The chapter is organized as follows. Section 2.2 reviews related work. Section 2.3 describes the architecture for a collaborative engineering design environment. Section 2.4 explains argumentation-based conflict resolution in collaborative engineering design. Section 2.5 describes the design and development of a software system implementing the described method. Section 2.6 presents an example to illustrate our method and system.

2.2 Related Work

2.2.1 Current State-of-the-art on Computer-aided Collaborative Engineering Design Systems

We will briefly review the current state-of-the-art on collaborative engineering design systems. A traditional Computer-Aided Design (CAD) system only allows a single user to do design while a collaborative CAD system allows multiple designers to work on a design together. Early research projects in collaborative CAD systems [1, 5-7] have successfully addressed some engineering design issues in collaborative environments. They were developed on local area networks, which are platform dependent, and they were not Web-enabled. It is hard to use them to support designers in locations thousands of miles away from each other to collaborate in heterogeneous platforms. There have also been research efforts toward enabling traditional CAD systems for collaborative design. For example, a Computer Supported Cooperative Work (CSCW) system [7] was developed using C++ and AutoCAD for collaborative design. It has a generic model of collaborative design. Another such system is DOME [5]. It was built by integrating existing single-user CAD systems using CORBA and C++.

The increasing power of the Internet makes collaborative CAD feasible. Recently, several Web-based CAD systems have been developed to allow multiple users from geographically distributed locations to share their design models over the Internet. They fall into three categories. The first category of Web-based CAD systems, including C-DeSS [8] and CDFMP [9], integrates Web-based multimedia tools, such as online chat and online meeting, with Web-based solid model displays so that designers from different locations can share their design ideas over the Internet. However, users cannot develop and edit their solid models online. The second category of Web-based CAD systems, including the Internet design studio [10], WCW [11], WebCAD [12], and NetFEATURE [13], allows multiple users to share their design over the Internet although multiple users can not develop their common models concurrently. The Web-based collaborative system for engineering design recently developed by us has the capabilities of both categories [14]. The third category of Web-based CAD systems, including CSM [14], CollabCAD [15], and Alibre Design [16], focuses on collaborative solid modeling.

All of the existing Web-based collaborative design systems provide very little or no support for detecting conflicts among requirements, exploring design alternatives, and identifying the best design through argumentation from multiple perspectives to resolve design conflicts. There is a clear need *to develop fundamental theoretic methodologies of conflict resolution and implement them with a Web-based collaborative engineering design system.*

2.2.2 Current State-of-the-art on Argumentation-based Conflict Resolution

Philosopher Stephen Toulmin developed a very influential model of argumentation [17] that has guided the development of software tools and systems for supporting the detection and resolution of conflicts in many knowledge domains. Argumentation is a process of arriving at conclusions through discussions and debates. Toulmin's work has promoted a more informal approach in dealing with argumentation than formal logic. In the area of engineering design, several argumentation-based conflict resolution methods and systems have been developed from Toulmin's model. The first of them, gIBIS (graphical IBIS), represents the design dialog as a graph [18]. While representing issues, positions, and arguments, gIBIS failed to support representation of goals (requirements) and outcomes. IBE [3] extended gIBIS by integrating a document editor. REMAP [19] (REpresentation and MAintenance of Process knowledge) extended gIBIS and IBE by providing the representation of goals, decisions, and design artifacts. As opposed to these systems, Sillince proposed a more general argumentation model [20]. His model is a logic model where dialogs are represented as recursive graphs and the rules of both rhetoric and logic were used to manage the dialog and to determine when the dialog has reached closure. Alexander [21] described the incorporation of Toulmin's approach into a software product (Teleologic DOORS) that represents the features of arguments in a visual hierarchy to aid the analysis of positions taken by proponents and opponents of particular design requirements. The biggest challenge with these systems is that the sizes of their argumentation networks are often too large to comprehend and therefore it is very difficult to use them to help make design decisions since they are qualitative and not computational. In addition, they cannot deal with uncertainties associated with argumentation from multiple perspectives. In a preliminary study, we developed a computational argumentation method for capturing and analyzing software design rationale [22]. Parsons and Jennings [23] proposed a framework, based upon a system of argumentation, which permits agents to negotiate to establish acceptable ways to solve problems. QuestMap [4] is a Computer Supported Collaborative

Argumentation (CSCA) tool developed to support legal argumentation by equipping the users with the language needed to construct and analyze arguments. The disadvantage of this tool is its lack of decision making capabilities. HERMES [24] was developed to aid decision makers reaching a decision, not only by efficiently structuring the discussion rationale but also by providing reasoning mechanisms that constantly update the discourse status in order to recommend the most backed-up alternative. Its disadvantage is that the weighting factor becomes very ineffective as it is not related to the entered position.

2.3 A Web-based Intelligent Collaborative Engineering Design Environment and Its Application Scenarios

A prototype Web-based intelligent collaborative system for engineering design has been developed by us. It extends a collaborative solid modeling tool from Alibre Co. [16] by adding an argumentation-based conflict resolution tool, a whiteboard, and a chat utility using a client-server architecture as shown in Figure. 2.1. On the client side, the system provides user interfaces for argumentation-based conflict resolution, whiteboards for design alternatives, and chat rooms for real-time information exchange. On the server side, it manages client communication and argumentation networks. Alibre Design is a collaborative solid modeling tool for creating 3D designs and 2D drawings. It allows engineering teams to work together concurrently over the Internet to create, visualize, review, and modify their designs and drawings.

In the collaborative design process, when a conflict is detected, an argumentation-based conflict resolution session will be initiated. A design issue concerning the conflict is raised first in the session. After multiple design alternatives are generated by the participants, arguments can be proposed by the collaborative designers to either support or oppose the design alternatives or arguments themselves. Our system can help identify the alternative that is most favored by all participants by considering all arguments to resolve the conflicts.

2.4 Argumentation-based Conflict Resolution in the Collaborative Engineering Design Environment

We have developed a computational argumentation method for collaborative engineering design based on our preliminary work on software design rationale capturing. The argumentation framework of this conflict resolution system is an extension of the informal IBIS model of argumentation using fuzzy logic. It will help achieve a consensus among stakeholders and identify the most favorable design alternative through argumentation by computing the favorability of individual design alternatives from all arguments in the argumentation network in an uncertain environment based on fuzzy logic.

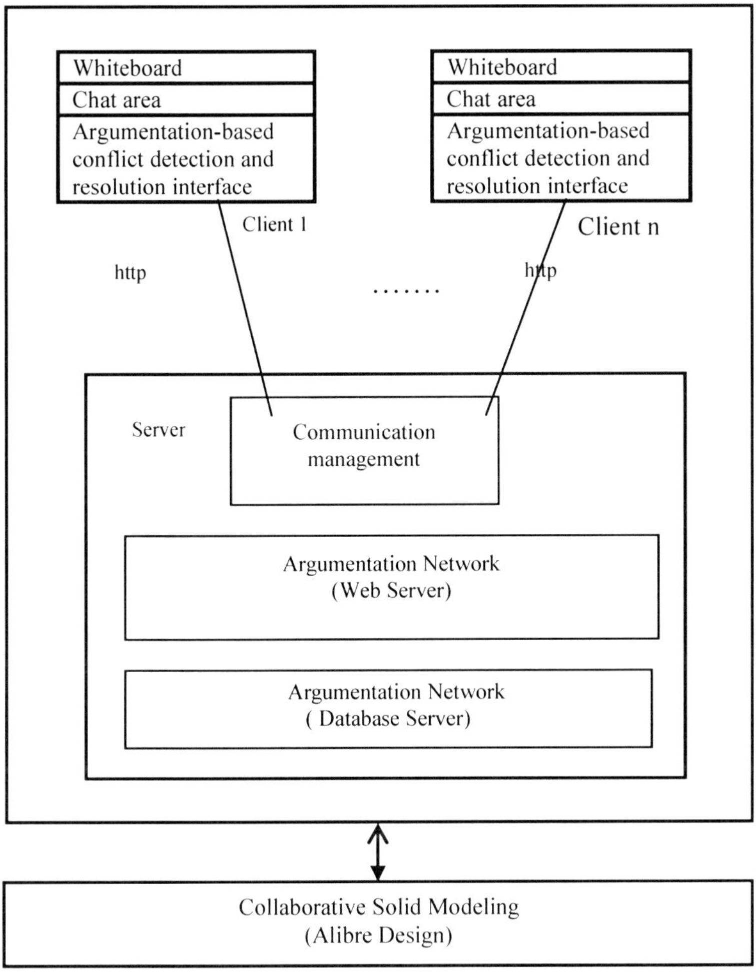

Figure 2.1. Architecture for a Web-based intelligent collaborative engineering design environment

The components of the design argumentation model for collaborative engineering design include stakeholders, requirements, conflicts, design issues, parts, alternatives, arguments, and decisions, as shown in Figure 2.2. We view collaborative design as the process of negotiating the resolution of design issues through dialogs between the stakeholders. A dialog for a given design issue is represented by the alternatives that are related to the design issue, and the arguments for or against each alternative. The resolution of a design issue is represented by a decision that selects an alternative which is most favored.

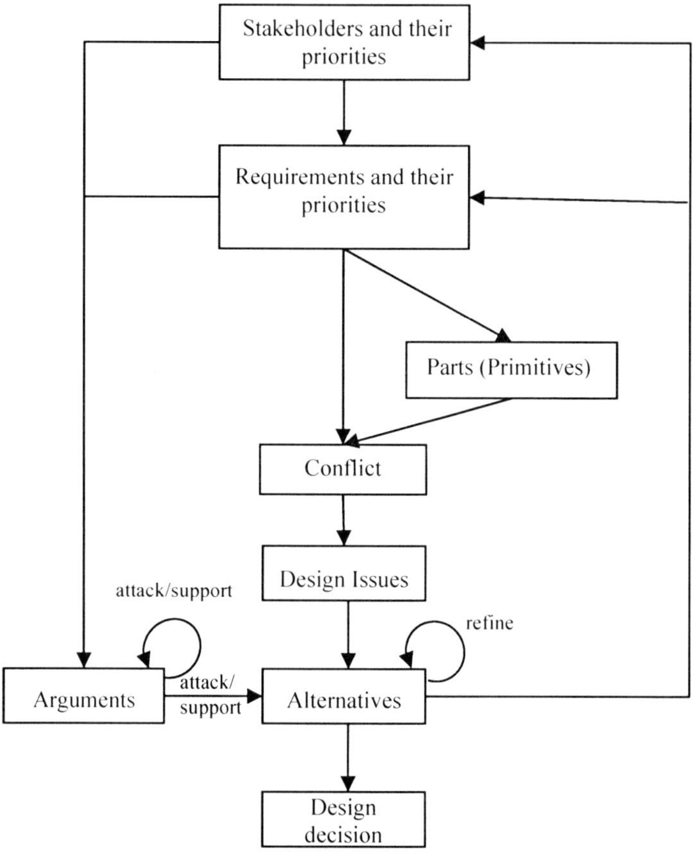

Figure 2.2. Framework for design argumentation

2.4.1 Structured Argumentation Through Dialog Graph

A design dialog for a design issue is captured as a weighted directed graph called a dialog graph [8], as shown in Figure 2.3. The nodes denoted by circles are *Positions.* A position is a statement or assertion that responds to an *issue,* which is a problem, concern, or question that requires discussion for the problem solving to proceed. The nodes denoted by rectangles are *Arguments*. Arguments are statements that support or attack Positions. Each Position may have one or more arguments that either *support* or *attack* it. Arcs represent a relationship (attack or support) from the originating argument node to the terminating argument or position node. The position node contains the name of the stakeholder posting the position and the text of the position. Each Argument node contains the name of the stakeholder posting the argument, the text of the argument and a weight value. The weight attached to an argument is the Argument Strength. It is the measure of an argument's degree of attack or support of either a position or another argument in the position dialog graph. The weight value is a real number between -1 and 1. A positive number denotes Support and a negative number denotes Attack while zero denotes Indecision. The strength of the argument is viewed as a fuzzy set and linguistic labels are used to represent the strength. It is easy to use linguistic labels, instead of real numbers, to denote the strength of an argument over another argument or a position. By doing so fuzzy inference can be used to evaluate a position. Both linguistic labels on arcs (branches) and strengths of arguments are given by participants. Since disagreements among participants are inevitable, how to objectively determine a position's overall favorability is a major research issue. A position node contains a label associated with it to give a measure of the strength of the position based on the strengths of the arguments under it. This measure represents the overall favorability of the position.

Let us use a simple example to illustrate the above concepts. Suppose that several designers in multiple locations collaborate to develop a speed reducer. They may have an *issue* of its gear design. Two design alternatives, which are represented as their *positions*, are proposed by participants. One focuses on cam and another focuses on linkage. Participants can debate about them by posting their arguments about their advantages and disadvantages to resolve their conflict. Another example will be given later to demonstrate how to apply the presented conflict resolution method.

2.4.2 Argument Reduction Through Fuzzy Inference

In Figure 2.3, we can see some arguments attached to other arguments, by a label to denote the degree of support or attack on the arc going between arguments, other than directly attached to the position. For example, A3 has Medium Attack (MA), and A1 and A5 have Strong Support (SS). Argument reduction is used to reduce the arguments which are not directly connected to the position, in order to have them directly connected to the position. For example, argument A3 which is posted as an argument that attacks argument A1, actually attacks the position P after argument reduction.

There are four General Argumentation Heuristic Rules that can be formulated as follows [2].

- General Argumentation Heuristic Rule 1: If argument B supports argument A and argument A supports position P, then argument B supports position P.
- General Argumentation Heuristic Rule 2: If argument B attacks argument A and argument A supports position P, then argument B attacks position P.
- General Argumentation Heuristic Rule 3: If argument B supports argument A and argument A attacks position P, then argument B attacks position P.
- General Argumentation Heuristic Rule 4: If argument B attacks argument A and argument A attacks position P, then argument B supports position P.

As the linguistic labels used are Strong Support (SS), Medium Support (MS), Indecisive (I), Medium Attack (MA) and Strong Attack (SA), the above four General Argumentation Heuristic Rules can be extended to obtain twenty-five Argumentation Heuristic Rules shown in Figure 2.4.

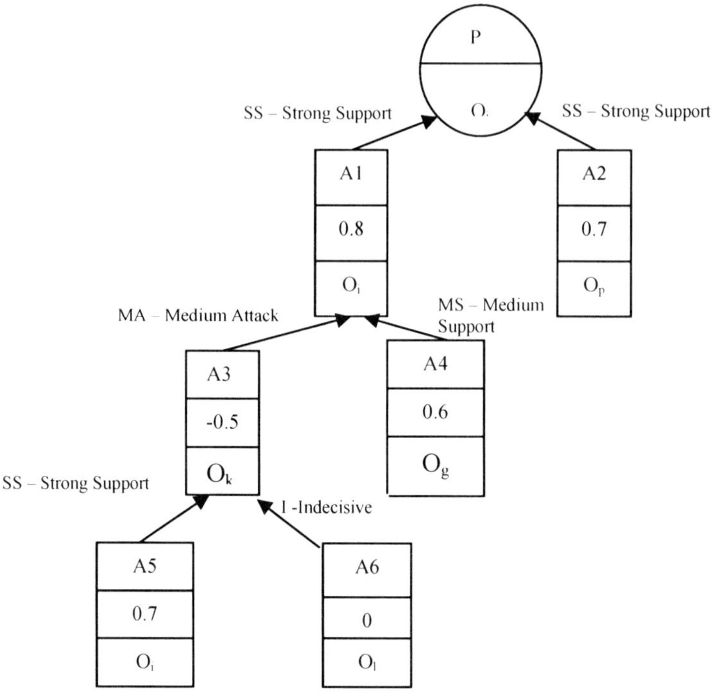

Figure 2.3. Position dialog graph

	SS	MS	I	MA	SA
SS	SS	MS	I	MA	SA
MS	MS	MS	I	MA	MA
I	I	I	I	I	I
MA	MA	MA	I	MS	MS
SA	SA	MA	I	MS	SS

SS: Strong Support MS: Medium Support

I: Indecisive MA: Medium Attack

SA: Strong Attack

Figure 2.4. Argumentation heuristic rules

Consider an instance where the strength of the level-1 argument is Strong Attack (SA) and that of the level-2 argument is Medium Support (MS), then the reduced strength of the level-2 argument will be Medium Attack (MA) as shown by the entry in column 3 and row 6 in Figure 2.4.

A fuzzy inference engine has been built to infer the reduced strengths of the arguments, as discussed later in this section. Using this fuzzy inference engine we can reduce a given Position Dialog Graph into one in which all the argument nodes are directly attached to the position node. Consider the example in Figure 2.3, where we have arguments occurring at level 3. The argument nodes at level 3 can be reduced and attached to the argument node at level 1. Their reduced strengths are computed using the fuzzy inference engine, as shown in Figure 2.5.

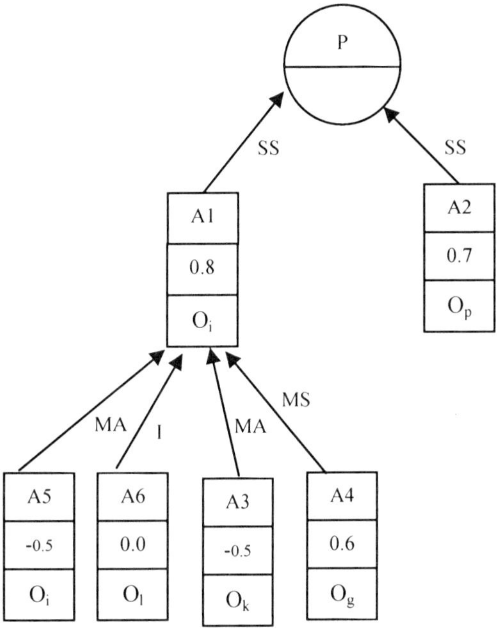

Figure 2.5. Position dialog graph after one level reduction

Now there is one level of arguments which are not directly attached to the position. Hence argument reduction has to be performed once again to have the reduced position dialog graph with all the arguments directly attached to the position. The arguments at level 2 are reduced using the fuzzy inference engine and attached directly to the position node, as shown in Figure 2.6.

In the procedure of argument reduction, the fuzzy inference engine takes in two inputs and generates one output. The inputs are the strengths of the argument to be reduced and the argument right above it. The output of the fuzzy inference engine is the strength of the argument after the argument reduction.

2.4.2.1 Linguistic Variable Through Fuzzy Membership Functions

Fuzzy membership functions are used to quantitatively characterize linguistic systems represented as fuzzy sets. The fuzzy membership function chosen for the system in our study is the piecewise linear trapezoidal function. Membership functions are defined by using a,b,c,d to denote the four vertices of the trapezoids.

Five membership functions have been defined for five fuzzy sets. The five fuzzy sets are Strong Attack (SA: a = -1, b = -1, c = -0.8, d = -0.5), Medium Attack (MA: a = -0.8, b = -0.6, c = -0.4, d = -0.2), Indecisive (I: a = -0.3, b = 0, c = 0, d = 0.3), Medium Support (MS: a = 0.2, b = 0.4, c = 0.6, d = 0.8) and Strong Support (SS: a = 0.5, b = 0.8, c = 1, d = 1). Figure 2.7 shows the five membership functions for the above five linguistic terms.

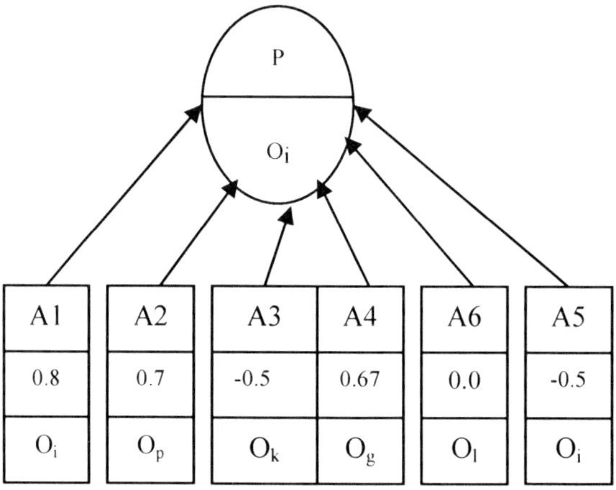

Figure 2.6. Position dialog graph after complete reduction

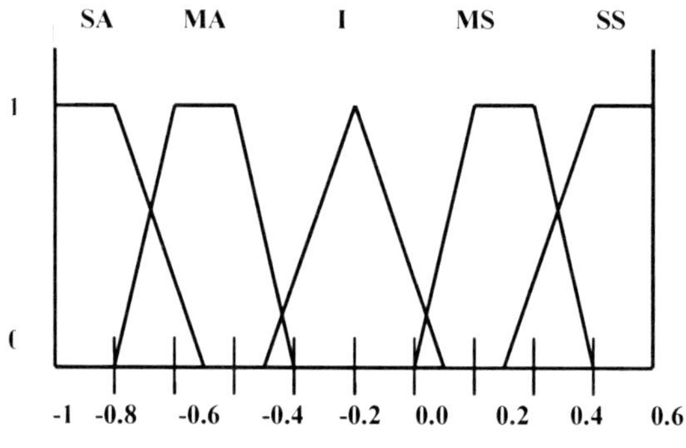

Figure 2.7. Five membership functions

2.4.2.2 Fuzzy Inference Rules

Fuzzy inference rules combine two or more input fuzzy sets and associate with them an output set. The input sets are combined by means of operators that are analogous to the usual logical conjunctives "and", "or", *etc.* The fuzzy rules, also known as argumentation rules, are given in Figure 2.4. The fuzzy or argumentation rules are stored and represented through the use of the Fuzzy Association Memory (FAM) matrix shown in Figure 2.8. There are two inputs X and Y for each rule.

Each input variable is one of five input sets, *i.e.*, "SS", "MS", "I", "MA", and "SA". The output variable Z is one of five output sets which are same as the five input sets. Each FAM matrix entry is a fuzzy set that is the output of the fuzzy rule. For example, the shaded part in Figure 2.8 represents the rule: "If X is Strong Support (SS) and Y is Strong Attack (SA), then the output Z is Strong Attack (SA)."

2.4.2.3 *Fuzzy System and Defuzzification*

The system associated with the FAM matrix is shown in Figure 2.8. In this case we have two input variables, *X* and *Y*, each with an associated fuzzy set from SS, MS, I, MA and SA. Figure 2.7 shows the membership functions for these sets.

x \ y	SS	MS	I	MA	SA
SS	SS	MS	I	MA	SA
MS	MS	MS	I	MA	MA
I	I	I	I	I	I
MA	MA	MA	I	MS	MS
SA	SA	MA	I	MS	SS

Figure 2.8. The Fuzzy Association Memory (FAM) matrix I

The membership functions for the fuzzy sets SS, MS, I, MA and SA are denoted by F_{SS}, F_{MS}, F_I, F_{MA} and F_{SA}, respectively. A value x of the input variable *X* then has membership degrees $F_{SS}(x)$, $F_{MS}(x)$, $F_I(x)$, $F_{MA}(x)$ and $F_{SA}(x)$ in respective fuzzy sets. For example, with the trapezoidal membership functions shown in Figure 2.7 and a value x = -0.7, we would have:

$F_{SS}(-0.7) = 0.0$
$F_{MS}(-0.7) = 0.0$
$F_I(-0.7) = 0.0$
$F_{MA}(-0.7) = 0.5$
$F_{SA}(-0.7) = 0.67$

Similarly, a value y of the input variable *Y* has membership degrees $F_{SS}(y)$, $F_{MS}(y)$, $F_I(y)$, $F_{MA}(y)$ and $F_{SA}(y)$. For example, the value y = 0.6 as shown in Figure 2.9 would result in

$F_{SS}(0.6) = 0.33$
$F_{MS}(0.6) = 1.0$
$F_I(0.6) = 0.0$
$F_{MA}(0.6) = 0.0$
$F_{SA}(0.6) = 0.0$

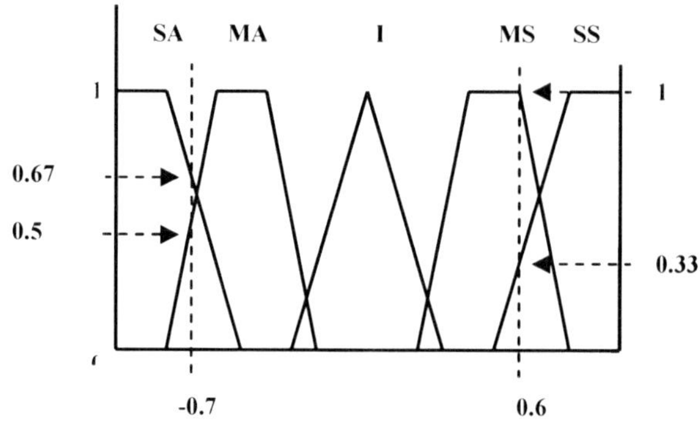

Figure 2.9. Membership degrees

Consider x = -0.7 and y = 0.6 as values of the input variables X and Y. A weight value for each entry in the FAM matrix is computed by taking the minimum value of the membership function associated with that entry. Now consider the FAM matrix entry corresponding to X as a member of the fuzzy set MA, and Y as a member of the fuzzy set SS. The weight w1 associated with the entry would be computed as:

$$
\begin{aligned}
w_1 &= \min [F_{MA}(-0.7), F_{SS}(0.6)] \\
&= \min [0.5, 0.33] \\
&= 0.33
\end{aligned}
$$

Only those FAM matrix entries which have nonzero membership-function values for both X and Y will have nonzero weights associated with them. The shaded entries in Figure 2.10 show the four activated rules for the values in the example. In addition to w_1, there are three more non-zero weights. They are

$$
\begin{aligned}
w_2 &= \min [F_{MA}(-0.7), F_{MS}(0.6)] \\
&= \min [0.5, 1.0] \\
&= 0.5 \\
w_3 &= \min [F_{SA}(-0.7), F_{SS}(0.6)] \\
&= \min [0.67, 0.33] \\
&= 0.33 \\
w_4 &= \min [F_{SA}(-0.7), F_{MS}(0.6)] \\
&= \min [0.67, 1.0] \\
&= 0.67
\end{aligned}
$$

The output variable Z also has five fuzzy sets associated with it, *i.e.* SS, MS, I, MA and SA. Specific values are assigned to these fuzzy sets, *i.e.* SS = 1, MS = 0.5, I = 0, MA = -0.5 and SA = -1. The system output is computed as follows:

$$
\text{Output} = \frac{(w1.MA + w2.MA + w3.SA + w4.MA)}{w1 + w2 + w3 + w4} = -0.59
$$

y \ x	SS	MS	I	MA	SA
SS	SS	MS	I	MA	SA
MS	MS	MS	I	MA	MA
I	I	I	I	I	I
MA	MA	MA	I	MS	MS
SA	SA	MA	I	MS	SS

Figure 2.10. The Fuzzy Association Memory(FAM) matrix II

2.4.3 Conflict Resolution by Computing Favorability of Positions (Design Alternatives)

The favorability of a position is a value indicating the strength of the position. It is calculated by taking the sum of the strengths of arguments obtained by performing reductions on the ones which are not directly connected to the position. Such a measure allows the participants in a design deliberation to compare positions objectively and quantitatively based upon the argument strengths.

To identify a good design concept, multiple design alternatives are usually developed and explored. These alternatives are known as positions. The designers would argue over each position by giving their arguments and respective weights. In order to resolve the conflicts, *i.e.*, to decide which is the best design alternative, the favorability is calculated for each position. The position with the highest favorability is the best design option.

At every point in the argumentation process, the designers can view the favorability values of various positions and can post their arguments accordingly. For example, a designer may observe that the favorability of a given position to which he is supporting is low. He may then decide to post a Strong Support (SS) on that position or a Strong Attack (SA) on an argument that has a Strong Attack (SA) on the position.

2.5 Design and Implementation

A Web-based intelligent collaborative engineering design system has been developed based on the above described method using Java on a client-server structure. Since whiteboard and chat utilities are commonly available for collaborative software systems today, we focus on design and implementation of intelligent argumentation for conflict resolution for the collaborative system.

The elements used for argumentation include Project, Issues, Positions and Arguments. The information has to be entered in text format, which can be viewed by every design member participating in the argumentation. If a conflicting issue has occurred in a new project, the designer has to first create a project and enter a detailed description of the project. Then he can add an issue under that project. If another conflicting issue occurs on the same project, the designer will need to retrieve the old project from the list of projects and then add an issue under the same project. Once an issue is created, the participating designers can enter their options *i.e.*, the positions to resolve the issue. The designers can then enter their opinions in the form of arguments to the positions.

At every stage in the argumentation process, the designers can view the result of the process, *i.e.*, they can view the position, it favorability, and the inputs by the other designers on the position. If the position with the highest favorability is the one the designer does not favor, he can then post an attack on that position or post a support on the position he favors (thus increasing the favorability of the position he supports).

The graphical user interface for the Web-based intelligent argumentation is shown in Figure 2.11. The Control Panel has five menu items: *Project, Issue, Position, Argument* and *Calculation/Clear*. Each menu item has submenus which perform unique actions on the respective argumentation elements.

As we discussed earlier, one of the drawbacks of the current systems developed in this field of research is that the sizes of their argumentation networks are often too large to comprehend and therefore it is very difficult to use them to help make design decisions. Hence in our system, we have represented the argumentation network in the form of a tree.

The basic argumentation elements are *project, issues, positions* and *arguments*. Project forms the root node, followed by issues, *i.e.*, the conflicting design issues that occur for a particular project. Under each issue are positions, *i.e.*, the design alternatives which address the issue. Arguments are under positions, and every argument can have any number of arguments. The tree structure is so designed that a designer at any time can work on any sub-tree of the argumentation tree. This helps the designer to concentrate on a specific part of the argumentation. The argumentation tree is not too large and as the fuzzy inference engine is used to resolve the conflicts, design decisions can be made without any difficulty.

2.6 An Application Example

UMR's Solar Car Team, a student design team which won the competitions in the American Solar Challenge in 2001 and 2003, is confronted with many challenging issues including resolving various design conflicts. One of the tasks of the team is to design a reliable latch mechanism that holds the base frame with the body of the solar car as shown in Figure 2.12. After the design team came up with two latch mechanisms as shown in Figures 2.13 and 2.14, from which the team needs to select the better design. Some obvious pros and cons of the two designs have been identified. While design 1 (Figure 2.13) is easier to be analyzed at the detail design stage and is also easier to be manufactured than design 2 (Figure 2.14), it is harder for the components to be assembled and needs extra work for the locking system. Solid models for design 1 and design 2 and their argumentation networks have been developed collaboratively using our collaborative design system, which incorporates Alibre Design, as shown in Figure 2.15 and Figure 2.16. Their comparison using the system is shown in Figure 2.17. An argumentation network has been developed to show resolution of conflicts, as shown in Figure 2.18. The argumentation network displayed by the system is shown in Figure 2.19. The design dialog reduction is done by the inference engine in the system. The reduced argumentation tree is shown in Figure 2.20 and the final result on favorability calculation is shown in Figure 2.21. It indicates that design 2 is favored by most participants based on the argumentation since its favorability is higher than that of design 1. This result of argumentation is concurred by the UMR Solar Car Design Team.

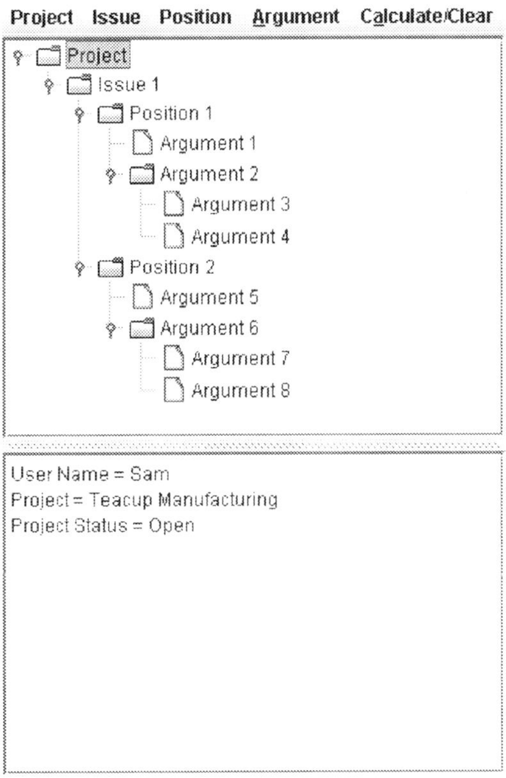

Figure 2.11. Conflict Resolution Window

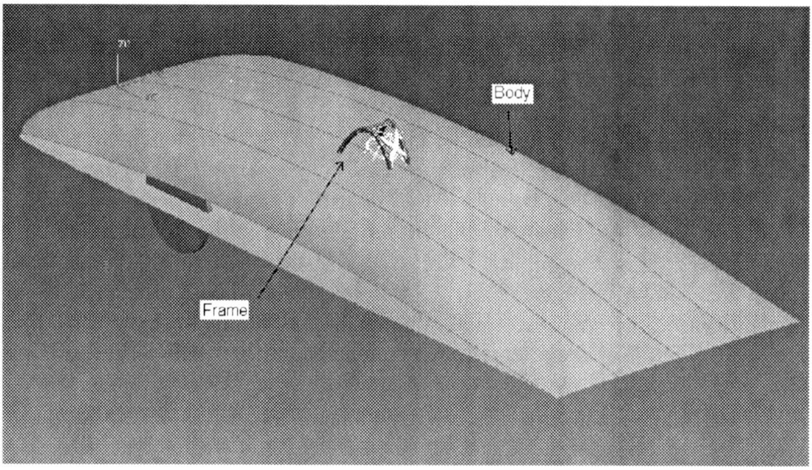

Figure 2.12. The solar car

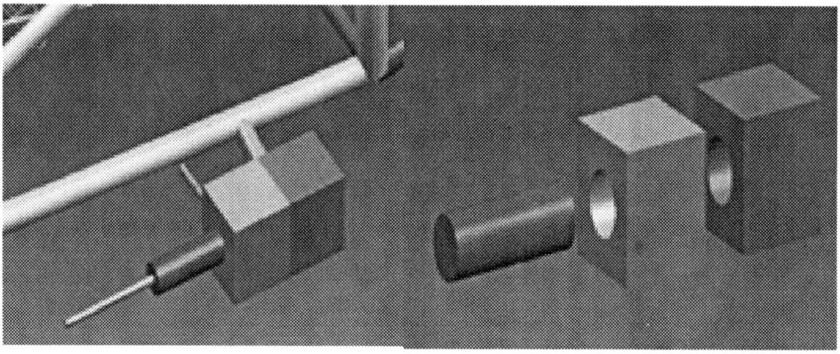

Figure 2.13. Design 1

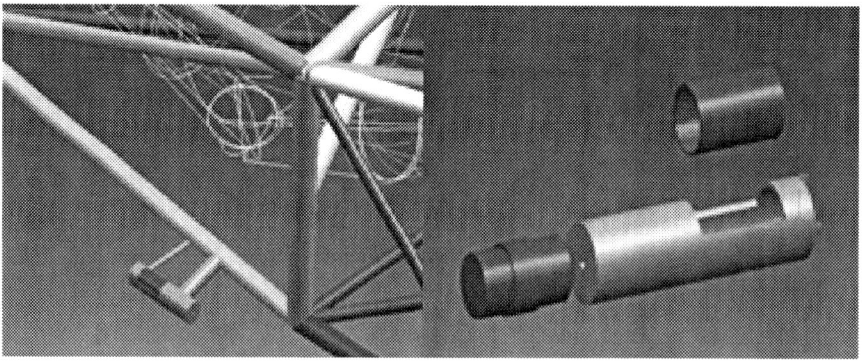

Figure 2.14. Design 2

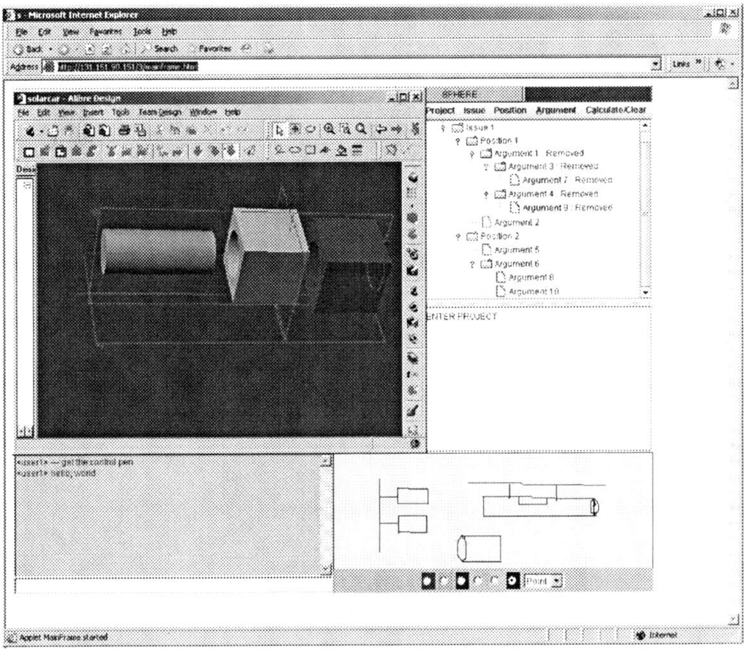

Figure 2.15. Collaborative design 1 for the Latch Mechanism

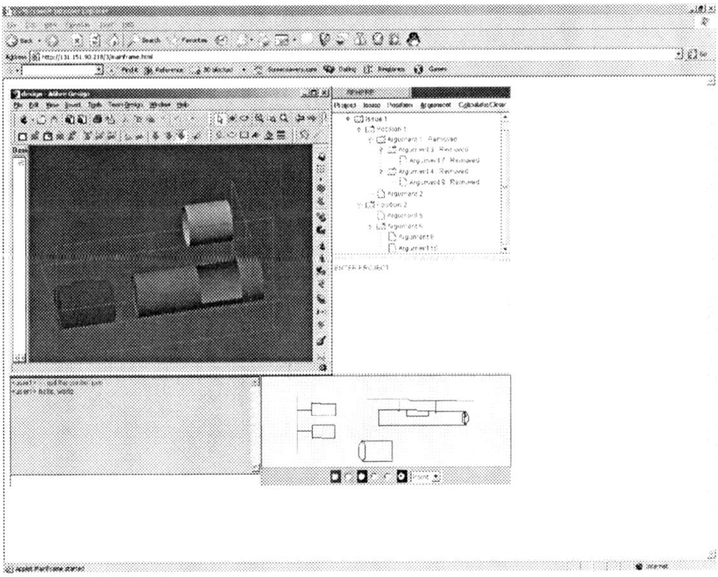

Figure 2.16. Collaborative design 2 for the Latch Mechanism

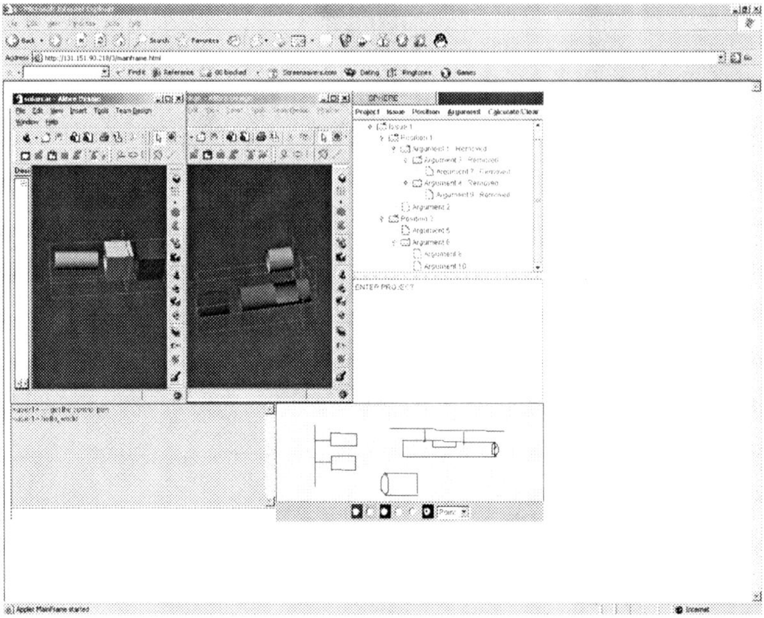

Figure 2.17. Comparisons of Design 1 and Design 2

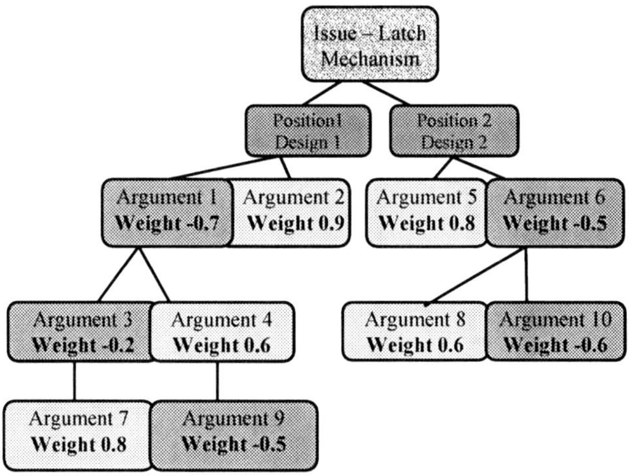

Figure 2.18. Argumentation tree

Argument 1 – The pin aligning will be a problem
Argument 2 – Design 1 is simpler and more cost-effective

Argument 3 – It is feasible to design an aligning pin and the locking can be designed easily
Argument 4 – The pin aligning is sensitive and will cause a lot of vibration
Argument 5 – A chamfer at both ends of the mating cylinder will allow smooth insertion
Argument 6 – Strength of the cylinders will depend on the material and dimensions and it is sensitive
Argument 7 – Manufacturing will be cost-effective
Argument 8 – The pin retraction will be a problem when removing the body from the frame
Argument 9 – If the two blocks are mated via a good design, then aligning will not be a problem
Argument 10 – The pin retraction should not be a problem with proper tolerance

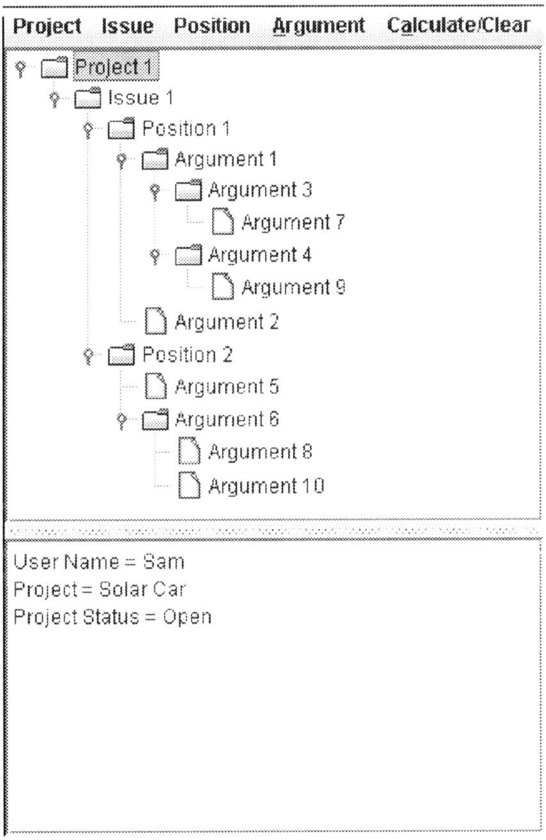

Figure 2.19. Argumentation network

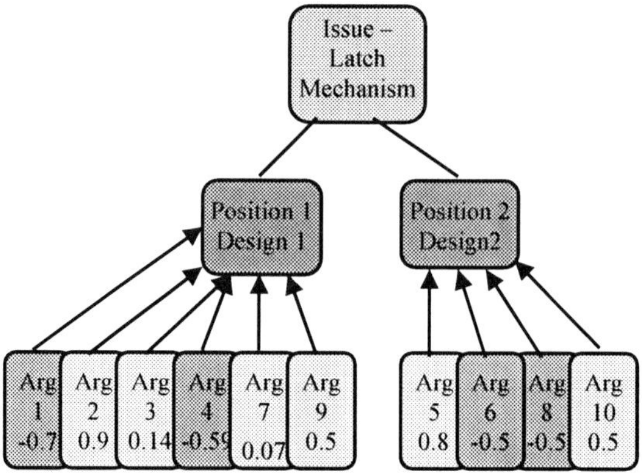

Figure 2.20. Reduced argumentation tree

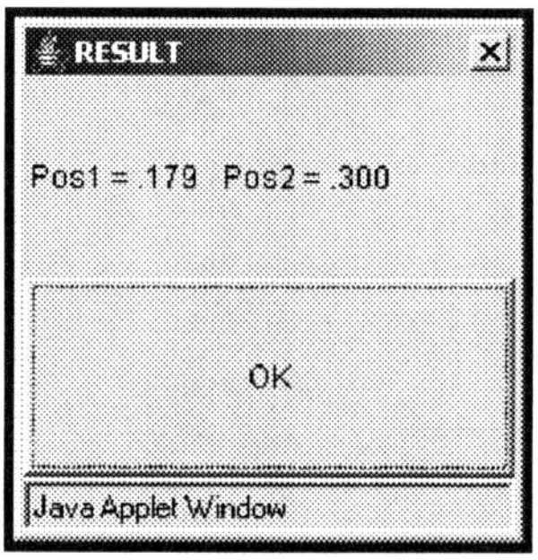

Figure 2.21. Favorabilty calculation result – solar car

2.7 Conclusions

An intelligent Web-based system has been developed using Java to facilitate collaborative engineering design by extending an existing collaborative solid modeling system to include an intelligent argumentation tool, a whiteboard, and a chat utility. It supports conflict resolution and decision making. The reduction of an argumentation hierarchy is based on fuzzy logic. The intelligent argumentation utility enhances conflict resolution capability in Web-based collaborative engineering design systems by capturing design rationale using argumentation hierarchies and providing intelligent aids to identify the most favored positions (design alternatives).

2.8 Acknowledgements

This research is supported by the Intelligent Systems Center (ISC) in the University of Missouri-Rolla. Man Zheng, Siddharth Shinde, and Yamini

Natarajan participated in and have contributed to this research project. Yan Sun has helped to edit the chapter.

2.9 References

[1] Sriram, R., 2002, *Distributed and Integrated Collaborative Engineering Design*, Sarven Publishers.

[2] Klein, M., 2003, "The dynamics of collaborative design: insights from complex systems and negotiation research," *Concurrent Engineering Research and Applications Journal*, 12(3).

[3] Lease, M., Lively, M. and Leggett, J., 1990, "Using an issue-based hypertext system to capture software life-cycle process," *Hypermedia*, 2(1).

[4] Morge, M., 2004, "Computer-supported collaborative argumentation," *CMNA IV. 4th Workshop on Computational Models of Natural Argument, ECAI 2004*, pp. 69–72.

[5] Pahng, G.-D. F., Seockhoon, B. and Wallace, D., 1998, "A Web-based collaborative design modeling environment," *Proceedings of the 7th IEEE International Workshops on Enabling Technologies: Infrastructure for Collaborative Enterprises (WET ICE '98)*, 17–19 June, pp. 161–167.

[6] Reddy, R., Srinivas, K., Jagannathan, V. and Karinthi, R., 1993, "Computer Support for concurrent engineering – guest editors' introduction," *IEEE Computer*, 26(1), pp. 12–16.

[7] Zhou, J. and Lin, G., 1999, "Implementation of collaborative design environment based on single user CAD systems," *Proceedings of the 3rd International Conference Knowledge-Based Intelligent Information Engineering Systems*, 31 Aug.–1 Sept., pp. 78–83.

[8] Klein, M., 1997, "Capturing geometry rationale for collaborative design enabling technologies," *Proceedings the 6th IEEE Workshops on Collaborative Enterprises*, 18–20 June, pp. 24–28.

[9] Zhang, H., Wu, H., Lu, J. and Chen, D., 2000, "Collaborative design system for performance," *Proceedings of Academia/Industry Working Conference on Research Challenges*, 27–29 April, pp. 59–63.

[10] Siddique, Z., 2004, "Internet design studio," *Proceedings of ASME 2004 Design Engineering Technical Conferences and Computers and Information in Engineering Conference*, Sept, 2004, Salt Lake City, Utah.

[11] Wang, L., Wong, B., Shen, W. and Lang, S., 2001, "A Web-based collaborative workspace using Java 3D," *Proceedings of the 6th International Conference on Computer Supported Cooperative Work in Design*, July, pp. 77–82.

[12] Peng, S., Tang, M. and Dong, J., 2001, "Collaborative model for concurrent product design," *Proceedings of the 6th International Conference on Computer Supported Cooperative Work in Design*, 12–14 July, pp. 212–217.

[13] Lee, J. Y., Han, S. B., Kim, H. and Park, S. B., 1999, "Network-centric feature-based modeling," *Proceedings of the 7th Pacific Conference on Computer Graphics and Applications*, 5–7 Oct.

[14] Chan, S., Ng, C. and Ng, V., 1999, "Real-time collaborative design of complex objects on the Web," *Proceedings of the 1999 IEEE International Conference on Systems, Man, and Cybernetics*, 2, pp. 120–125.

[15] CollabCAD, 2005, "CollabCAD software," http://www.collabcad.com.

[16] Alibre, 2005, http://www.alibre.com.

[17] Toulmin, S. E., 1958, *The Uses of Argument*. Cambridge, UK: University Press.

[18] Conklin, J. and Begeman, M. L., 1988, "gIBIS: A hypertext tool for exploratory policy discussion," *ACM Transactions on Information Systems (TOIS)*, 6(4), pp. 303–331.

[19] Ramesh, B. and Dhar, V., 1992, "Supporting systems development by capturing deliberations during requirements engineering," *IEEE Transactions on Software Engineering*, 18(6), pp. 498–510.

[20] Sillence, J., 1997, "Intelligent argumentation systems: requirements, models, research agenda, and applications," in *Encyclopedia of Library and Information Science* (Allen Kent, Editor), Marcel Dekker, New York, pp. 176–217.

[21] Alexander, I., 2003, "Modeling argumentation: toulmin-style," Retrieved April 15, 2005. http://easyweb.easynet.co.uk/~iany/consultancy/papers.htm.

[22] Sigman, S. and Liu, X. F., 2003, "A computational argumentation methodology for capturing and analyzing design rationale arising from multiple perspectives," *Information and Software Technology*, 45, pp. 113–122.

[23] Parsons, S. and Jennings, N. R., 1996, "Negotiation through argumentation – a preliminary report," *Proceedings of the 2nd International Conference on Multi Agent Systems*, pp. 267–274.

[24] Karacapilidis, N. and Papadias, D., 1998, "HERMES: supporting argumentative discourse in multi-agent decision making," *Proceedings of the 15th National Conference on Artificial Intelligence (AAAI-98)*, Madison, WI, AAAI/MIT Press, pp. 827–832.

3

A Shared VE for Collaborative Product Development in Manufacturing Enterprises

G. Chryssolouris, M. Pappas, V. Karabatsou, D. Mavrikios and K. Alexopoulos

Laboratory for Manufacturing Systems and Automation
Department of Mechanical Engineering and Aeronautics
University of Patras, Greece

This chapter describes the development of an integrated Virtual Environment (VE) for collaborative product design. The objective of this approach is to enable real-time collaboration among multiple users, at different sites, in the same VE. The concept includes the use of Virtual Reality (VR) technology for the development of a working display environment that provides all collaborative users with navigation, immersion and interaction capabilities in real time. The scope of this work is to provide an efficient, robust collaboration tool for the real time validation of a manufacturing product, from the early stages of the conceptual design up to the latest stages of the production chain. In order to demonstrate the benefits of virtual collaboration that a shared environment can offer to manufacturing, a pilot application, based on the requirements of a "real life" manufacturing company, has been developed and presented.

3.1 Introduction

Today's global business environment in the manufacturing industry is characterized by competitive pressures and sophisticated customers, who demand innovative and speedy solutions. Understanding and optimizing design processes is the cornerstone of success in these fast-changing environments. Short time to market, while maintaining a high product quality, has become the main success factors.

Manufacturing companies need to innovate, both by designing new products and by enhancing the quality of the existing ones [1]. Usually, during product design, all the persons involved share a great number of drawings-files and assembly models. Often, different components or sub-assemblies of the product are designed by different groups of designers at geographically different locations. Companies are frequently out-sourcing engineering activities, performed internally, in order to accelerate the design and the product development process [2]. Nowadays 50–80% of all the components manufactured by original equipment manufacturers are out-sourced to external suppliers [3]. This often creates problems due to the lack of distributed and collaborative design and manufacturing systems, which would effectively disseminate product design knowledge. These problems are typically resolved through meetings or via e-mails and phone discussions. Colleagues are not capable of collaborating and exchanging their ideas easily, if they work in different places or particularly in different countries. An operating shared VE could solve this problem by eliminating unnecessary meetings, repetitive e-mails and costly product mistakes and delays. The use of such a system aims at identifying, quickly and efficiently, both feasible and optimal designs through collaboration among product development partners at different locations.

The main goal of the present work is the conceptualization, design and development of a shared VE for supporting real-time collaboration onto the same virtual product. The proposed shared VE not only does provide collaboration capabilities among multiple users, but also immersion and interaction with products under evaluation. Collaboration features related to users, roles, events, projects and files management have also been developed into a Web-based platform in order to support the simulation features, which are provided by the shared environment and which are related to product design verification.

3.2 Background

In the past decade many research approaches and applications focused on the use of VR for overcoming the complexity of product design and manufacture [4]. A lot of them also included human simulations in order to perform ergonomic analysis of virtual products or assembly processes [5–7]. On the other hand, various Web-based manufacturing systems have been developed for supporting collaborative activities, in different life-cycle phases of product development. These include marketing, design, process planning, production, distribution, service, *etc.* Distributed product development life-cycle activities in a globally integrated environment are associated with the use of internet as well as Web technologies. Many product development software systems, such as Computer-Aided Design (CAD), Computer-Aided Manufacturing (CAM), database management and intelligent knowledge-based, have also been integrated, through Web technologies, into these Web-based collaboration systems [8].

An asynchronous collaborative system has been presented [9], called Immersive Discussion Tool (IDT), which emphasizes on the elaboration and transformations of a problem space and underlines the role that unstructured verbal and graphic communication can play in design processes. A prototyped system called cPAD has been developed [10, 11] to enable designers to visualize product assembly models and to perform real time geometric modifications, based on polygonized representations of assembly models. The Detailed Virtual Design

System (DVDS) for shape modeling in a multi-modal, multi-sensory VE has been presented [12], enabling collaborative design and design among multiple designers, both in the same site and in remote site VEs. An Internet-based VR collaborative environment, called Virtual-based Collaborative Environment (VRCE) developed with the use of Vnet, Java and VRML [13], demonstrates the feasibility of collaborative design for small to medium size companies that focus on a narrow range of low cost products. A Web-enabled Product Data Management (PDM) system which facilitates various collaborative design activities [14] has been developed providing 3D visualization capabilities as well. Another tool for dynamic data sharing in collaborative design, has been developed [15], ensuring that experts use it as a common space to define and share design entities.

Further to the research approaches to the field of a Web-based collaborative product design, a few commercial tools are available to support such functionalities. OneSpace.net (http://www.cocreate.com/) is a lightweight Web collaboration tool that supports online team collaboration for project development. It combines architecture for Web services with popular concepts, such as organized projects, secure messaging, presence awareness and real time online meetings. IBM's Product Lifecycle Management (PLM) Express Portfolio has been designed specifically for medium-sized companies that design or manufacture products. This system mainly focuses on business processes and also allows design engineers to share 3D data, created with diverse authoring tools and thus, product development can be managed. It includes CATIA Version 5 collaborative product design software and SMARTEAM for product data and release management (http://www.ibm.com/). Matrix10 is designed to support deployments of all sizes. It includes PLM business process applications that cover a wide range of processes, namely product planning, development and sourcing and program management. Moreover, it allows diverse design disciplines to be synchronized around design activities and changes, by reducing the critical errors and cost associated ones with poor collaboration (http://www.matrixone.com/). eDrawings Professional (http://www.solidworks.com/) is an email-enabled communication tool that eases the review of 2D and 3D product design data across extended product development teams.

Despite the investment made in the last years, both in research and in industrial applications, the global market still lacks in collaboration tools, capable of providing VR techniques with the possibility of product design evaluation. Most collaborative tools are more related to a PLM and less to shared VEs. Thus, the development of a lightweight collaborative VE, supporting the validation and dissemination of product designs as well as the immersive interaction of multiple users with the virtual prototypes, comprises the goals of this research work.

3.3 Building the Shared VE

The widespread commercial VR software tool Division MockUp2000i2 (dV/MockUp), which is provided by PTC (http://www.ptc.com/), was used as a basis of building up the distributed and collaborative VE of the present work. The dV/MockUp is a high performance digital mock-up tool used for visualizing, analysing, and interacting with 3D CAD models in real-time, in an immersive and interactive way, by providing functionalities for geometry input and graphics rendering, interfaces to VR peripheral devises, and digital humans (mannequins) library. The software tool includes:

- a Database, which stores the entities of the virtual world together with their attributes
- Actors that manipulate the entities present in the Database and constitute the VR engine of the platform
- a Core Application, which provides functionality for loading, processing and saving the objects of the virtual world, and
- a Virtual Product Manager that provides the user with a desktop GUI for the control of the VE.

The tool is event-based, allowing users to create and edit real-time behaviors, constraints, animations and part assembly/disassembly sequences. Ergonomics evaluation of a product's design can also be performed into the VE, by using Division Safework mannequin tool, which is added-on to the dV/MockUp.

The backbone of the proposed framework is the functionalities that have been implemented into the pilot VE, which enable distributed users to visualize, simulate, modify and analyze (in terms of ergonomics) the virtual prototype (product), during a collaborative design evaluation scenario. The users are able to create new VE as well as to open and modify the existing ones. All the required materials for the synthesis of the VE (geometries, materials, textures, *etc.*) should be stored locally in each distributed station before joining a collaborative session. All collaborative distributed users can work simultaneously on the same environment, through a master-client interface, either in desktop mode (Figure 3.1) or in immersive mode, by using VR peripheral devices.

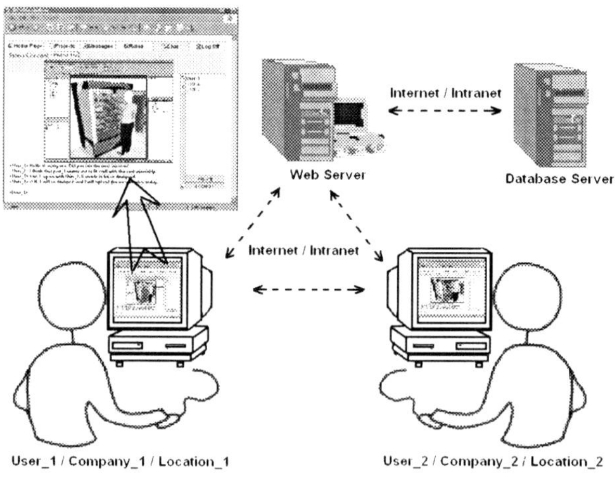

Figure 3.1. Collaborative design using the proposed shared VE

3.4 Virtual Environment Functionality

The key features of the shared VE have been implemented so as for the requirements of a typical industrial virtual collaborative scenario to be covered. These functions have been implemented into the dV/MockUp to allow the visualization and functional simulation of products as well as the users' immersion and interaction within the VE. The basic functions that have been implemented in the VE are described in the following sections.

3.4.1 Virtual Prototyping Function

In order to create a realistic VE, several functions, related to the appearance of the product, as well as to its environment, have been implemented in order to enable users to change the material or the texture of a part, the transparency level of a part or a sub-assembly, or even the lighting of the environment (Figures 3.2 and 3.3).

Figure 3.2. Visualization of the virtual prototyping function related to the transition of one's part transparency level

Figure 3.3. Visualization of the virtual prototyping function related to the mutation of the environmental lighting

3.4.2 Behavioral Simulation Function

The behavioral simulation controls the functional characteristics of the virtual systems, involved in the process performance. Based on the Event/Action mechanism of dV/MockUp, developers can model complex behaviors in the VE (assembly joint constraints, part movement restrictions, *etc.*), in order for the virtual objects to 'behave' in a real-life like manner. The Event/Action mechanism is the dV/MockUp's way of modeling the real world. When an event, such a collision occurs, the actions defined by the user, can be spawned. An example of modeling the real-life functionality of the refrigerator's door is presented in Figure 3.4. The result of this modeling is the opening (or closing) of the door, once the user has picked the handle of the door.

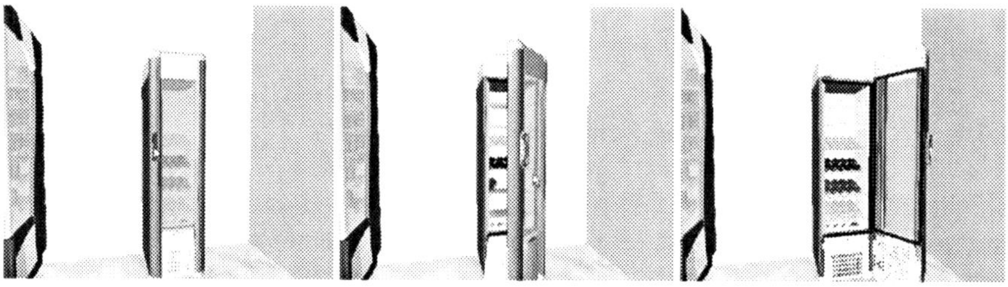

Figure 3.4. Visualization of the behavioral simulation function related to the assembly joint constraints

3.4.3 Assembly Support Function

This function allows for the accurate assembly execution within the VE. During an assembly process, the part to be assembled is released from the user's hand, so as to be assembled in its final position, as soon as a good positional and rotational orientation has been achieved (magnet concept). This orientation is very close to the exact final mounting position. The field of the 'magnet' can be adjusted to account for the various levels of fitting precision and is enabled while the part to be assembled is approaching its corresponding sub-assembly. A red transparent cube appears once the 'magnet' field has been enabled (Figure 3.5). The size of this cube determines the sensitivity of this function as well.

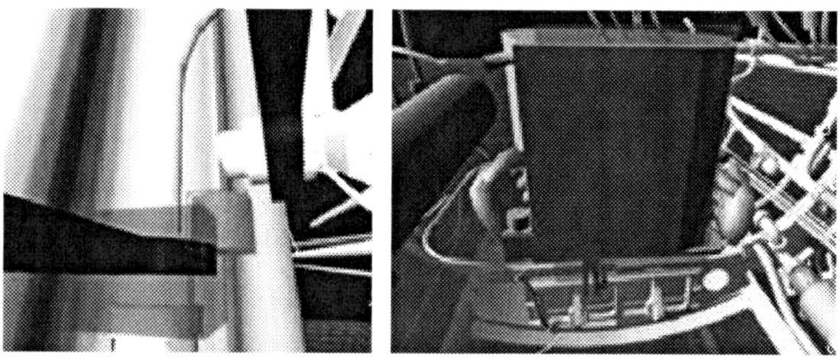

Figure 3.5. Assembly support function (magnet concept) in desktop and immersive mode

3.4.4 Collision Detection Function

Dynamic clash detection is provided within the simulation environment among static parts and either moving parts or the user's hands. In this way, visual and acoustic alerts enable the user to verify the feasibility of a process, in terms of reachability of picking and mounting locations and handlability of parts. Based on the collision detection function, an advanced mechanism has been implemented to support the manipulation of objects, enabling the immersive interaction with components and tools, in a way similar to that in the real world [16]. The case specific gesture modeling enables the realistic handling of objects in accordance with their shape and function (Figure 3.6).

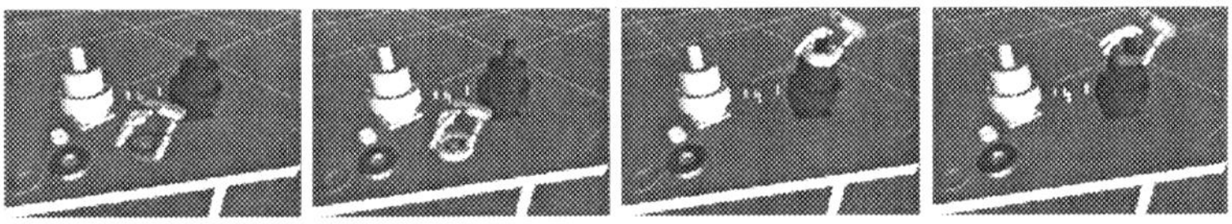

Figure 3.6. Visualization of the assembly support function (magnet concept)

3.5 Pilot Application

In order to demonstrate the benefits of incorporating VR technology in collaborative manufacturing, a pilot VE has been developed based on the needs and requirements of a manufacturing industry that produces commercial refrigerators. The aim of this pilot environment was to help a customer (mini-market owner) to decide, with the help of the product designer, which products should be the most suitable for his needs, having also taken into account the mini-market's layout. Thus, a virtual representation of the mini-market has been created, into which three different types of refrigerators were included (Figure 3.7). Several functionalities were implemented in this pilot VE, in order to support the collaboration between the designer and the customer, during the evaluation of several alternative product designs and layouts. A number of combinations of different colors and textures of these three refrigerator types, have been evaluated in order for the customer to make the final decision. Moreover, the refrigerators were evaluated both in terms of their capacity and their ergonomics (*i.e.*, reachability tests, kids/adults field of view, *etc.*), in order for the position of the goods (*i.e.*, refreshments) on the refrigerators' shelves to be decided (Figure 3.8). Another requirement of the customer to be enabled to test alternative designs of the refrigerators' door handle. This requirement was taken into account during the development of the pilot VE. Thus, the option of introducing several 3D objects to the VE by selecting them, from a virtual database, was also incorporated into the pilot VE. Finally, the functional simulation of the refrigerators' door opening/closing was of great importance during the evaluation of several layouts. Many alternative layouts were rejected due to the collision of one refrigerator's door, while being opened, with other contiguous space objects (*i.e.*, another refrigerator).

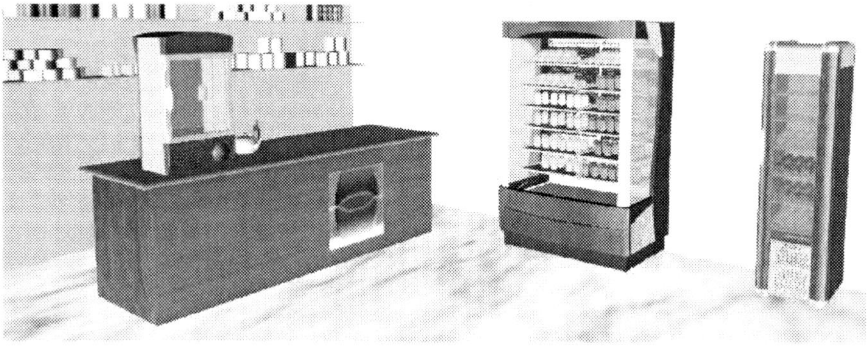

Figure 3.7. The mini-market layout, including the three different type of refrigerators

Several other collaborative sessions, of different scenarios, have also been performed in order for all the implemented functionalities of the collaborative VE to be evaluated. These scenarios fulfiled the major needs for collaboration in design phase of several types of users (*i.e.*, designers, managers, marketists, suppliers, customers, *etc.*). Immersion capability is also available for realistic human interaction.

Figure 3.8. Ergonomic analysis of the final product

During a multi-user collaborative session, each user has his/her own copy of the Graphical User Interface (GUI), which provides a rendered 3D view of the virtual product (Figure 3.1). All users can interact with the virtual product at any time, without any restrictions to the number of simultaneous interactions. Any change prerformed by a user is immediately visible by all the others. Real-time chat capability supports the users' communication during a collaborative session (Figure 3.9). Moreover, a user can be optionally represented by an animated digital figure, called avatar, in case of making use of VR peripheral devices. Any number of users can join a collaborative session by using Transmission Control Protocol/Internet Protocol (TCP/IP) over Local or Wide Area Network (LAN or WAN). In order for one to make use of the developed shared environment there is no enforcement on the use of specific Operational System or VR peripherals.

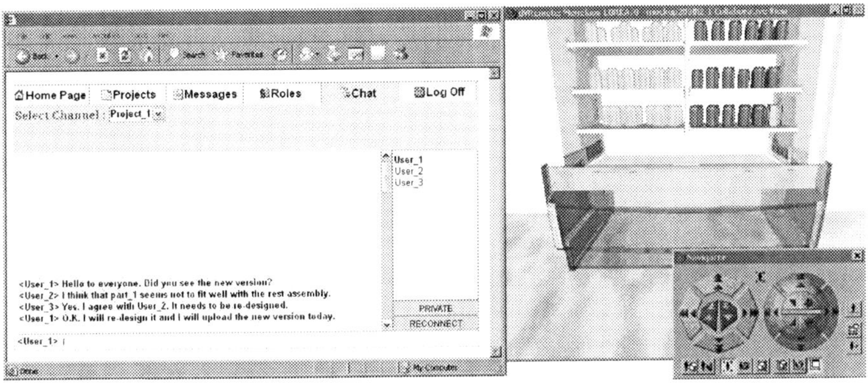

Figure 3.9. Real-time collaboration capability of the shared VE

The present shared VE provides an advanced environment in the network, as a common virtual design space, in which people can simultaneously work during the product life cycle. The developed pilot environment enables:

- The cooperation among distributed designers and manufactures during the refrigerator's designing stage.
- The real-time multi-user interaction in the same virtual prototype/design
- The effective and efficient sharing and evaluation of design and manufacturing data through internet.
- The ergonomic evaluation of the products with the use of mannequins that represent different user populations.
- Activities in many a session within a common virtual space (*e.g.*, conceptual design, assembly execution, ergonomic analysis, *etc.*).
- The advanced product demonstration by using VR (a virtual showroom).

3.6 Conclusions and Future Research

A shared virtual collaborative environment for the evaluation of a manufacturing product design has been developed and presented in this chapter. Providing a multi-user real-time collaboration as well as VR-based product verification, this environment could be used as an efficient tool by designers, engineers and managers.

The shared VE allows multiple users to work in a collaborative and distributed way, by decreasing considerably the time required for the designing phase to be completed. This work focuses on improving team productivity, providing the infrastructure necessary to make the engineering teams efficient, even if they are dispersed over different sites, without changing the existing design environment. The benefits of using the proposed shared VE include:

- Multi-user visualization, immersion and interaction.
- Real-time collaboration on the same virtual design.
- Simultaneous review and maintenance of alternative virtual designs.
- Evaluation of ergonomics by using digital human simulation.

Future research will focus on elaborating current functionality of the VE with tools for collaborative decision making support. The aim is to develop functionality for a systematic quantified assessment of alternative designs and plans. Thus, metrics and techniques for getting measures out of collaborative product simulation sessions should be identified. Intelligent reasoning, based on the quantified performance measures, the decision policy and the estimation weights, will be provided as output to support decision making, by taking under consideration the special conditions and requirements of team work. Thus, any future work will focus mainly on two directions:

- Quantified validation of design / plans.
- Intelligent reasoning on alternative solutions.

In this way, the VE for collaborative design, will be capable not only of team reviewing designs and plans, but also of "suggesting" proper solutions on design or re-conceptualization problems, based on collaborative interactions and testing, which can happen in VEs. Moreover, in terms of user-to-system interactions, Augmented Reality interfaces will be further investigated in order for the potential of running the simulation "on-top" of already set-up real working environments to be identified.

3.7 Acknowledgements

This work was partially supported by the Greek National research project e-MERIT/EB-26, funded by the General Secretariat of Research and Technology (GSRT).

3.8 References

[1] Chryssolouris, G., 2006, *Manufacturing Systems: Theory and Practice*, 2nd Edition. (Springer-Verlag: New York).

[2] Park, H. and Cutkosky, M. R., 1999, "Framework for modeling dependencies in collaborative engineering processes," *Research in Engineering Design - Theory, Applications, and Concurrent Engineering*, 11, pp. 84–102.

[3] Rezayat, M., 2000, "The enterprise - Web portal for life cycle support," *Computer Aided Design*, 32(2), pp. 85–96.

[4] Chryssolouris, G., Mavrikios, D., Fragos, D., Karabatsou, V. and Pistiolis, K., 2002, "A novel virtual experimentation approach to planning and training for manufacturing processes-the virtual machine shop," *International Journal of Computer Integrated Manufacturing*, 15(3), pp. 214–221.

[5] Chryssolouris, G., Karabatsou, V. and Kapetanaki, G., 2001, "Virtual Reality and Human Simulation for Manufacturing," *Proceedings of the 34th International CIRP Seminar on Manufacturing Systems*, Athens, Greece, pp. 393–398.

[6] Chryssolouris, G., Mavrikios, D., Fragos, D., Karabatsou, V. and Alexopoulos, K., 2004, "A hybrid approach to the verification and analysis of assembly and maintenance processes using virtual reality and digital mannequin technologies," *Virtual Reality and Augmented Reality Applications in Manufacturing*, Nee A. Y. C. and Ong S. K. (Eds.), Springer-Verlag, London, pp. 97–110.

[7] Chryssolouris, G., Mavrikios, D., Fragos, D. and Karabatsou, V., 2004, "Verification of human factors in manufacturing process design. A virtual experimentation approach," *Methods and Tools for Co-operative and Integrated Design*, Tichkiewitch S. and Brissaud D. (Eds.), Kluwer Academic Publishers, pp. 463–474.

[8] Yang, H. and Xue, D., 2003, "Recent research on developing Web-based manufacturing systems: a review," *International Journal of Product Research*, 41(15), pp. 3601–3629.

[9] Craig, D. L. and Craig, Z., 2002, "Support for collaborative design reasoning in shared virtual spaces," *Automation in Construction*, 11(2), pp. 249–259.

[10] Shyamsundar, N. and Gadh, R., 2001, "Internet-based collaborative product design with assembly features and virtual designspaces," *Computer Aided Design*, 33, pp. 637–651.

[11] Shyamsundar, N. and Gadh, R., 2002, "Collaborative virtual prototyping of product assemblies over the Internet," *Computer Aided Design*, 34, pp. 755–768.

[12] Arangarasan, R. and Gadh, R., 2000, "Geometric modeling and collaborative design in a multi-modal multi-sensory virtual environment," *Proceeding of the ASME 2000 Design Engineering Technical Conferences and Computers and Information in Engineering Conference*, Baltimore, Maryland, pp. 10–13.

[13] Kan, H. Y., Duffy, V. G. and Su, C. J., 2001, "An internet virtual reality collaborative environment for effective product design," *Computers in Industry*, 45, pp. 197–213.

[14] Xu, X. W. and Liu, T., 2003, "A web-enabled PDM system in a collaborative design environment," *Robotics and Computer-Integrated Manufacturing*, 19(4), pp. 315–328.

[15] Noel, F. and Brissaud, D., 2003, "Dynamic data sharing in a collaborative design environment," *International Journal of Computer Integrated Manufacturing*, 16(7–8), pp. 546–556.

[16] Pappas, M., Fragos, D., Alexopoulos, K. and Karabatsou, V., 2003, "Development of a three-finger grasping technique on a VR glove," *Proceedings of the 2nd Virtual Concept Conference*, Biarritz, France, pp. 279–283.

4

A 'Plug-and-Play' Computing Environment for an Extended Enterprise

F. Mervyn

G. W. Woodruff School of Mechanical Engineering, Georgia Institute of Technology, USA

A. Senthil Kumar and A. Y. C. Nee

Department of Mechanical Engineering, National University of Singapore, Singapore

With the emergence of the extended enterprise where different companies are involved in the product development process, the successful implementation of collaborative product design and manufacturing across the extended enterprise has become a difficult task. This chapter deals with the issue of developing an integrated computing environment for facilitating collaborative product design and manufacturing across the extended enterprise. An application development framework is presented that is geared towards a 'plug-and-play' computing environment. The framework describes how design and manufacturing applications can be developed independently, yet be seamlessly integrated simply by plugging the application into common computing environments.

4.1 Introduction

Faced with a rapidly changing global environment, product development enterprises today are reformulating their strategies to be globally competitive. One strategy that enterprises have adopted is to concentrate on their core competencies and build closer relationships with their partners. The resulting organization of geographically distributed companies working together to realize a product is known as the extended enterprise and is the new unit of business competition [1]. Facilitating collaborative product design and manufacturing across an extended enterprise is a difficult task that requires various cultural and technical issues to be resolved. The aim of this chapter is to address one such technical issue – the development of an integrated computing environment to support the collaboration across an extended enterprise, by facilitating information exchange between product designers and manufacturing process designers, and coordinating their activities. The necessary information is critical to make rapid trade-off decisions and collaboratively arrive at the optimal design and manufacturing processes of the product.

The rest of this chapter is organized as follows. Section 4.2 discusses the related research in developing integrated computing environments. Section 4.3 proposes an approach to develop design and manufacturing applications that is geared towards a 'plug-and-play' capability. Section 4.4 presents an illustrative example and Section 4.5 concludes the chapter.

4.2 Related Research

An integrated computing environment enables collaborative product design and manufacturing by providing the necessary mechanisms for exchanging information and coordinating information flow.

Early efforts in developing integrated computing environments concentrated on the integration of the various standalone computer-aided systems used in the design and manufacture of a product. Standalone systems are applications in which the entire functionality of the application is hosted on a single computer. Cutkosky, *et al.*, [2] presented a notable work in this regard based on an agent approach. Agents were used to encapsulate the standalone applications and agent interaction was based on shared concepts and terminology for communicating knowledge across disciplines. Sriram, *et al.*, [3] proposed the use of the blackboard architecture for facilitating communication and coordination between different standalone computer-aided systems. The blackboard was implemented as an object oriented database. The use of a central repository as a product master model was another approach described by Hoffman and Joan-Arinyo [4] to create an integrated computing environment. The clients of the master model are domain-specific standalone applications that can deposit and retrieve information from the master model. The master model repository provides mechanisms for maintaining the consistency of the deposited information structures.

Another approach is the use of standard file formats such as STEP and IGES located at central databases. Roy and Kodkani [5] proposed the use of a translator to convert CAD models into VRML-based models, which can then be viewed over the WWW. The VRML models are stored in an existing product data repository. The translator resides on a main central server and can be accessed remotely by a designer. Xie, *et al.*, [6] developed an integrated CAD (Computer-Aided Design) / CAPP (Computer-Aided Process Planning) / CAM (Computer-Aided Manufacturing) system for sheet metal product development platform based on an information integration framework where the geometry of the product was represented in STEP files. The information integration framework was developed using Pro/INTRALINK.

As seen from the literature, various external communication and coordination mechanisms need to be developed to integrate the standalone computer-aided systems. Therefore, before a company collaborates with a new partner these interfaces to the external mechanisms have to be developed. Companies normally employ the services of systems integrators to develop the mechanisms and the required interfaces to the mechanisms. While this is a feasible solution, it is an expensive solution, as each pair of systems requires a dedicated solution [7]. Further it is time consuming and delays the effective exchange of information between new partners.

Another problem with the integration of standalone systems to support collaborative design and manufacturing is the loss of associated information under design changes. This problem can be illustrated with the following example. In this example, we assume two systems as part of an integrated computing environment, a CAD system and a CAPP system. The part shown in Figure 4.1(a) has been created using the CAD system and sent to the CAPP system as a STEP file. When the part is loaded into the CAPP system, it creates an internal representation of the model to carry out machining operations on the model. In this internal representation, each geometric entity is identified by a tag. The CAPP system then identifies the three faces with the face tags, 38, 42 and 52 as shown in Figure 4.1(a) as a machining feature and determines the tools required to machine the feature. If the product designer then changes the part as shown in Figure 4.1(b), the CAPP system has to retrieve a new STEP file. When the CAPP system loads the new STEP file, the system will not be able to recognize this as a modified part and will create new tags for the geometric entities of the altered part. As can be seen in Figure 4.1(b), the three faces of the machining feature are now referred to by the tags, 372, 456 and 516. This new reference to the geometric entities results in a need for the CAPP system to recognize the three faces again as a machining feature. In collaborative product design and manufacturing, various design changes occur to accommodate the requirements of the different domains involved in product development. Such a problem results in inefficient systems that need to restart their tasks each time a change occurs.

To solve the problems associated with integrating standalone systems, research efforts progressed towards developing distributed collaborative systems. As opposed to standalone applications, in distributed systems, the functionalities of the system are hosted on different computers. Several researchers have proposed developing distributed collaborative systems based on the use of a central geometric modeling server. Han and Requicha [8] discussed an approach that provides product and process design applications with a transparent access to diverse solid modelers located at a central server. A feature-based design system, an automatic feature recognizer and a graphics rendering system were developed around the modeling server. The central geometric modeling server stores the boundary representation model of a designed part. When a design change occurs, the design system communicates the change to the feature recognition system, which can then access the new data from the modeling server. Shyamsundar and Gadh [9-10] proposed a client-server based architecture for collaborative virtual prototyping of product assemblies over the Internet. A polygonized representation of the part was used for visualization and an Internet-centric, compact assembly representation was also developed. In their system, design changes are not automatically transmitted to users working on the model. However, assembly features are tagged and if a designer attempts to modify that face, the designer receives a warning. Bidarra, *et al.*, [11] developed a web-based collaborative feature modeling system known as webSPIFF. It is based on a client-server architecture where the server coordinates the collaborative session, maintains the shared model and makes use of a multiple-view feature modeling kernel [12]. The multiple-view feature modeling kernel provides different users with different views of the product model. All views are kept consistent by feature conversion.

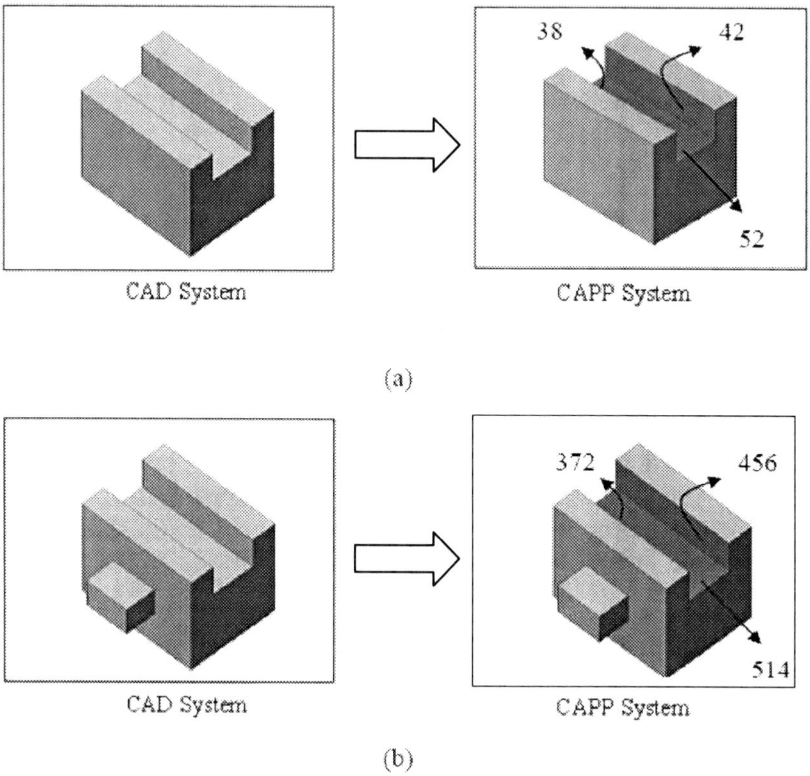

Figure 4.1. Loss of associated information under design changes

The efforts in developing distributed collaborative systems have solved several important problems. The use of Web-based or simple application clients to access functions hosted on a server removes the need for enterprises to maintain expensive standalone systems at their sites. This allows enterprises to collaborate with one another without purchasing compatible systems and without the need for customized communication and coordination mechanism to be developed. Users can just use a Web browser or download a simple application client to carry out their tasks, exchange information and coordinate their activities. Further, the use of a central geometric modeling kernel ensures that references to geometric entities are consistent under design changes, solving the problem of associated information loss under design changes. However, present efforts in developing collaborative systems have mainly concentrated on collaborative part and assembly modeling. Several important issues in an overall computing environment for collaborative product design and manufacturing have yet to be addressed. One such issue is the integration of new applications into the overall distributed collaborative computing environment. The problem to be addressed here is, "If a new application is developed, how can it be seamlessly integrated into the overall computing environment without developing new communication and coordination mechanisms?" This requires the development of collaborative systems that are geared towards 'plug-and-play' capability. The computing environment should allow advanced domain-specific applications to be developed independently, yet be integrated simply by plugging the application into computing environments. This chapter presents such an approach.

4.3 Application Develoment Framework

The proposed framework for developing an integrated computing environment for collaborative product design and manufacturing is based on the architecture as shown in Figure 4.2. Central to this architecture is the use of a common manufacturing application middleware. Middleware is systems software that resides between applications and the underlying operating systems, network protocols and hardware [13]. The essential role of middleware is to manage the complexity and heterogeneity of distributed infrastructures and thereby provide a simpler programming environment for distributed application developers [14].

Early efforts in middleware development dealt mainly with connectivity issues, *i.e.*, how programs on different computers can connect to one another. These middleware technologies that deal with connectivity are referred to as distribution middleware [13]. Examples of distribution middleware include OMG's CORBA (Common Object Request Broker Architecture), Sun's Java RMI (Remote Method Invocation) and Microsoft's DCOM (Distributed Component Object Model). Distribution middleware technologies are at a mature stage today. Middleware technologies have since progressed to dealing with other issues in developing distributed systems. One such group of middleware technologies deals with domain specific issues. These middleware technologies, referred to as domain specific middleware, concentrate on providing domain specific services that applications can access in a transparent and integrated manner. This chapter envisions that domain specific middleware technologies will be instrumental in the future development of product and process design applications.

The architecture shown in Figure 4.2 presents a paradigm where product and process design applications access various services provided by the manufacturing application middleware to exchange information and be coordinated in a seamless manner. The current implementation of the framework proposes the use of two middleware services, geometric modeling services and process data exchange services.

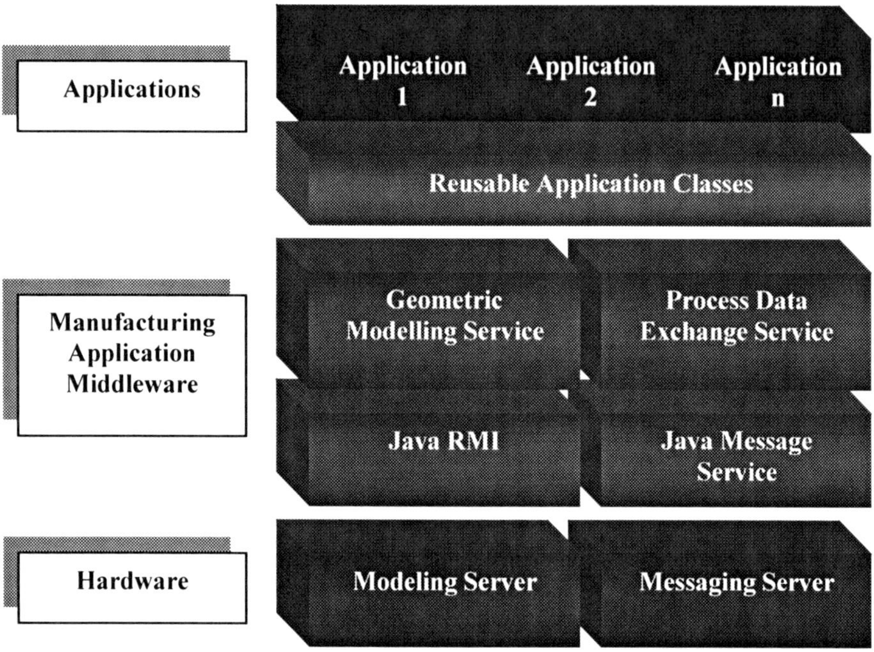

Figure 4.2. Architecture for developing an integrated computing environment for collaborative product design and manufacturing

Geometric modeling services refer to the ability to create and manipulate geometric models, and access geometric data. Geometric modeling services are proposed for three reasons. Firstly, many of the currently developed CAD, CAPP and CAM applications do not develop their own geometric modeling capabilities. Many of these applications are developed based on external geometric modeling kernels. These external geometric modeling kernels provide functionality to build, manipulate, view and interrogate geometric models. It is therefore sensible to provide geometric modeling services as a common service. Secondly, many applications use geometric modeling kernels only to extract necessary information from product data to carry out their own tasks. Providing the ability to access geometric data from a common service would remove the reliance of these applications on geometric modeling kernels just for information extraction. Thirdly, providing a common geometric modeling service where applications create and access geometric data provides a unique opportunity to manage the concurrent authoring and processing of geometric data by product and process design applications. In the proposed architecture, the geometric modeling services are deployed on a central geometric modeling server.

Process data is data generated by the process design applications. This includes feedback to upstream applications and data for downstream applications to carry out their tasks. A data exchange service that facilitates the exchange of process data ensuring applications receive data in a timely manner is an important service for effective collaboration. The process data exchange services are deployed on a central messaging server.

The product and process design applications that access the middleware services are not the conventional CAD, CAPP and CAM systems currently used by enterprises. These are new applications that are developed to conform to the middleware-based architecture. In order to facilitate the development of these applications, the framework provides various reusable application development classes. These reusable application development classes make application development easier by already implementing object-oriented classes for interfacing with the middleware services, storing of data and visualization of geometric models. The following sections discuss the implementation of the middleware services.

4.3.1 Geometric Modeling Middleware Services

The geometric modeling middleware services are hosted on a central geometric modeling server. The geometric modeling server, shown in Figure 4.3, was implemented in Java and consists of the following components: (i) Java RMI interfaces, (ii) Implementation classes, (iii) Java Native Interface (JNI), (iv) Parasolid Modeling Kernel and (v) Apache HTTP Server. Application clients access the services offered by the geometric modeling server through Java RMI. In the present system, two main RMI interfaces have been implemented: Modeling Functions and the Applications Relationship Manager (ARM). The Modeling Functions interface allows application clients to create and manipulate geometric models. The ARM interface allows applications to build relationships with the geometric models. The ARM serves as the mechanism for synchronising all the different product and process applications when a design change is made.

4.3.1.1 Modeling Functions

The Modeling Functions Interface declares the methods that application clients can invoke to make function calls to a geometric modeling kernel. In the developed system, the Parasolid modeling kernel has been utilized to perform the modeling operations. As the Parasolid modeling kernel is written in the C programming language, a Java Native Interface (JNI) is needed to utilize the modeling functions of Parasolid. The result of an application client invoking one of these methods is the creation or modification of a geometric model. Data of the created or modified geometric model is then written to a geometric data XML file and stored in the Apache HTTP server for application clients to access.

The details of the sequence of activities when any of the methods is invoked are as follows:

1. An application client invokes one of the methods of the Modeling Functions RMI interface.
2. The Modeling Functions Implementation classes invoke the necessary Parasolid functions, resulting in the creation or modification of a geometric model.

3. Parasolid generates the necessary information to describe the geometric model based on a boundary representation. A boundary representation describes a geometric model by defining the model's boundary as a set of geometric entities including faces, edges and vertices. An example of a boundary representation is shown in Figure 4.4. The geometric model and the constituent geometric entities are identified by tags.
4. As the boundary representation data is not sufficient for application clients to view a solid model of the created part, the Modeling Function Implementation classes invoke the Parasolid function to tessellate the model into triangles. The tessellated triangles can then be rendered on the application client's screen to provide a solid view of the geometric model. An example of a tessellated model is shown in Figure 4.5.
5. The information on the tessellated triangles and the geometric entities are then written to a Geometric Data XML file and stored in the Apache HTTP server. Application clients can then access this data easily from the Apache HTTP server, visualize the geometric model and carry out further operations. Section 4.3.1.2 discusses the geometric data XML file in greater detail.

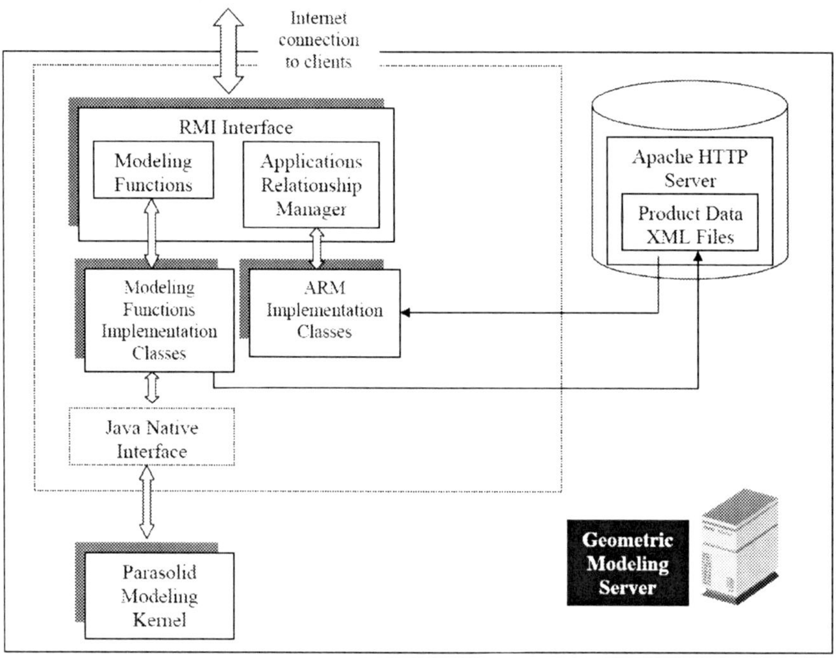

Figure 4.3. Architecture of geometric modeling server

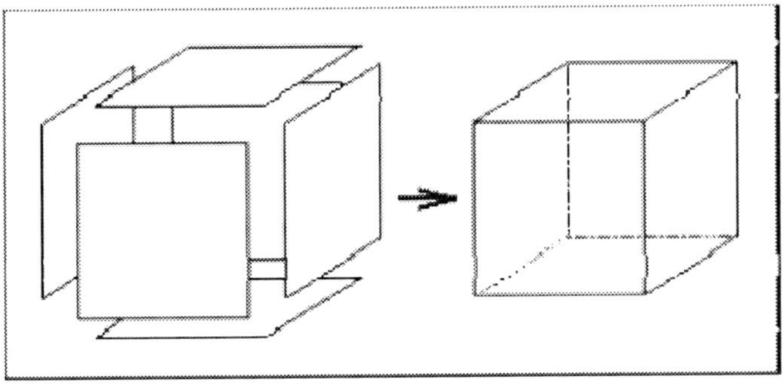

Figure 4.4. A cube represented by its boundary

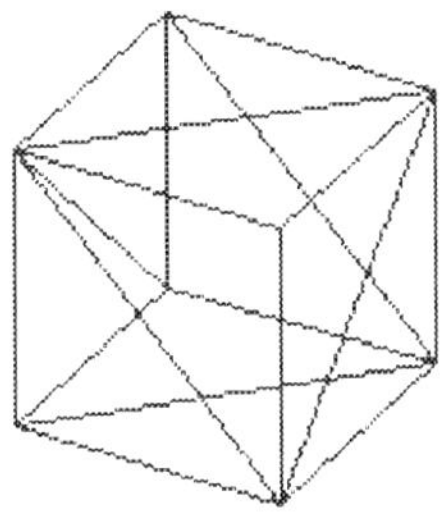

Figure 4.5. Example of a tessellated model

4.3.1.2 Geometric Data XML File

XML is an industry standard markup language used for representing data in a platform independent manner. When representing data using XML, a Document Type Definition (DTD) has to be specified first. This would govern the data structure contained in the XML file. The structure of the DTD of the Geometric Data XML file is shown in Figure 4.6.

Tags in XML follow a hierarchical structure. The root tag of an XML file is always <DOCUMENT>. In the geometric data DTD, each body, identified by a <BODYTAG>, is divided into faces. A <FACETAG> is present to identify the various faces of the body. <FACETYPE> provides information on the type of the face, for example, cylindrical, plane and spherical. <SNAPPOINT> refers to the vertices of each face. Each face is further divided into elemental triangles known as facets. The <FACET> tag contains the coordinates of the vertices of each triangle.

Figure 4.7 shows a solid model of a cube and a portion of the corresponding Geometric Data XML file. From the data, it can be seen that the <BODYTAG> of the part is 119. The highlighted face has a <FACETAG> of 179 and a <FACETYPE> of plane. The face has been divided into two facets and the corresponding vertices of the first facet can be seen in the figure. In Ref [15], we provide an alternative representation where facet information is compressed using the Edgebreaker algorithm [16].

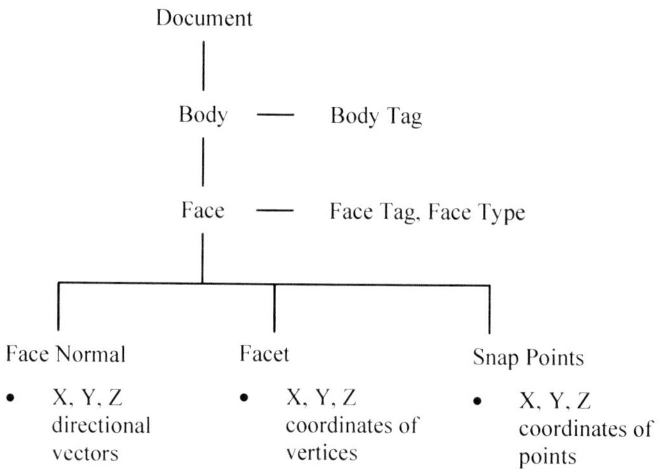

Figure 4.6. Document Type Definition (DTD) of the geometric data XML file

4.3.1.3 Application Relationship Manager (ARM)

The ARM serves two main roles. Firstly, it serves as the mechanism for synchronization by propagating design changes to all affected applications. Secondly, it aids collaborative decision-making by creating relationships between the requirements of the different domains involved in collaborative product design and manufacturing. The ARM works by allowing downstream applications to create relationships with the geometric entities of a geometric model. For example, if an assembly modeling system uses a face of the geometric model as a mating face with another model, it creates a relationship with that face. The use of geometric entities provides a general means to link downstream decisions to the product model. When a design change is made, all applications that have created relationships with the model are notified. In this way, the ARM serves as the synchronisation mechanism. However, before a design change can be implemented, the ARM allows the part designer to determine which applications will be affected by implementing the change simply by viewing all the different applications that created relationships with the model. In an interactive environment, product and process designers can discuss the design change before it is implemented. For example, if the product designer wants to make a change to a face, which the assembly modeling system created a relationship with, he/she can discuss with the engineer who created the relationship without involving the other people. In an automated environment, the downstream applications that are affected can be notified of the change and the affected application can then deal with the change automatically. If the change cannot be dealt with, then the proposed change is not acceptable. In this way, the ARM aids collaborative decision-making.

In order to facilitate the creation of relationships and notification of changes, the ARM RMI Interface declares the following methods.

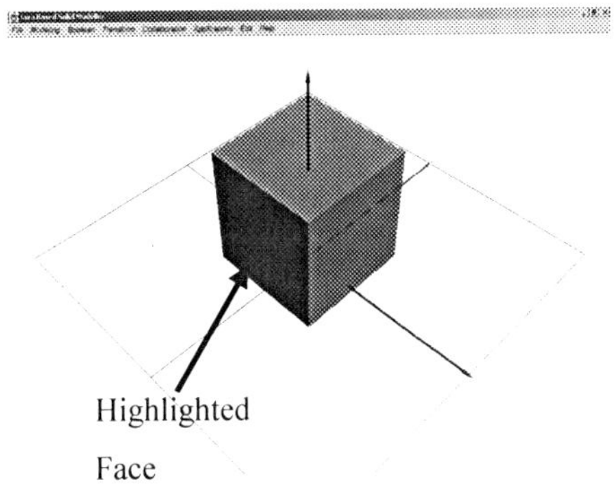

<BODY>
<BODYTAG>119</BODYTAG>
- <FACE>
<FACETAG>179</FACETAG>
<FACETYPE>PLANE</FACETYPE>
+ <NORMAL>
+ <SNAPPOINT>
+ <SNAPPOINT>
+ <SNAPPOINT>
+ <SNAPPOINT>
- <FACET>
<X1>-0.25</X1>
<Y1>0.25</Y1>
<Z1>0.5</Z1>
<X2>-0.25</X2>
<Y2>-0.25</Y2>
<Z2>0.5</Z2>
<X3>0.25</X3>
<Y3>-0.25</Y3>
<Z3>0.5</Z3>
</FACET>
+ <FACET>
</FACE>

Figure 4.7. Example of a geometric data XML file

- *public void deposit_model (int bodytag)*: This method is to be called by an application to make a geometric model ready for creating relationships. The method is normally called by a design client. The input to the method is the body tag of the geometric model. The method subsequently retrieves data from the appropriate geometric data XML file and updates data structures created for managing relationships. The data structure for managing relationships was implemented in an object-oriented manner and is as shown in Figure 4.8.

  The ARM_Models class contains a list of the various geometric models that relationships can be built on. The information on the geometric models is stored in a Geometric_Model class. In this class, each model is identified by its body tag. The Geometric_Model class in turn contains a list of the different faces that make up the geometric model. It is on these faces that relationships are created. The information on each face is stored in a Face class. The Face class contains information of the tag used to identify the face and the status of the face. The status of the face could either be "changed" or "unchanged". The Face class also contains information on the different relationships that have been created on that face. The information on relationship is stored in a Client_Relationship class. The Client_Relationship class stores information on the URL of the client that made the relationship, the client type, the type of relationship and any comments that an application client would like other applications to take note of. The ARM uses the URL of the client to notify design changes. Restrictions are not made on application clients to specify certain values for client type and type of relationship. Typical values of client type could be 'process planning client' or 'assembly modeling client'. Type of relationship refers to how the application relates to the face. For example, an assembly modeling client could describe the type of relationship as a 'mating face'.

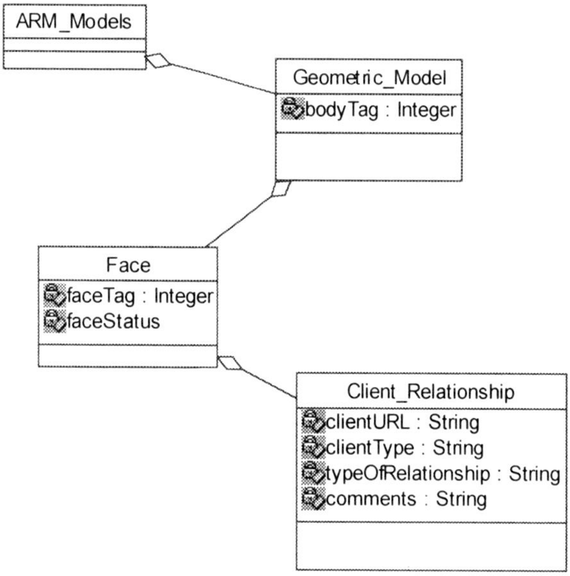

Figure 4.8. Data structure for managing relationships

- *public boolean create_relationship (RelationshipInfo info) and public boolean delete_relationship (RelationshipInfo info)*: These methods are called by application clients to create and delete relationships with faces of the geometric model. The input to this method is a class RelationshipInfo, which contains the information required to create a relationship. The RelationshipInfo class is shown in Figure 4.9.
When the method is called, it subsequently checks if the model has been deposited for creating relationships. If so, it will update the ARM data structure for managing relationships with the information from the RelationshipInfo class. If a relationship is created successfully, the method returns a TRUE Boolean value to the calling method. If a relationship could not be made, it returns FALSE.

- *public RelationshipInfo[ ] query_relationship (int facetag):* This method allows an application client to query information on the client relationships that have been made on a face. This will allow application clients to know which domains will be affected by a design change on a face. In the comments attribute in the RelationshipInfo class, clients could restrict other domains from suggesting changes to be made to that face. The method returns an array of RelationshipInfo objects to the calling method.
- public void transmit_design_change (int bodytag): This method is called when a change is made to the geometric model. The method subsequently determines the faces that have been affected and notifies clients that created relationships with affected faces.

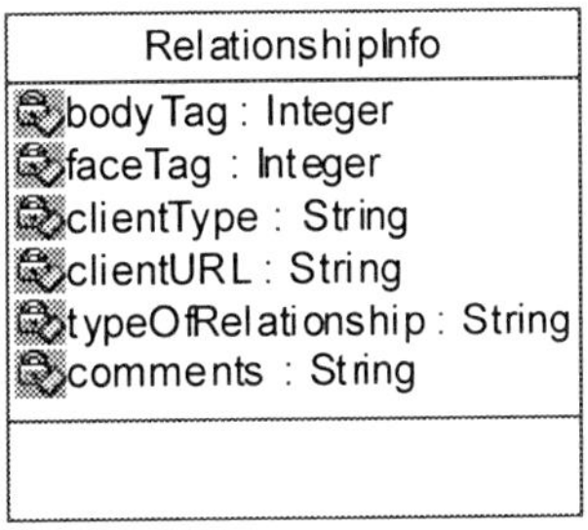

Figure 4.9. RelationshipInfo class

4.3.2 Process Data Exchange Middleware Services

The key role of the process data exchange middleware services is to facilitate the exchange of data from a process design application to all related applications. While the geometric modeling middleware services integrate product design with downstream domains, the process data exchange middleware services aim to integrate the different process design domains. An example would be the ability for a process planning system to send information to a shop floor execution system. The process data exchange middleware services also allow process design applications to send feedback information to the product design domain.

In the current system, a messaging approach has been adopted for the exchange of process data. Messages provide a dynamic means to exchange information. When a process design application completes its tasks and generates the necessary information, it can immediately send the information to the related applications. This provides a triggering mechanism for other applications to react to this information, either by beginning the application's tasks or by evaluating previous decisions based on the new information. Further, through the use of messages, only necessary information has to be sent to the related applications. This provides a means to decouple private data from data to be shared. The process data exchange services were developed based on the Java Message Service (JMS) specification. The JMS specification allows Java applications to create, send, receive and read messages. JMS prescribes a set of rules and semantics that govern messaging, including a programming model, a message structure and an Application

Programming Interface (API). In the JMS programming model, JMS clients exchange messages through the use of a JMS message service. Clients that produce messages send the messages to the message service, which then sends the message to the message consumer. Although JMS describes a message structure, the message structure adopted in this work is based on the Simple Object Access Protocol (SOAP). SOAP is an XML-based protocol that facilitates the exchange of structured data. The information from process design applications is therefore to be structured using XML files. An example of fixture design information models being exchanged using this approach is presented in [17].

The framework proposed in this section also provides classes at the client end to deal with the incoming messages. These are implemented as part of the reusable application development classes and will be discussed in the next section.

4.3.3 Reusable Application Classes

The reusable application development classes facilitate the development of applications that conform to the overall architecture proposed in the framework by providing the necessary interfaces to communicate with the manufacturing application middleware services. Application developers develop their applications using these classes as a basis. In the present implementation, three groups of reusable Java classes have been developed: (i) XML data parsing and storage classes (ii) Geometric model visualization classes and (iii) Middleware interface classes.

There are two groups of XML data parsing and storage classes. One group deals with parsing the Geometric Data XML file stored in the Apache HTTP server of the geometric modeling middleware services and storing the parsed data in developed data structures. The other deals with parsing incoming XML messages from the process data exchange middleware services and storing the data in developed data structures. Application developers use these classes to retrieve and store data in their applications. They can then develop the application logic using these data structures.

The geometric model visualization classes are used to visualize the geometric model using Java 3D. The middleware interface classes facilitate communication with the geometric modeling server and the messaging server. The necessary classes for invoking the Modeling Functions remote methods and the ARM remote methods are provided for in this group of classes.

4.4 Illustrative Case Study

This section describes how the proposed framework achieves the 'plug-and-play' capability for collaborative product design and manufacturing based on a scenario where several companies collaborate to design and manufacture a product. Four companies are involved in this scenario. Company A produces product A, which is made up of three parts. Company A designs Part 1, but outsources the manufacturing to Company B. Company A purchases Part 2 from Company C and Part 3 from Company D. This scenario is depicted in Figure 4.10.

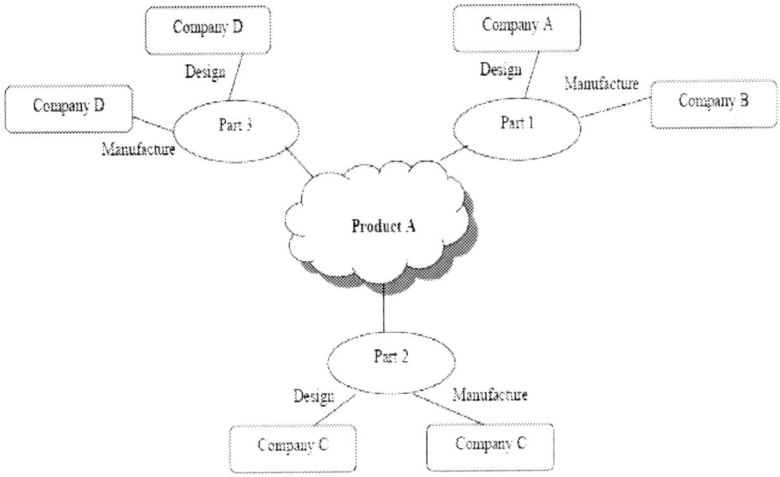

Figure 4.10. Example Scenario

In such a scenario, if every company creates its own legacy system for product and process design, huge compatibility problems would arise. Synchronised integration of activities would be a difficult task.

However, using the proposed framework, it is shown how companies can develop their own customised applications, yet exchange information seamlessly with the other companies. It is assumed that all companies have developed applications based on the reusable application development classes and the logic of the applications is also based on the middleware services. In this scenario, company A has created a product design client [18], company B a fixture design client [19] and companies C and D have created assembly evaluation clients [20]. The overall integrated computing environment for the design of Part 1 and the required manufacturing processes is shown in Figure 4.11. Company A, as the designer of Part 1, hosts the geometric modeling server and the messaging server at its site.

As the product designer designs Part 1 using the developed product design client, the evolving data of the geometric model is written to the Geometric Data XML file and stored in the Apache HTTP server. When the product design is complete, the product designer deposits the model in the ARM. The designed part and the corresponding information model set up for relationship management are shown in Figure 4.12.

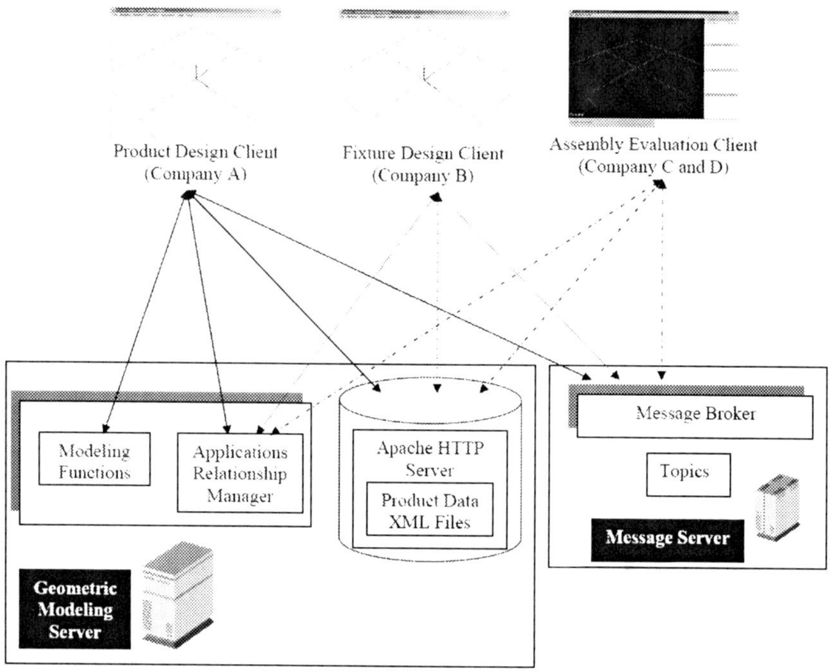

Figure 4.11. Integrated computing environment for collaborative product design and manufacturing

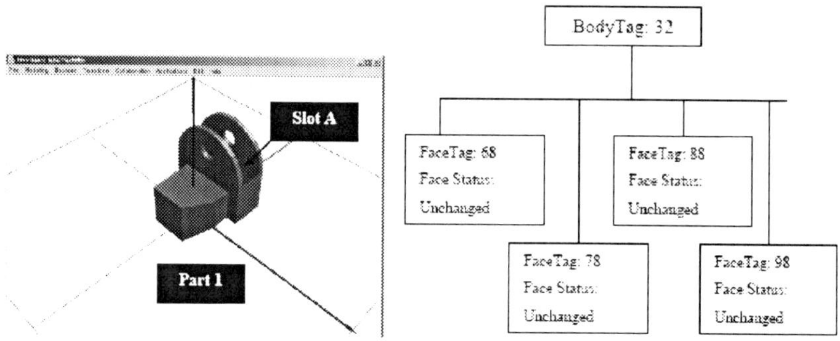

Figure 4.12. Designed part and corresponding information model

Company B manufactures Part 1 by first casting and then machining the different features. The fixture designer of Company B can now easily access the geometric model data of Part 1 by using the developed fixture design client. The fixture design client can be plugged into the geometric modeling server simply by providing the URL for the Apache HTTP server and the name of the part. The reusable application development classes would then obtain the geometric model data and visualize the part. If Company B works with another company, all that is needed to plug into their geometric modeling server is to specify the new URL for the Apache HTTP server. This allows different companies to work with one another without purchasing systems from the same vendor and building customised point-to-point communication mechanisms. However, it should be noted that security mechanisms have to be in place to prevent unauthorised usage of the data. Security mechanisms are beyond the scope of this chapter. Figure 4.13 shows the fixture design application being used to design a fixture. The fixture designer decides to use faces with the tags '68', '78' and '88' as locating faces and creates relationships with these faces. An example relationship is as follows:

Client Type = "Fixture Design Client";
Type of Relationship = "Locating Face";
Comments = "Face used to locate workpiece";

The fixture design application can also generate the necessary information to provide feedback to the product designer. The fixture design application can be plugged into the messaging server by simply providing the URL of the messaging server. The information can then be sent to the product designer who can then take the necessary action to deal with the incoming message.

Concurrently, Company C can also access the geometric model data of Part 1 to check the assemblability of Part 1 with Part 2 based on their assembly evaluation application. It is assumed that Company C deems the assembly as feasible and creates relationships with the geometric model of Part 1. The relationship is created with the following information:

Client Type = "Assembly Evaluation Client";
Type of Association = "Assembly";
Comments = "Face part of a feature used for assembly. Do not change";

Company D also loads Part 1 into their assembly evaluation client (Figure 14) to check the assembly of Part 3 with Part 1. It deems that assembly is not feasible and requires a slot to be included as shown in Figure 4.15. Company D can immediately determine which companies or domains would be affected if the change is suggested by querying the ARM for a list of relationships that were created with the affected faces. In this case, the fixture design of Company B would be affected. We assume that the design change is critical and Company A makes the change to the product model. The ARM then determines all clients that made relationships with the affected faces and propagates the change. The affected applications can then retrieve the geometric data of the modified part and deal with the changes. A point to note there is that the tags used to reference the geometric entities are consistent throughout the modifications. Necessary action can then be taken by the applications to deal with the change. An example of the fixture design system adaptively dealing with design changes can be found in [21-23].

In this scenario, it was assumed that the application clients were developed by the companies. These applications could also have been developed by commercial vendors.

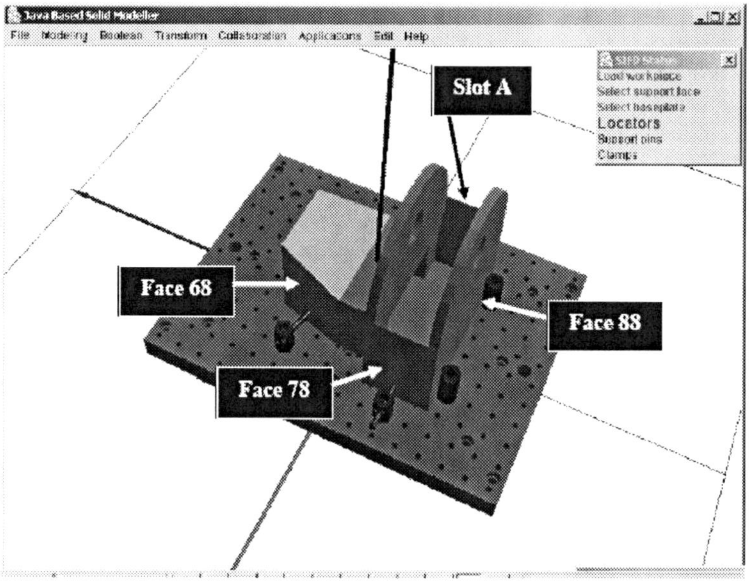

Figure 4.13. Fixture design client of Company B

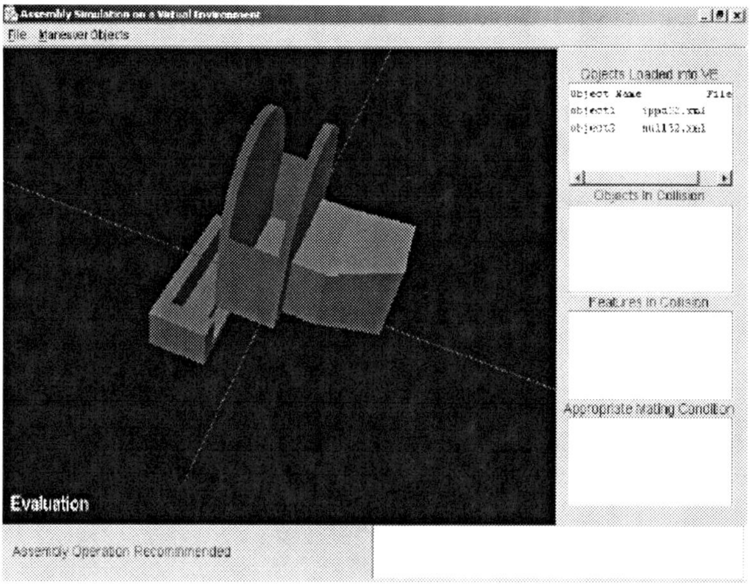

Figure 4.14. Assembly evaluation client of Company D

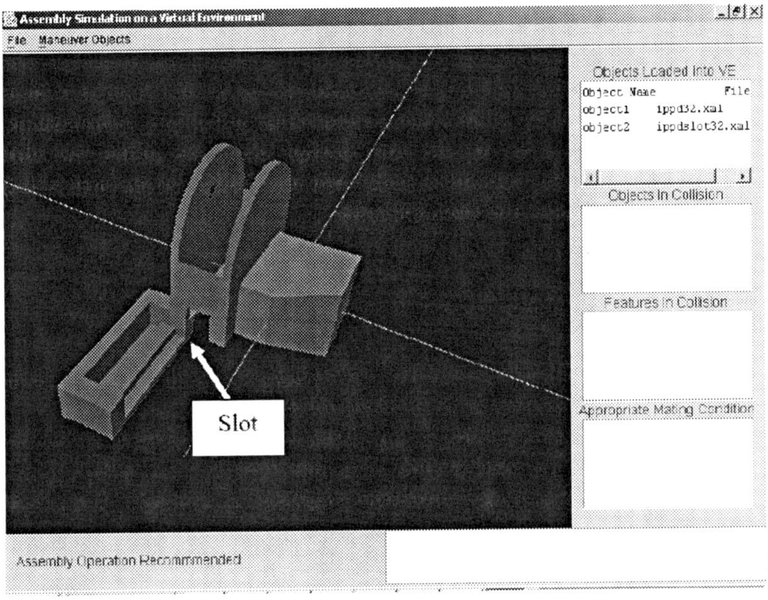

Figure 4.15. Company D's suggestion to include slot A

4.5 Conclusions

In today's business environment where companies collaborate with an array of different partners, developing applications to be integrated is a tall order, as the required interfaces with other applications are not known during the development of the application. An attempt has been made in this chapter to solve this problem by developing a 'plug-and-play' computing environment for collaborative design and manufacturing. An application development framework has been proposed to achieve this goal. Applications developed using the framework can be plugged into common computing environments and seamlessly exchange information. The framework is based on the use of a common manufacturing application middleware. The design of the middleware has further solved important problems faced in the development of integrated computing environments for collaborative design and manufacturing.

Firstly, it has solved the problem of losing associated information under design changes when standard file formats are used as an information exchange mechanism. The loss of associated information under design changes is a big hindrance to product and process design applications concurrently performing activities. In the proposed middleware approach, all applications have a consistent reference to geometric entities as data is obtained from a central modeling server. This consistent reference to geometric entities will allow applications to deal with changes in an intelligent manner without having to redo the process design.

Another problem in implementing collaborative design and manufacturing is how to consider the different requirements of the different domains involved in product development during the design of a product. The ARM has proposed a solution to this problem by allowing applications to create relationships with the geometric model and notifying all affected applications when a change is made. As the different applications carry out tasks concurrently and create relationships with the model, it becomes easy to see how design decisions affect the different domains. Through this way a product design can be made to be optimal taking into account the requirements of different domains.

4.6 References

[1] Dyer, J. H., 2000, *Collaborative Advantage: Winning Through Extended Enterprise Supplier Networks*, Oxford University Press.

[2] Cutkosky, M. R., Engelmore, R. S., Fikes, R. E., Genesereth, M. R., Gruber, T. R., Mark, W. S., Tenenbaum, J.M. and Weber, J.C., 1993, "PACT: an experiment in integrating concurrent engineering systems," *Computer*, 26, pp. 28–37.

[3] Sriram, D., Logcher, R., Wong, A. and Ahmed, S., 1990, "An object-oriented framework for collaborative engineering design," *Computer-Aided Cooperative Product Development*, (Edited by Sriram, D., Logcher, R. and Fukuda, S.), pp. 51–92.

[4] Hoffman, C. M. and Joan-Arinyo, R., 1998, "CAD and the product master model," *Computer Aided Design*, 30, pp. 905–919.

[5] Roy, U. and Kodkani, S. S., 1999, "Product modeling within the framework of the World Wide Web," *IIE Transactions*, 31, pp. 667–677.

[6] Xie, S. Q., Tu, P. L., Aitchison, D., Dunlop, R. and Zhou, Z. D., 2001, "A WWW-based integrated product development platform for sheet metal parts intelligent concurrent design and manufacturing," *International Journal of Production Research*, 39, pp. 3829–3852.

[7] Ray, S. R. and Jones, A. T., 2003, "Manufacturing interoperability," *Proceedings of the International Conference on Concurrent Engineering: Research and Applications*, Portugal, July.

[8] Han, J. H. and Requicha, A. A. G., 1998, "Modeler-independent feature recognition in a distributed environment," *Computer Aided Design*, 30, pp. 453–463.

[9] Shyamsundar, N. and Gadh, R., 2001, "Internet-based collaborative product design with assembly features and virtual design spaces," *Computer Aided Design*, 33, pp. 637–651.

[10] Shyamsundar, N. and Gadh, R., 2002, "Collaborative virtual prototyping of product assemblies over the Internet," *Computer Aided Design*, 34, pp. 755–768.

[11] Bidarra, R., van den Berg, E. and Bronsvoort, W.F, 2001, "Collaborative Modeling with Features," *CD-ROM Proceedings of the ASME Computers and Information in Engineering Conference*, 9-12 September, Pittsburgh, PA, ASME, N.Y.

[12] de Kraker, K. J., Dohmen, M. and Bronsvoort, W. F., 1997, "Maintaining multiple views in feature modeling," *Proceedings of the 4th Symposium on Solid Modeling and Applications*, ACM Press, pp. 123–130.

[13] Schantz, R. E. and Schmidt, D. C., 2001, "Middleware for distributed systems: evolving the common structure for network-centric applications," *Encyclopedia of Software Engineering*, (Edited by Marciniak, J. and Telecki, G.), Wiley and Sons.

[14] Campbell, A. T., Coulson, G. and Kounavis, M., 1999, "Managing complexity: middleware explained," *IEEE IT Professional*, October.

[15] Mervyn, F., Senthil kumar, A., Bok, S. H., Nee, A. Y. C., 2004, "Developing distributed applications for integrated product and process design," *Computer Aided Design*, 36, pp. 679–689.

[16] Rossignac, J., 1999, "Edgebreaker: Connectivity compression for triangle meshes," *IEEE Transactions on Visualization and Computer Graphics*, 5(1), pp. 47–61.

[17] Mervyn, F., Senthil kumar, A., Nee, A. Y. C., "Fixture design information support for integrated product and process design," to be appeared in *International Journal of Production Research.*

[18] Ratnapu, K. K., 2001, *Web Based CAD System*, M.Eng. Thesis, National University of Singapore.

[19] Mervyn, F., Senthil kumar, A., Bok, S. H. and Nee, A. Y. C., 2003, "Development of an Internet-enabled interactive fixture design system," *Computer Aided Design*, 35, pp. 945–957.

[20] Mervyn, F., 2001, *Development of a Virtual Assembly Evaluation Environment*, B.Eng Thesis, National University of Singapore.

[21] Mervyn, F., Senthil kumar, A., Nee, A. Y. C., 2004, "Design change synchronization in a distributed environment for integrated product and process design," *Computer-Aided Design and Applications*, 1(1–4), pp. 43–53.

[22] Mervyn, F., Senthil kumar, A. and Nee, A. Y. C., 2005, "An adaptive modular fixture design system for integrated product and process design," *Proceedings of the 2005 IEEE Conference on Automation Science and Engineering*, August 1–2, Edmonton, Canada.

[23] Mervyn, F., 2004, *Integrated Product and Process Design across a Global Extended Enterprise: An Implementation in Fixture Design*, PhD Dissertation, Department of Mechanical Engineering, National University of Singapore.

5

Cooperative Design in Building Construction

Yuhua Luo

Department of Mathematics and Computer Science, University of Balearic Islands, Spain

The chapter describes a state-of-the-art information technology system for cooperative design in building construction. The system is a high level cooperative tool for an architectural design team without geographic limitation on their locations. It provides on-line, real time interaction between long distance participants for their multi-discipline design projects. The team members can edit, modify and verify the design on-line or off-line cooperatively. Advanced concurrent control guarantees the mutual exclusion in the cooperative design process. The system has the capability to handle the design in 3D down to any level of details for construction design projects.

5.1 Introduction

The design phase in construction process is critical for the quality and cost of the constructed building. A substantial amount of cost waste in building construction is due to the errors in the design. The design phase is a complex process shared by many specialists, such as architects, structural engineers, air conditioning engineers, energy supply designers, *etc*. Different specialists usually use different CAD tools that produce very different design results. The whole design process is an iteration of decomposing and integrating designs of different specialist teams at different levels of scales and details. This iteration process has a very high possibility of error occurance. It is extremely costly to correct the errors after going to the site construction operations. There were no sophisticated information technology tools that could support this iteration process to produce error free global designs cooperatively [19]. The computer supported cooperative work in many industrial areas is limited to cooperative visualization via communication networks where the user interaction is kept to minimum. The existing CSCW (Computer-Supported Collaborative Work) solutions are very far from what demands in the building construction design.

A cooperative working system to support this multiple design iteration became an obvious solution. Targeting at developing an innovative system towards a complete solution, a European project was launched. At In addition to supporting the multiple iterations in design, the system was developed to support many other cooperative working tasks towards a new way of working in AEC (Architecture, Engineering and Construction) industry.

To have a solid base for the system development, a deep investigation on the actual situation in the design phase in Portugal, Spain, Italy and the UK was performed. Some key lacking elements for cooperative design were identified. At the same time, a close insight into the field of 3D computer graphics, database technology, telecommunication, and computer supported cooperative work was carried out. This allowed the developers to see the possibility and at what degree the current ICT technology could provide a solution to the needs in AEC practice. A team of computer scientists, architects and engineers started to work closely together in the project. The strategy is to find a solution to bridge the gap between the current situation and the future way of working. They believed that not all the problems in this traditional industry could be solved at once. But certainly some key elements for the most critical cooperative working scenarios can be developed. This leads to the birth of a cooperative design system which is the first time attempt to provide a high level cooperative working tool for an architectural design team geographically spread. The developed system, for the first time, can support real time, long distance, multiple location simultaneous interaction cooperative work.

One of the major objectives of the system is to make early integration to explore design conflicts, errors at early stages long before the construction begins. Another objective is to let different CAD tools from all the disciplines talk together towards an error-free design. A neutral 3D data format [7]: VRML is selected for the system. It can accept designs from any architectural or engineering design tools that can output VRML. For user convenience, DXF or 3DS format are also accepted. The system is called M3D which stands for Multi-site cooperative 3D design for architecture [4].

The system provides the on-line and off-line cooperative working capability and information sharing to a group of long distance participants. The team members can integrate their own design to form a global design by on-line cooperative working sessions or off-line individual work. They can use a rich set of editing and integration operations to verify the design and make modification on 3D design objects together. They work in a virtual design room together despite of the geographic distance. The concurrent control guarantees the mutual exclusion in the cooperative design process. The system also facilitates the decomposition and reorganization of the design for any number of iterations during the whole design phase.

M3D differs from simple application sharing which does not provide concurrent control and authorization of the cooperative design object. Application sharing requires all the partners use exactly the same CAD tool. In contrary, M3D aims at providing communication among a wide range of CAD tools [7]. M3D has a Web-based database [6, 8, 9] storing all the project information of all the phases along the building lifetime. It has a direct interface with the integration tool – the M3D Editor [5].

There have been efforts in the society to use virtual reality for visualization of architectural design. However, in these applications the 3D models are roughly made for visualization purpose only. Making these models itself is an extra work, not a natural product of the design phase. Therefore, due to the lack of detail of the 3D models, the communication support to transmit large scale design models, the design objects in these applications often look too simple and too naive to reach the real need of the industry. It is not acceptable for the design projects without extra cost to create these models.

M3D applies virtual reality technique to the architectural design at a higher level. M3D system [16, 17] introduces the 3D design technology for the whole design process from early conceptual design until detail design. Therefore, it overcomes the problems of the existing VR application for the AEC industry. The designs are in 3D by nature with all the details necessary for conflict and error detection and construction. The visualization is only a by-product since the buildings and their supporting elements are already constructed virtually from the very early beginning of the design. As a solution to identified problems in the architectural production, M3D proposes the re-engineering of current business processes towards a new business model in the AEC industry. The system supports the cooperative design and integration in the following aspect:

- Accepting design in 3D from all the multi-disciplinary specialties in the project
- Providing the capability of holding on-line cooperative working meetings
- Automatic verification for 3D design
- Integration with outputs from other CAD tools
- Information storage and retrieval for AEC projects.

5.2 System Architecture and Components

The major function of the system is for cooperative work for an architecture design team including on-line and off-line working. To provide all the team members the capability of initializing an on-line cooperative working meeting, a peer to peer and layered structure has been designed.

Figure 5.1 shows the M3D system architecture. Each column in the figure represents one cooperative member's site in the global system. All the sites have the same resources and same right. Therefore, any of the members can initiate a cooperative working session if necessary using their own set of applications. The number of members of the system is flexible. Any number of the members can join a cooperative working session if necessary.

The system contains three major components

- The cooperative 3D editor
- The cooperative support platform
- The integrated design project database

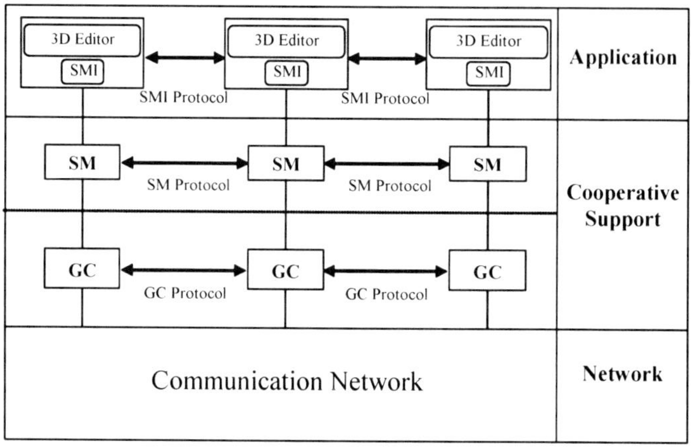

Figure 5.1. The system architecture of the M3D system

The cooperative 3D Editor is where all the on-line and off-line cooperative work happens. It supports cooperative editing in three dimension and real time. Individual components modification or the integration of the design can be realized by a cooperative working session. The cooperative support platform governs all the on-line working sessions and group operations. The project database is where all the project information is stored. All the members can access to the database in the on-line working session or off-line independent design works.

5.2.1 The Cooperative 3D Editor

The cooperative 3D Editor is the central tool for on-line cooperative design working sessions. It can also be a stand-alone single user tool. The editor has to satisfy a highest requirement than the normal visualization tool because it has to support the on-line cooperative modification from long distance locations.

The Editor supports on-line modification of the design. The changes of the design will immediately appear to all the participants in the session. Modifying the position, orientation and scale for an object can be realized by interactive manipulation or exact numerical specification. The shape of any single object can also be modified. Common functions in a single interactive system, such as undo and clipboard operations, can be performed in the cooperative 3D Editor. Undo can be performed on modification, insertion and deletion.

The major cooperative editing operations are: scene tree editing; geometric transformations; object editing; light management and material editing.

The design is organized as a VRML scene tree inside the cooperative editor. There is a window specially showing the current design tree by names of the components. This graphical textual tree can be edited by dragging its nodes around. This makes the decomposition of the design easier. The objects can be selected from the tree using their names. This is proved to be very useful since there are usually large amount of objects in one design scene. Furthermore, it allows the user to select a group of objects by selecting their parent node on the tree. See the left side of the screen on Figure 5.2.

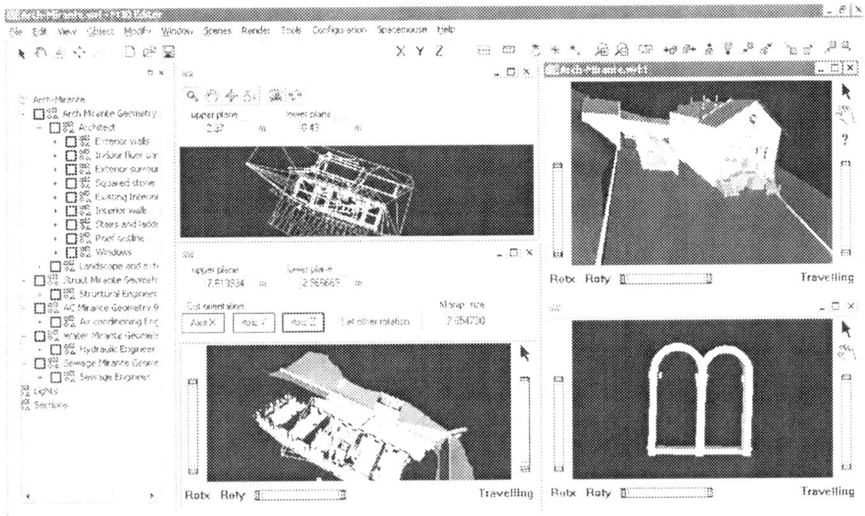

Figure 5.2. A snapshot of an on-line cooperative working session

Geometric transformations of any object in the design can be performed interactively or numerically. A group of manipulators are provided for the interactive transformation. Dialog box is provided for numerical specifications.
The object editing option in the editor allows the user to isolate a particular part of the design and make modifications on its geometry other than applying transformations. It can also produce 2D drawings from the projection of the 3D structure.

The object editing option in the editor allows the user to isolate a particular part of the design and make modifications on its geometry other than applying transformations. It can also produce 2D drawings from the projection of the 3D structure.

The editor includes an important module for error detection in the AEC design called Automatic Design Verifier (ADV). It is particularly developed for the error checking and design verification of the architecture and building construction projects.

The module integrates the 3D design of all the specialties, such as architectural, structural, water and sewage, *etc.* within the same project to check for possible geo-metrical and topological inconsistencies. "Inconsistency" here means an undesired geometric and/or topological condition. For example, a sewage pipe run intersects a pile foundation is a case of inconsistency.

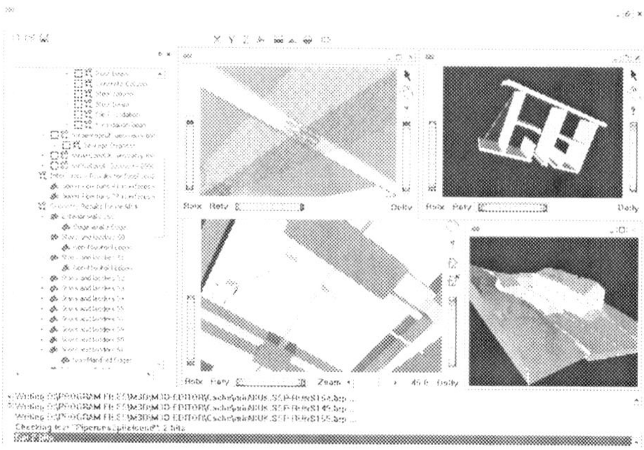

Figure 5.3. The automatic design verifier

5.2.2 The Cooperative Support Platform

The 3D Editor tool is layered on top of an IP compatible cooperative support platform, as shown in Figure 5.1. The cooperative support platform adopted a distributed architecture and a peer-to-peer network model. This module has two layers one on top of the other, the Group Communication (GC) and Session Management (SM). This platform provides a set of necessary communication services. The layer structure hides the point-to-multipoint con-figuration from the applications – the 3D Editor and others. An important task of the platform is to ensure a source ordering delivery of multicast messages over TCP/IP. The session control mechanisms keep the applications free from specific functions necessary for their inclusion into cooperative working environments. These specific functions include: communication services multiplexing, support of quality of service, consistency control, admission of new members into a running session, managing early members leave, invocation of new distributed applications, handling of exception events or failures and definition of roles within the group.

5.2.3 The Integrated Design Project Database

As a complete solution to the problem we identified in the architectural design and production process, M3D provides an integrated architectural information storage and retrieval system - an integrated design project database.

The database uses a client/server architecture. The database architecture supports multi-site access. The database users can connect to the database server by any communication network. There are two ways to access the database: via the 3D Editor and via a usual Web browser. The editor uses the data access Application Programming Interface (API) to access the database server. It connects to the client network library. The client network library uses a networked inter-process communication method to communicate with the server network libraries on the server system. The browser application uses a similar method to communicate with the database server.

With the integrated design project database, all the teams can work with a unique version of the same design document. All the results of an individual design or the decision of a collaborative design working meeting will be stored in the database. All the authorized users can access to the design project data any time and anywhere which is essential for cooperative working teams spreading geographically and may cross different time zones.

5.3 Considerations and Implementation for Collaborative Design

To develop an innovative cooperative working system, substantial amount of studies have been made to include all the important considerations for a multidiciplinary design team. The system has to be interoperative and multi-disciplinary. The on-line cooperative working meetings are the most important form of cooperation. It is technically very demanding and difficult to reach. It has to have a rich set of tools for on-line cooperative working, easy to manage, but powerful to include all the basic operations. It has to support design error detection in all the stages of the design particularly at early design stage. The following sections will explain the basic considerations and the results of implementation

5.3.1 Interoperative and Multi-disciplinary

The architecture design is a complicated process involving design from multi-disciplinary specialists. As a high priority in system design, the system has to be interoperative and multi-dsciplinary. To provide interoperability and cross-disciplinary interaction, the M3D system has been designed based on four conditions. First condition is the adoption of data inputs from a wide range of specialists independent of the CAD tools they use [1, 5, 6, 7]. The only requirement for all the design from the M3D system is that they have to be in 3D and they can generate the acceptable formats of the system such as: VRML, 3DS, DXF[7]. The second condition is a harmonic organization of design data from different disciplines. To satisfy this condition, M3D uses a hierarchy – a tree structure to store and manipulate designs from multi-disciplines. Each specialty is a branch of the whole design tree. The whole tree is a harmonic body of the global design project. There is no limit in space or subbranches for the growing of the whole design or a particular design specialty. The third condition is the accessibility of the global design. M3D stores all the project information in a Web accessible database. Any specialty team can access to the global design from anywhere and any time with secutiry control. The fouth condition is using the 3D geometry as a common base for integration and error detection. Independent of the discipline, an air-conditioning engineer or a structure engineer has its own way of design, its own CAD tool. But they have one thing in common, all designed objects are in 3D and can occupy space. 3D has been chosen as the basic format to integrate designs from different specialties which is intuitive, good for visualization and design error checking.

Figures 5.3 and 5.4 show some of the examples from live projects of cooperative design using the M3D system [18]. Multi-disciplinary design teams are located in different cities. During the system development trial, the architecture, structure, and water, sewage teams are in different locations in Lisbon, Portugal while the electricity design team is at Barcelona, and the air conditioning designer is located in Palma de Mallorca, Spain. The architect of the projects located in Lisbon started by designing the architecture part in 3D. After the 3D architecture design was finished, the geometry was stored into the project database. Other specialists groups in other cities can obtain the most recent version of the design by accessing the project database. In a conventional design project, they would take the final drawings of the paper project and interpret them, and insert their specialist geometry into the paper drawings. This would demand a series of steps. They have to spend a substantial time to study the architecture project and digest it in detail, since the majority of the participants and specialists do not know the project. In addition, a supplementary drawing work has to be done since in the original project there were only paper drawings. Furthermore, an interpretation of the representations of the original project, by technicians from several countries has to be done. This usually involves different rules in different countries. Successive steps are drawing, checking, changing, redrawing, re-checking, *etc.* for each specialist team. The process can be very time consuming and errors, misuse of design versions can occur.

Using the M3D system, the situation is completely different. The users used their own CAD design tools and input their designs to the M3D system. By using the M3D Editor [5] and M3D database the architecture project becomes very intuitive and easier for other cooperative teams to understand since it is already in 3D format.

The M3D Editor provides many ways of viewing the 3D design including navigating into the building itself, separating a part from the whole building to view it in detail etc. The supplementary drawing is not necessary since it is easy to have any clipping plane with arbitrary orientation to see the architecture. The interpretation is no longer difficult since the 3D architectural design is already there with all the material, lighting available for any kind of interpretation. If there is any doubt about any part, the team member can just go into it and look at it in detail in 3D. The M3D Editor can provide convenient navigation, show and hide any part, and even make a direct measure to find the physical distance between elements in the design.

Coordination of a building and construction design project including all the specialties is usually a very complicated task due to the incompatibility of the design tools each team uses, the geographic distance between them, and a tremendous number of iterations necessary during different phases of the design. The M3D system [16, 17] makes the coordination of all the individual specialty design and integration work a lot easier. All the teams design their special part using their own CAD design tools, the outputs are presented in VRML format. During the integration phase, all the teams use the M3D Editor to insert their own design into the global design either on-line or off-line. Obvious errors are easily visible and the correction of them becomes straightforward. The tool supports integration and decomposition of the design. It involves only the dragging of the sub-trees in the global design tree. In addition, the database in the system stores all the design data with version control support. The most obvious advantage is that for the first time, we have a common 3D working space for all the cooperative specialties without the constraint of geography and time.

5.3.2 The on-line Cooperative Working

The M3D system supports both off-line and on-line integration [3, 11, 12, 16, 17]. A multi-operator conference through the long distance network between several participants in the project can be held regardless where they are in the world. Figure 5.4 shows the network connection configuration of such on-line conference [14, 15]. The project was divided into a developer group and a user group. The developer group consisted of computer scientists, software engineers who designed and developed the system. The user group was formed by architects, structure engineers, air conditioning, power supply, sewage engineers who are the representatives of the users that will use the M3D system. The system is new and innovative. The developer group has to take the opinions of the user group seriously to design, develop and improve the system. All the components were tested by the user group during its development. After the components integrated, the project undertook a final testing phase. The user group in the project made intensive international connectivity tests and on-line co-operative design work using the system as a new design tool. The trials were held among three locations: Lisbon, Palma and Barcelona.

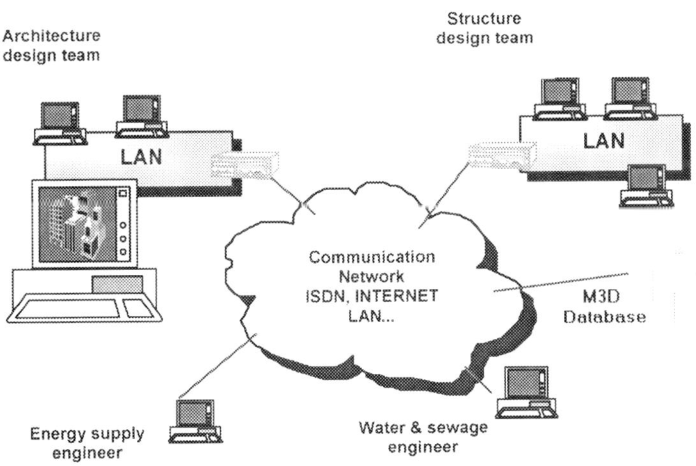

Figure 5.4. Network connection among cooperative participants

During the cooperative working session [11, 12, 13], the system was tested in different aspects. For example, the cooperative on-line viewing and organization of the designs were tested intensively. The partners inserted their new specialty design into the system from their own locations while others could see them immediately. The geometry of the air conditioning, structure, water sewage was inserted to the global design one by one. Each inserted part becomes a part of the global design tree. Each part can be shown or hidden. The regrouping of the design is very easy. The tool supports operations not only on one design project. The participants can load different design projects onto the common virtual working space by opening new windows. Within the design, any element can be selected for manipulation and modification. An annotation can be attached to any object for future reference or other off-line users. The annotation can have hyperlinks linking to any further documents. The on-line interaction seems not be affected by the distance unless a completely new project has to be loaded from the long distance database. The system has been designed to minimize the data traffic during the working session. Voice and text communication is also provided which are very useful for supplementing the on-line editing tool. See Figure 5.2 for a snapshot of the user interface.

The on-line cooperative working is a strong support for a design team. Many conflicts, errors, misunderstandings can be discovered and resolved. The tool is effective for the integration of different specialties in a design project. The 3D display, simultaneous multiple user viewing and manipulation provide a very intuitive way for common discussion and decision making to support the cooperative design work. The possibility to manipulate the geometry such as dividing them into elements, making changes to the geometry and display of the clipping sections in real time and many more other features show that the system is an innovative tool for the architecture design sector. The experiments also showed that the strong support of the database during and after the cooperative on-line working session is essential. A new concept of the architecture design business process is being formed by using the M3D system.

5.3.3 Design Error Detection During Integration

Because of the complexity of the architecture, the occurrence of conflicts, errors or omissions during the design is high. These errors are mostly detected only in the construction phase, which is far too late and has an enormous impact in terms of costs and deadlines. One of the major purposes of the M3D system is to detect any possible error in the design especially the incompatibility among different specialties. In addition to using M3D Editor to integrate the specialist design, obvious error can show up visibly. As mentioned above, the system also provides a detector that automatically checks incompatibilities of the geometry, according to the settings of the users [10].

The M3D ADV [10] supports the operations of intersection, subtraction, addition, bounding, growing between building elements. The system can check if the geometries of different specialties fulfill the rules established by the designers, or if within the same specialty the legislative, geometric or use rules are fulfilled. The possibility of debating and checking in group and real time makes it a very good way to validate the integration. Throughout the design verification process, it is possible to use the ADV to find out some hidden errors in a design project even it is in its final stage. Figure 5.3 is a scenario of using the ADV for design verification to find out the inconsistency during the integration.

The following presents a typical on-line cooperative design verification session. All the team members are connected for an on-line working session according to their previous communication. The architect opens a design file of the architecture from his local machine and shows to other participants. He can use voice communication or text description in the text chat window to explain his design. All the team members in other locations can see the design in 3D immediately. The structure engineer inserts the structure design from his own machine located in another city. His structure design will be integrated into the global design. Other team members can see the two designs identified by different colors in their own Editor window. The sewage designer does the same, loading his sewage design from his own file system located in the third location. A global design is formed and appears on all the participants' screen. The structure engineer then starts the ADV [10] from his M3D system by clicking on one of the menus. The architect types in a sentence: "Growing the interior wall by 3 meters, where do they intersect the roof outline?" The ADV system forms a statement in IDL through analyzing the sentence and performing the necessary operation [10]. After the calculation, the result shows up from the engineer's machine. At the same time, the architect and the sewage designer from long distance can see the result from their own location. The possible error parts are shown with distinguished color. Many errors appear which could not be discovered only by human eyes. The team members led by the architect can then discuss with the errors and make some decisions or on-line corrections. The concluded design will be stored into the project database for the next scale design or for final construction. If important modification has to be made, the corresponding designers of the related parts may go back to their own CAD tools to modify the design formally. The new version of the designs with modification and error correction will be stored into the project database.

5.4 System Evaluation

A series of five real life design projects [18] have been performed using the system for the evaluation during the system development. Figures 5.5-5.9 show some of them. All the projects required different specialties and the design teams were located in the three European cities. With conventional design model, the teams would have to travel a lot and caused a lot of communication time using email, fax or telephone. Using the M3D system, the users had regular on-line cooperative working sessions weekly. They have successfully finished these projects. All the connection time, user interactions, and system performance were recorded to provide to the developers further improvement. The users are satisfactory to have the cooperative working tool. They appreciate very much the M3D Editor, the ADV and the database. They found the tool very useful in their daily practice and have been using it from then on for other design projects. They do not find current tools that have the same functionality as M3D.

Figure 5.5. A historical house for reconstruction

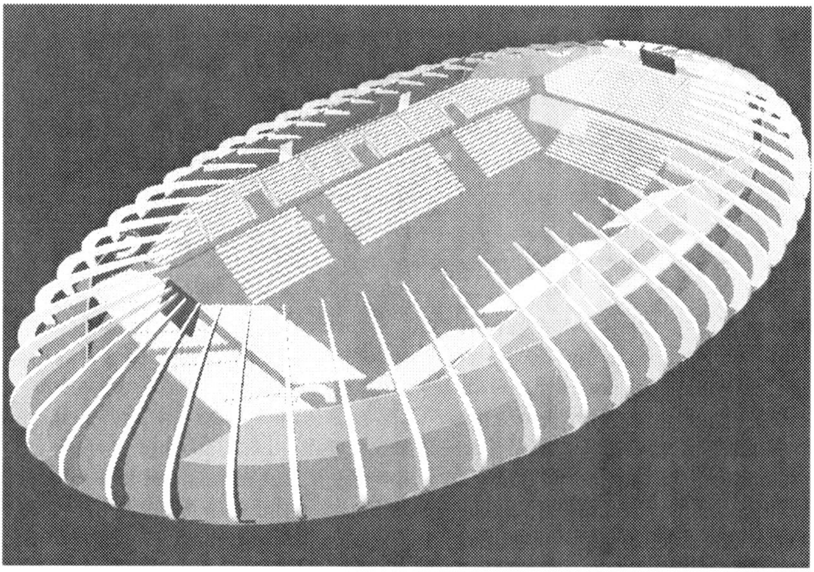

Figure 5.6. Live project 1

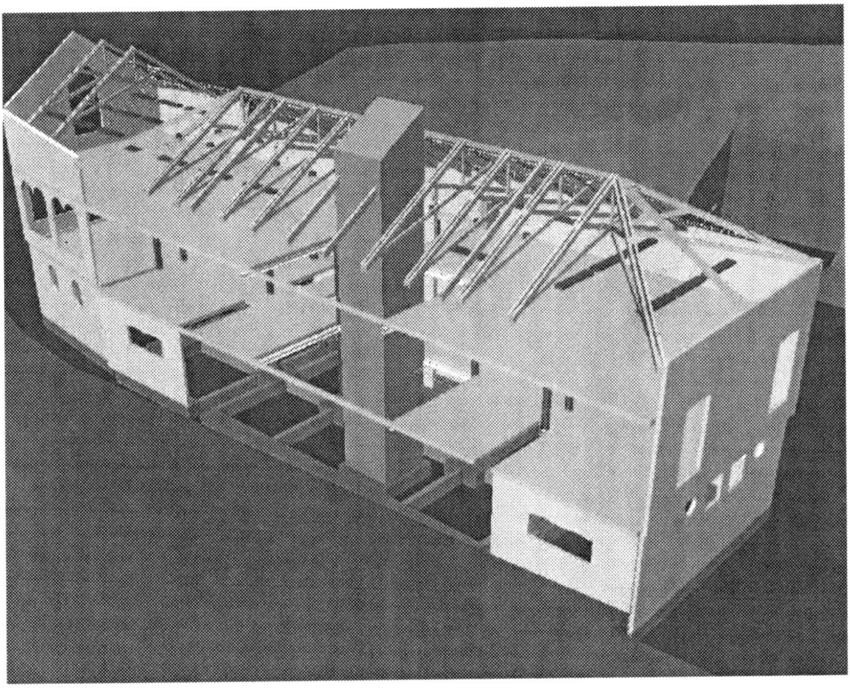

Figure 5.7. Live project 2

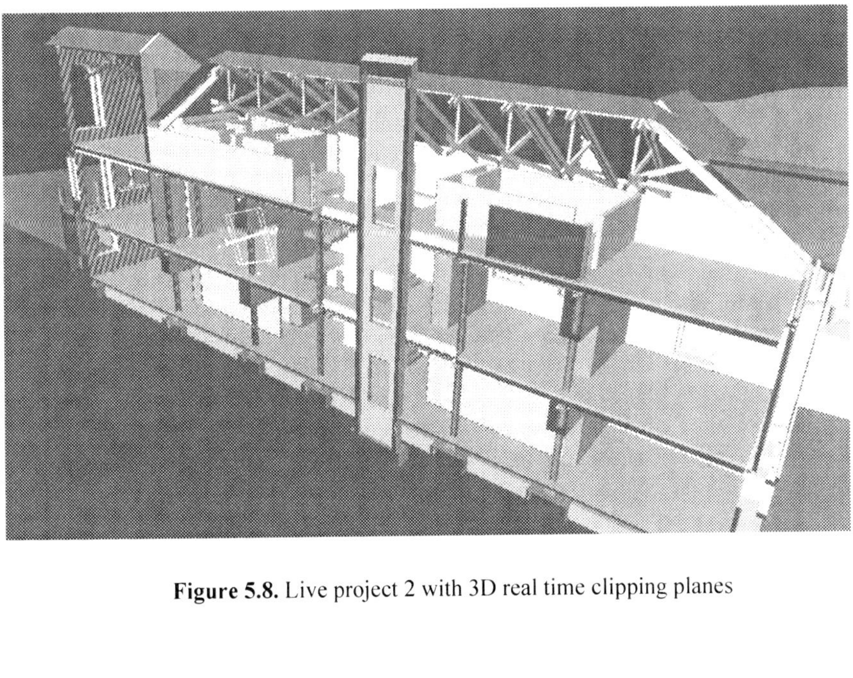

Figure 5.8. Live project 2 with 3D real time clipping planes

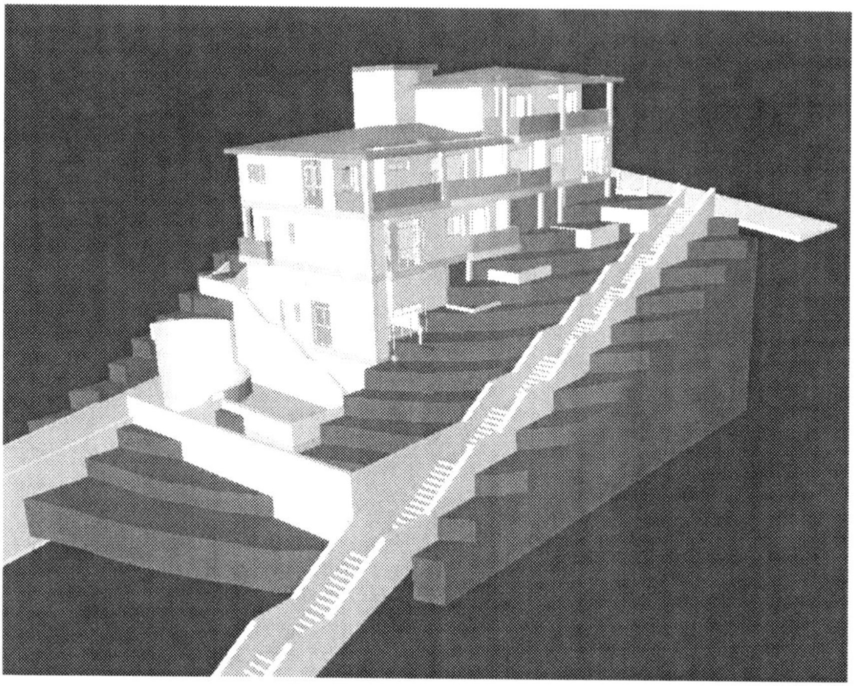

Figure 5.9. Live project 3, a residential building on complicated terrain

5.5 Conclusions

The M3D system supports a new business model in AEC industry. It has the capability to integrate the complete architectural production processes: design, construction, monitoring and maintenance. The key element in this new business model is the integration of multi-discipline 3D design for early error detection and correction. This can solve a substantial amount of problems in current practical business model.

The system provides the capacity of on-line cooperative work of the whole design team among several locations. It resolves conflicts, reduces errors, misunderstanding, and redundant work and therefore the project time and cost. The construction phase will be the most benefited. We can predict that by using the M3D technology, there will be a great reduction of errors caused by poor coordination and poor integration of multi-disciplinary design. The principal proposed by M3D can be applied to other industries for their cooperative work such as mechanical engineering, aerospace engineering, etc.

5.6 Acknowledgements

This work was directed by the author of this chapter, funded by the European Union, the Spanish CICYT, and each individual partner. The author would like to thank all the members in the project whose names are shown in the project deliverables listed in the references.

5.7 References

[1] Luo, Y., Sánchez, D., Alves, S., Dias, M., Marques, R., Almeida, A., Silva, J., Manuel, J., Tummers. B. and Galli, Ed. R., 1999, "M3D technical specifications," *ESPRIT Project No. 26287 M3D*, Deliverable 1.2.

[2] Tummers, B., Josquin, M., Galli, R., Dias, M., Almeida, A., Pons, B. and Silva, J., 1999, "Definition of database access, data structure and data exchange tools," *ESPRIT Project No. 26287 M3D*, Deliverable 1.3.

[3] Almeida, A., Dias, M., Marques, R., Estevens, F., Silva, J., Fonseca, M., Galli, R., Luo, Y., Pons, B. and Gonzàlez, A., 1999, "Definition, test and management of network model," *ESPRIT Project No. 26287 M3D*, Deliverable 1.4.

[4] Moat, M., Laves, S., Marques, R., Gamito, M., Silva, J., Fonseca, J. M., Vilar, M., Galli, R., Sanchez, D. and Luo, Y., 1999, "M3D prototype," *ESPRIT Project No. 26287 M3D*, Deliverable 1.5. April.

[5] Sánchez, D., Bennasar, T., Fornés, J., Galli, R., Huéscar, J., Serra, J. C., Luo, Y., Gamito, M., Dias, M., Albiol, M. and Arias, J., 2000, "The multi-site 3d editing tool for architectural production integration," *ESPRIT Project No. 26287 M3D*, Deliverable 2.1. April.

[6] Tummers, B., Smulders, A., Josquin, M., Galli, R., Huéscar, J., Luo, Y., Serra, J. C., Pons, B. and Ramos, R., 1999, "Report on database access modules," *ESPRIT Project No. 26287 M3D*, Deliverable 2.2. October.

[7] Dias, M., Bastos, R., Santos, R., Coelho, D., Bennasar, A., Fornés, J., Galli, R., Huescar, J. and Tummers, B., 2000, "Report on data exchange tool," *ESPRIT Project No. 26287 M3D*, Deliverable 2.3. April.

[8] Tummers, B., Smulders, A., Josquin, M., Serra, J. C., Huéscar, J., Galli, R., Luo, Y., Pons, B. and Ramos, R., 1999, "Report on the development of M3D database part 1," *ESPRIT Project No. 26287 M3D*, Deliverable 3.1.1. October.

[9] Tummers, B., Smulders, A., Galli, R., Huéscar, J., Serra, J. C., Luo, Y. and Ramos, R., 2000, "Report on the development of M3D database part 2." *ESPRIT Project No. 26287 M3D*, Deliverable 3.1.2. April.

[10] Gamito, M., Dias, M., Lope, A., Santos, P. and Lopes, P., 2000, "Report on development of interference detection and basic complexity reduction," *ESPRIT Project No. 26287 M3D*, Deliverable 3.2.2. April.

[11] Almeida, A., Dias, M., Alves, S. and Galli, R., 2000, "Cooperative support report and prototype," *ESPRIT Project No. 26287 M3D*, Deliverable 4.1. April.

[12] Dias, M., Alves, S., Almeida, A., Monteiro, L., Pinela, P., Silvestre, R. and Salas, A., 2000, "Conference management for 3D editing," *ESPRIT Project No. 26287 M3D*, Deliverable 4.3. April.

[13] Almeida, A., Marques, R., Silva, J., Fonseca, M., Galli, R., Luo, Y. and Pons, B., 1999, "Report on National connectivity." *ESPRIT Project No. 26287 M3D*, Deliverable 5.1.1. April.

[14] Dias, M., Alves, S., Moita, M., Galli, R., Pons, B. and Tummers, B., 1999, "Report on international connectivity," *ESPRIT Project No. 26287 M3D*, Deliverable 5.1.2. September.

[15] Dias, M., Ramos, R., Gayá, J., Galli, R., Josquin, M., Smulders, A., Tummers, B., Silva, J., Henriques, B., Fonseca, J. M. and Pons, B., 2000, "User trial plan and protocol," *ESPRIT Project No. 26287 M3D*, Deliverable 5.1.3. April.

[16] Luo, Y., Galli, R., Sánchez, D., Bennasar, A., Fornés, J., Serra, J.C., Huéscar, J. M., Gayà, J., Dias, M., Gamito, M. A., Tummer, B., Smulders, A., Silva, J., Fonseca, J. M. and Villar, M., 2000, "Alpha release of M3D application," *ESPRIT Project No. 26287 (M3D)*, Deliverable 5.2.1, November.

[17] Luo, Y., Galli, R., Sánchez, D., Bennasar, A., Fornés, J., Serra, J. C., Huéscar, J., Gayà, J., Dias, M., Gamito, M. A., Tummer, B., Smulders, A., Silva, J., Fonseca, J. M., Villar, M., Albiol, M., Arias, J., Pons, B. and Castañeda, D., 2001, "Beta release of M3D application," *ESPRIT Project No. 26287 (M3D)*, Deliverable 5.2.2, March.

[18] Castañeda, D., Silva, J., Fonseca, J. M., Torres, I., Villar, M. and Albiol, M., 2001, "Report on the user trial evaluation," *ESPRIT Project No. 26287 (M3D),* Deliverable 5.3. March.

[19] Luo, Y. and Dias, J. M., 2004, "Development of a cooperative integration system for AEC design," In *Cooperative Design, Visualization, and Engineering*, Luo, Y. (Eds.), LNCS 3190, Springer-Verlag, Berlin Heidelberg, pp. 1–11.

6

A Fine-grain and Feature-oriented Product Database for Collaborative Engineering

Y.–S. Ma, S.–H. Tang and G. Chen

School of Mechanical and Aerospace Engineering
Nanyang Technical University, Singapore

Traditionally, product databases are either purely geometric or meta-linked to Computer-Aided Design (CAD) files. The first type lacks feature semantics and hence is too rigid for collaborative engineering. The second type is dependent on CAD files which are system sensitive and has too large information grain size that makes information sharing and engineering collaboration difficult. This chapter introduces a fine-grain and feature-oriented product database design. It is ideal to support Web-enabled collaborative engineering services. For this purpose, a four-layer information integration infrastructure is proposed. A solid modeler is incorporated to provide low-level geometrical modeling services. The novelty of this research includes three aspects: (1) a generic feature definition for different applications in the form of EXPRESS-schemas; (2) the integration of a solid modeler with feature-oriented database by mapping from EXPRESS-defined feature model to the runtime solid modeler data structure as well as to the targeted database schema; and (3) modeler-based generic algorithms for feature validation and manipulation via the database. A modeler-supported history-independent approach is developed for feature model re-evaluation.

6.1 Introduction

Due to the stiff competition and rapid changes of globalization, shortening time-to-market has become the critical success factor for many companies [1, 2]. As a result, Concurrent and Collaborative Engineering (CCE) has become a norm. CCE has been recognized as the systematic approach to achieve the integrated, concurrent design of products and their related processes, including manufacturing and support [3], via collaborations across virtual project teams of different business partners.

In a CCE environment, many engineers with diverse skills, expertise, temperament and personalities are responsible for different tasks. The vast amount of knowledge and information involved in product development is certainly more than any individual can manage. Many computer-aided software tools have been incorporated into the product development process, which include Computer-Aided Design (CAD), Computer-Aided Process Planning (CAPP), Computer-Aided Engineering (CAE), and Computer-Aided Manufacturing (CAM) tools. However, information sharing among these applications has not been very well handled so far. Currently, almost all the existing CAx applications, which include individual installations, project Web portals, groupware tools and PDM (Product Data Management) systems, are based on files as their repositories. File-based approach has large information grain-size that results in data redundancy, storage space waste and potential conflicts [4]. Therefore, such design is no longer adequate for web-based CCE environment. It can be appreciated that, instead of managing the information via each application system in the separated data formats, a database management system (DBMS - Database Management System) can be used to manage all the product information concurrently, and at the same time in a consistent manner in order to eliminate the duplicated data. A DBMS can also provide shared user-access to databases and the mechanisms to ensure the security and integrity of the stored data.

Some research work has been carried out in product DBMS. CAD*I, a research project by ESPRIT (European Strategic Program for Research and development in Information Technology) was among the first to use DBMS to realize the data exchange among different CAD systems [5]. Similar research work includes [6], [7] and [8]. However, in these product databases, only geometric data can be managed. This means high-level feature information (semantic information) is lost. Therefore, it cannot support complete information integration.

Currently, most of the CAx systems are feature-based because features are a very useful data structure that associates engineering semantics with tedious geometrical data entities. Therefore, feature information must be represented such that engineering meaning is fully shared among CAx applications. To represent high-level feature information in database, Hoffman *et al.*, [9]proposed the concept of product master model to integrate CAD systems with downstream applications for different feature views in the product life cycle. Wang, *et al.*, [10, 11] put forward a collaborative feature-based design system to integrate different CAx systems with database support. However, these proposed databases lack geometrical engine to support model validation.

A geometrical modeling kernel, which is also referred to as a modeling engine, provides lower-level geometrical modeling service. Therefore, it can be integrated with database to support feature management operations, such as saving, restoring and updating, and hence product model integrity and consistency can be maintained. In the previous work [12, 13], a four-layer information integration infrastructure is proposed based on the architecture of a feature-oriented database. Ideally, it will enable information sharing among CAx applications by using the unified feature model [14] in the EPM (Entire Product Model), and allows the manipulation of application-specific information with sub-models. However, the method to provide low-level geometrical modeling services remains as a major question for research.

Martino *et al.* [15] proposed an intermediate geometry modeler to integrate design and other engineering processes with a combined approach of "design-by-feature" and "feature recognition". Bidarra [16, 17] and Bronsvoort [18, 19] proposed a semantic feature model by incorporating ACIS into webSPIFF, a web-based collaborative system. However, the above-mentioned research has little discussion on the integration of solid modeler with database, and it is not clear whether they have managed product data in files or with a database. Kim *et al.*, [20] described an interface (OpenDIS) for the integration of a geometrical modeling kernel (OpenCascade) and a STEP database (ObjectStore). However, their work cannot ensure full information integration because STEP cannot cover feature information for different feature-based CAx applications.

Traditionally, feature information cannot be exchanged among different applications. More recently, researchers, such as Bhandarkar *et al.*, [21], Dereli *et al.*, [22] and Fu *et al.*, [23], proposed different algorithms to identify useful feature information from the exchanged part models. Although feature extraction [24] and identification can partially recognize some feature information, information loss still occurs because these approaches depend on pure geometric data. For example, feature relationships (constraints) cannot be recovered from the geometric data model.

In order to enable higher-level feature information sharing among different applications, many researchers [25-27] proposed to use design information as the input and derive downstream application feature models by feature conversion. However, their works support only one-way link which means they can only convert from design features to other application features. In [28, 29], a multi-view feature modeling approach that can support multi-way feature conversion by feature links, is proposed. Separately, an "associative feature" definition was developed in [30, 31] for establishing built-in links among related geometric entities of an application-specific and multi-facet feature while self-validation methods were defined for keeping feature validation and consistency. Compared with one-way feature conversion approach, these multi-facet feature representations are promising for supporting multi-view product modeling.

The concept of unified feature model was first proposed by Geelink, *et al.*, [32]. The interactive definitions for design and process planning features were focused. However, the constraints defined were limited within one application feature model. Therefore, different application views could not be integrated in their model. Chen *et al.*, [14] proposed a new unified feature modeling scheme by introducing inter-application links for higher-level feature information sharing among different CAx applications. The unified feature model is essentially a generic semantic feature model for different CAx applications covering three-level relations among geometric and non-geometric entities. The unified feature model includes a knowledge-based model by incorporating rules and the necessary reasoning functions [33, 34].

This chapter focuses on the investigation of mechanisms to integrate a solid modeler with a feature-oriented database, such that multi-application information sharing can be realized over the Web. This chapter consists of seven sections. After this introduction, Section 6.2 gives a generic definition of features with the consideration of unification of applications. Section 6.3 investigates the mapping mechanisms between the proposed feature type, consisting of properties and methods, and a solid modeler data structures. Section 6.4 explores the integration of the solid modeler and database with key algorithms, *e.g.*, feature validation, constraint solving. Section 6.5 describes the method for solid modeler-supported feature model evaluation. A case study is presented in Section 6.6. Section 6.7 gives the conclusions.

6.2 Generic Feature Model

To consider integrating a solid modeler with the feature-oriented database, the mapping method between the database schemas and the feature definitions based on the solid modeler entities is critical. A unified feature model allows different applications to define different features with a set of well-defined generic types [14]. It is essential that each feature type has well-defined semantics [16]. The semantic attributes specified in each feature definition have to be associated with the structured elements of the given feature type. Such elements include feature shape representation with parameters, constraints that all feature instances should satisfy, and the non-geometric attributes to be used for embedded semantic properties, such as classifications, names, labels, and relations. All types of constraints are used for capturing design intent in the context of a product model. A generic feature representation schema is described in Figure 6.1. Note that the original information model is described in EXPRESS-G. Details for the convention of EXPRESS-G are shown in Figure 6.2 [35].

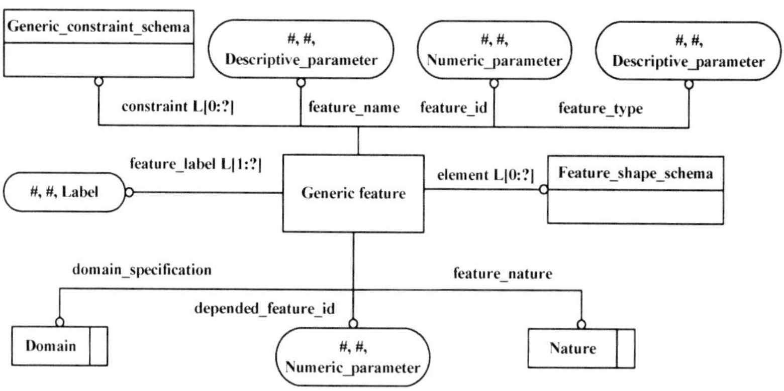

Figure 6.1. Generic feature representation schema

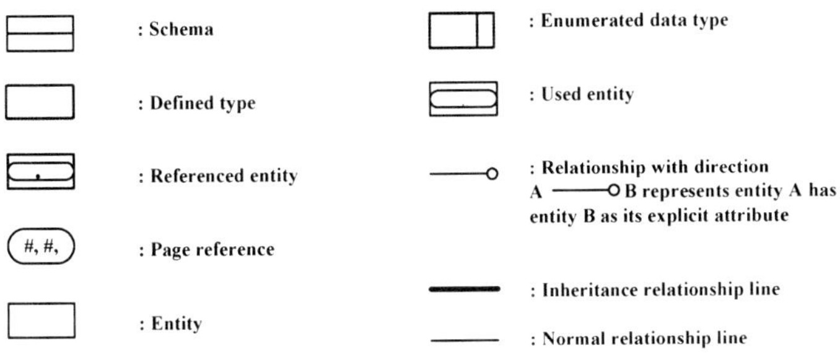

Figure 6.2. Convention of EXPRESS-G

6.2.1 Feature Shape Representation

To represent the shape of a feature means defining feature geometrical and topological constraints or relations with parameters and associating these parameters with feature manipulation (creation, modification and deletion) functions. The parameters are used to provide user interfaces to create and modify features in the modeling operations.

6.2.2 Constraint Definition

Constraints must be explicitly defined in the feature model to specify relationships among features, geometric or topological entities. Such constrints provide invariant characteristics of a feature type in the product model. Constraints may have various types (*e.g.*, geometric constraints, tolerance constraints and others). In generic feature definition, constraints are regarded as attributes attached to a set of associated entities, *e.g.*, geometric and non-geometric entities or even features. Although different types of constraints may have different attributes, they fall into a few common types, which can be generalized as shown in Figure 6.3.

Constraint_ID: It is the identifier of a constraint instance.

Constraint_name: It specifies the name of a constraint instance.

Owner_ID: It uniquely identifies which feature a constraint belongs to.

Constraint_expression: It represents the relationship between the constrained elements and reference elements.

Constrained_entity_ID list: It is used to specify a list of constrained entities with reference to the referenced entities.

Referenced_entity_ID list: It can be used to uniquely identify other related reference entities.

Constraint_strength: It has an enumeration data type, which may include several levels, such as *required, strong, medium or weak*. It represents the extent that the constraint needs to be imposed when constraints conflict with each other.

Constraint_sense: It is used to specify the direction between constrained entities and referenced entities. It has the select data type which maybe *directed* and *undirected.* A constraint is directed if all members of a set or list of constrained entities are constrained with respect to one or more referenced entities. A constraint is undirected if there are no referenced entities and the constraint is required to hold between all possible pairs of a set of constrained entities. Stated differently, in the undirected constraint, there is no difference between constrained entities and referenced entities. For example, if a directed constraint is applied to two lines (line1 and line2), which requires line2 to be parallel with reference to line1, it implies that line1 existed in the model before line2 was created. The corresponding undirected constraint would simply assert that line1 and line2 are parallel, with no implied precedence in their order of creation.

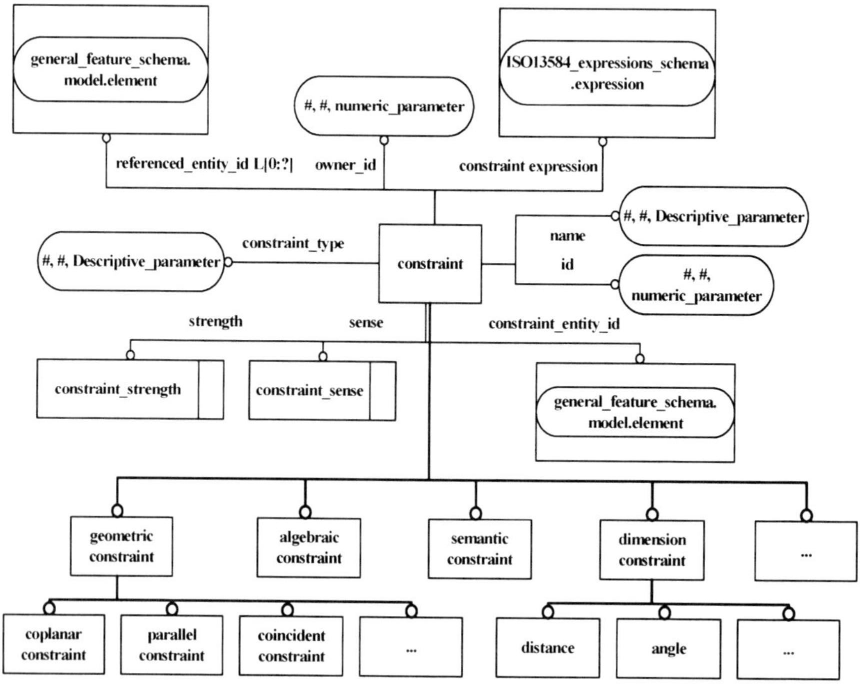

Figure 6.3. Constraint representation schema

Constraint solving functions: They are responsible for solving constraint according to constraint types.

Other manipulation functions: These functions may include attributes access functions, behavior control functions, *etc.*

6.2.3 Other Feature Properties

Other feature properties can be defined as follows:

General feature attributes- *Feature_name* and *feature_id*

General feature attributes such as *feature_name* and *feature_id* shall be realized with the instantiation of a specific feature according to the *application_specific* feature definition. These attributes are necessary when searching for the relevant feature properties during feature modeling operations.

Depended_feature_id_list

To maintain feature relationship, *depended_feature* shall be explicitly defined in feature definition. Feature dependency relation definition is described by Biddara [16, 17] as "feature *f1* directly depends on feature *f2* whenever *f1* is attached, positioned or, in some other way, constrained relative to *f2*". *Depended_feature_id_list* plays an important role in maintaining feature dependency graph, and furthermore, feature relations during feature modeling operations.

Feature label

A feature label is attached as an attribute to every face of a particular feature instance. In a feature, its member face labels are defined as a list of strings in the definition, to record feature face elements. Then the face corresponding to the label is referred to as the owner.

Domain specification

Domain specification has the ENUMERATION data type, which represents the application scope such as *design*, *manufacturing*, *assembly* and others. By specifying the different domains, multi-views can be supported with certain filtering and synchronizing mechanisms.

Nature

The nature of a feature also has ENUMERATION data type. It could be either positive or negative. A positive value means the instances of the feature are created by adding material. A negative value means forming a feature instance is realized by subtracting material.

6.2.4 Member Functions

Four groups of member functions are required to support the generic feature class. *Attribute access functions* shall be defined to manage a feature's attributes. Some functions are common to all types of features, *e.g.*, *backup()*. Others are feature-specific such as *findOwner()*, *findConstraint()*, *getParameter()*, *setParameter()*, *etc.* Object technology with a proper polymorphism design can be applied well here.

Modeling operation functions (*e.g.*, *splitOwner()*, *mergeOwner()*) are used to control the behaviors of feature during a modeling operation, *e.g.*, splitting, merging, or translation.

Feature evaluation and validation functions are responsible for feature model modification. Feature validation functions are used to validate feature geometry and solving constraints after each feature modeling operation. These functions will be discussed in detail in Section 6.4.

In order to persistently manage product and process information, which includes feature information, geometrical data and other information, *saving and restoring functions* of the database, which are the interactions between the run-time feature model and the database, must be defined in individual feature classes because these functions have to organize information for different applications according to the functional requirements. Details will be explained in Section 6.4.

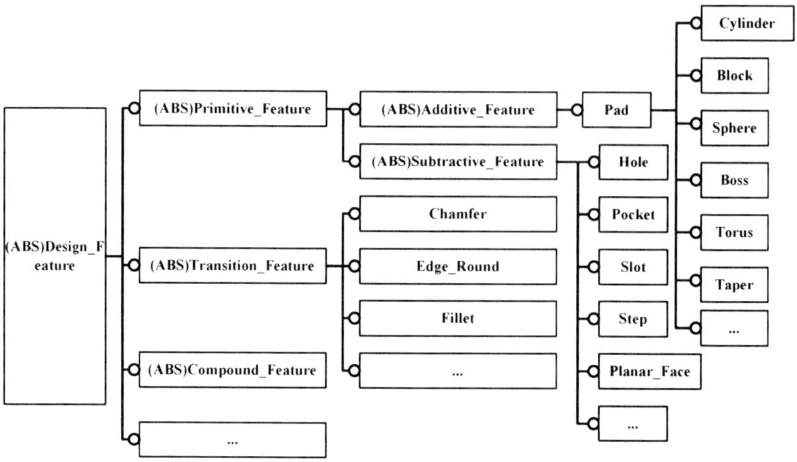

Figure 6.4. Design feature representation schema

6.2.5 Application-specific Feature Model

Application-specific feature model can be defined on the basis of generic feature model. As shown in Figure 6.4, the design feature type has three subtypes: primitive feature, transition feature and compound feature. The primitive feature type is separated into two subtypes, additive and subtractive features. Additive feature is represented as "pad", which covers all instance features formed by adding material such as cylinder, taper, sphere, boss, block, torus and so on. Subtractive feature type represents all features such as hole, pocket, and slot that are formed by subtracting material. The transition feature type includes chamfer, edge_round and fillet, which are always associated with other primitive features. The compound feature type is a union of several primitive features. For each specific design feature type, it has predefined explicit geometry, topology, parameterization and constraints specifications. For example, a design feature slot can be defined as shown in Figure 6.5.

6.3 Mapping Mechanisms

To provide lower-level geometrical modeling services, a geometrical modeling kernel is required. In this work, ACIS, a commercial package, is incorporated into the proposed system. An EXPRESS-defined and extended STEP feature model, which includes geometrical and generic feature representation schemas, is mapped to the data representation schemas in ACIS such that the proposed system will have the required fine grain functionality. On the other hand, this feature model would also need to be mapped to the target database schema so that it can be interfaced with a consistent repository.

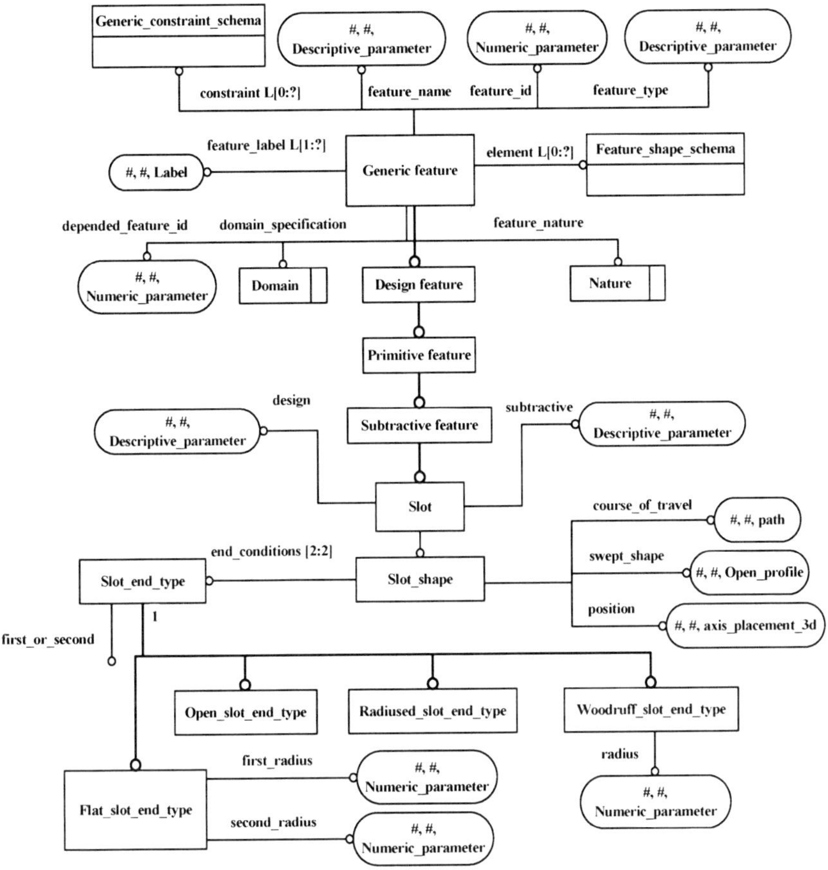

Figure 6.5. Slot feature definition in EXPRESS-G

6.3.1 Mapping from Extended EXPRESS Model to ACIS Workform Format

6.3.1.1 Geometry Mapping

In this research, in order to explicitly maintain feature shape and associative relations in the product model, a cellular model is adopted. Cellular model represents a part as a connected set of volumetric quasi-disjoint cells [36]. By cellular decomposition of space, cells are never volumetrically overlapped. As each cell lies either entirely inside or outside a shape volume, a feature shape can be represented explicitly as one cell or a set of connected cells in the part. The cellular model-based geometrical representation schema adopted in this research is shown in Figure 6.6. Basically, there are three types of topological entities for cellular topology, which are *CELL*, *CSHELL* and *CFACE*. *CELL* has two subtypes, namely

CELL2D and *CELL3D*. A *CELL2D* contains a list of *CFACE*s, each of which points to faces that are double-sided and both-outside. A *CELL3D* contains a list of *CSHELL*s. A *CSHELL* represents a connected set of *CFACE*s that bound the 3D region of the cell. A CELL is attached to the normal ACIS topology in the *LUMP* level (which represents a bounded, connected region in space, whether the set is 3D, 2D, 1D, or a combination of dimensions). Each CFACE has a pointer to a face in the lump and use it in *FORWARD* or *REVERSE* sense.

As cellular model is directly supported in an ACIS, cellular husk is adopted. Therefore, geometry mapping is one-to-one straight forward.

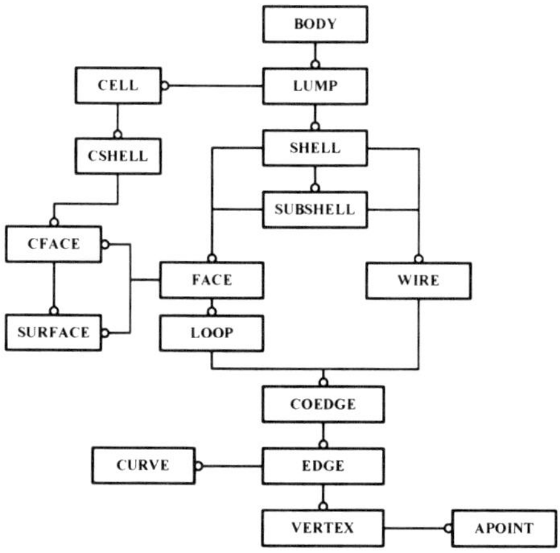

Figure 6.6. Partial geometrical representation schema according to cellular topology [36]

6.3.1.2 *Generic Feature Definition under ACIS Framework*

ACIS provides ENTITY-ATTRIBUTE architecture [36], under which we can specify user-defined attributes (features, constraints or others). The following rules are developed and used by the authors for defining features, constraints and other attributes in ACIS:

Use simple attributes to represent properties such as the material of a body or color of a face.

Use complex attributes to represent properties such as features, dimensions, tolerance, or constraints.

Use bridging attributes to link an ENTITY with some application-specific and parametric variables, such as dimensions.

Use instruction attributes placed on entities to force certain behavior.

Attributes of features and constraints may have various data types, *e.g.*, string, integer or ENTITY pointer.

Aggregating data type has been defined as ENTITY_LIST. The ENTITY_LIST is a variable length associative array of ENTITY pointers and provides common functions for the manipulation of itsmembers, *e.g.*, *add* ENTITY, *look up* ENTITY and *[]* operator for accessing list member by position.

Enumeration data type can be simulated by defining a string as the enumeration member or simply using an integer data type.

Selecting data type which can be simulated by using an abstract class and defining specific types of the abstract class.

On the basis of the above proposed mapping rules, a generic feature definition is created as shown in Figure 6.7.

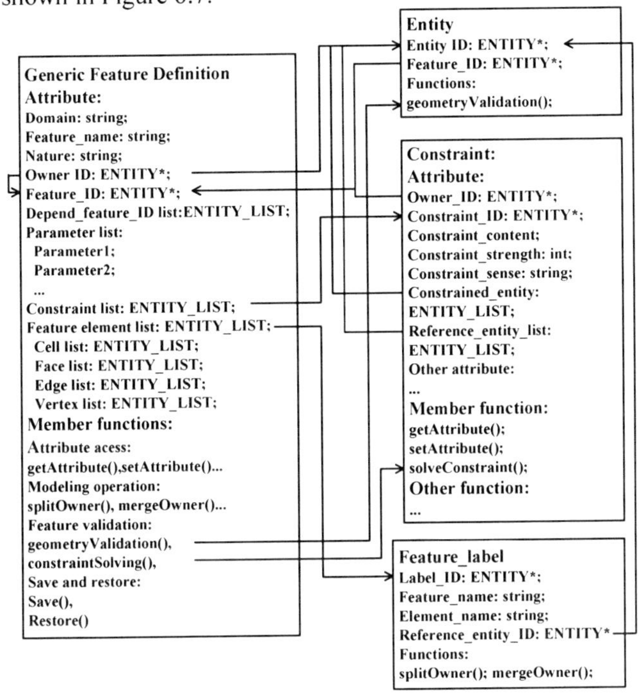

Figure 6.7. Generic feature definition with ACIS entities

6.3.2 Database Representation Schema

According to the mapping mechanisms proposed in [12], a geometrical representation schema as well as generic feature representation schema in the database has been developed. For details, please refer to [12].

6.4 The Integration of Solid Modeler and Database

The solid modeler has been tightly integrated in four layers in order to manage product and process information (see Figure 6.8). First, its API functions are called constantly which are encapsulated within the feature manipulation methods during the collaboration sessions between the end users and the application server. Second, all the geometrical entities are manipulated and their run-time consistency maintained through the solid modeler's implicit runtime data structure module.

Third, it also provides runtime functional support directly to the end users via commands dynamically. Fourth, the solid modeler has also to support the repository operations via the DB manager.

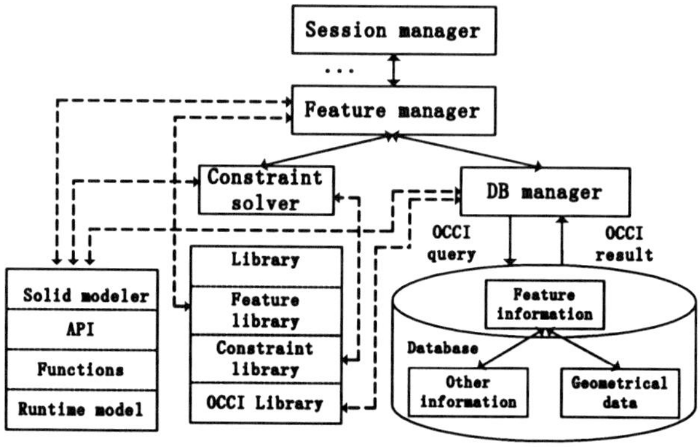

Figure 6.8. Partial integration diagram of a solid modeler and the feature-oriented database

This chapter focuses on the forth layer. In the proposed architecture of the web-based feature modeling system [12], database (DB) manager is responsible for managing the geometrical entities via the solid modeler runtime model and manipulating the data elements to be stored and extracted in the database for different applications. With the support of a solid modeler, the database manager can provide data manipulation functions such as *save*, *restore* and *validate* functions. These functions are fundamental to support different applications. In the following sub-sections, feature validation methods together with the generic *save* and *restore* algorithms are explained. In order to manage the connection between the DB manager and the database during saving and restoring processes, OCCI (Oracle C++ Call Interface) [37] is adopted as the bridge (see Figure 6.8).

6.4.1 Feature Model Re-evaluation and Constraint Solving

Once feature operations are specified via User Interfaces (UIs), the product model needs to be modified and updated. This process is achieved through feature evaluation. The geometrical model has to be managed to ensure the consistency. Here, the run-time product model should be generated via the integrated solid modeler and managed based on the database records. All feature evaluation operations call solid modeler APIs to realize the geometrical procedures while the rest of the functions are implemented separately. In this way, the bottom-level geometrical operations are readily looked after by the solid modeler; hence, the development effort is significantly reduced. Details of feature model re-evaluation will be explained in Section 6.5.

Theoretically, feature validation functions include two kinds: those dealing with the geometry, and those dealing with constraints. With the incorporation of a solid modeler, geometry validation functions are not really necessary under the proposed design because the solid modeler is responsible for manipulating and validating feature geometry. On the other hand, constraint-solving functions need to call specific algorithms defined in the individual constraint sub-classes to solve different kinds of constraints according to their types. Globally, all the constraints are maintained by the Constraint Manager in a constraint graph for EPM (Entire Product Model), which contains sub-graphs for specific application views. Constraint manager solves constraints by calling the corresponding solvers according to different constraint types. For example, SkyBlue algorithm [38] can be used to solve local algebraic constraints in design domain; Degrees of Freedom analysis algorithm [39] can be used to solve geometrical constraints in design domain. If conflict of intra-application constraints occurs, local constraints solver can determine automatically which constraint should be satisfy first according to the value of *constraint_strengh*, which is an attribute of constraint defined in Section 6.2. Inter-application constraints can also be solved under the control of constraint manager according to the value of *domain_strength*. For the definition of *domain_strength*, also refer to Section 6.2. The value of *domain_strength*, which regulates priority sequence of different domains, can be predefined, or is set by an authorized user. Any conflict of inter-application constraints will be detected by constraint manager after which the constraints solver can trigger the corresponding applications to reevaluate the product model according to *domain_strength*. Only when all constraints are checked and feature geometry is validated, does feature validation finish.

6.4.2 Save Algorithm

To elaborate, during the saving process, the solid modeler has to extract all the information from its runtime data structure and then save them into the database after a format conversion according to the mapping relations and the database mapping schema described in [12]. The *Save* algorithm can be expressed in the steps as follows (see Figure 6.9):

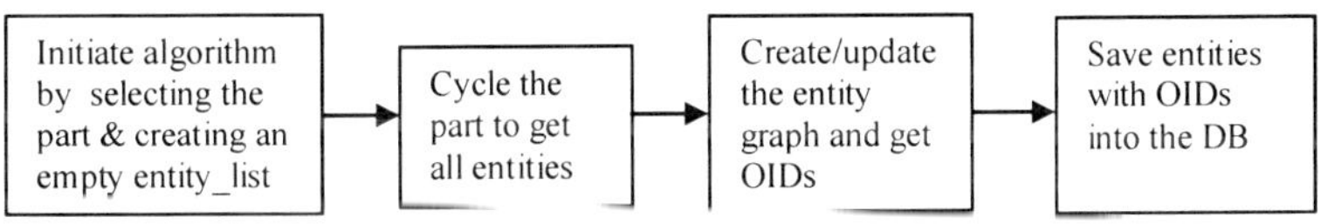

Figure 6.9. Save algorithm

- Select the part to be saved. Create an empty entity list and add the part attributes to be saved to the list;
- Cycle all entities (features, topological entities, such as solids, shells, faces, and geometrical entities, such as lines, planes, curves, and surfaces) from the part and add them to a graph map so that object pointers can be fixed as unique database Object Identifiers (OID). ACIS API functions, *e.g.*, *api_get_xxxx()*, are used to get all saved ENTITIES;

- Use such object pointers to call *save* functions of the specific class (e.g. *point.save()*, *vertex.save()* or *feature.save()*) to save part data to the database.

6.4.3 Restore Algorithm

In a reverse way, the uploading process is triggered when the product model is being established during the session initiation from the database.

Restore algorithm has the following steps (see Figure 6.10):

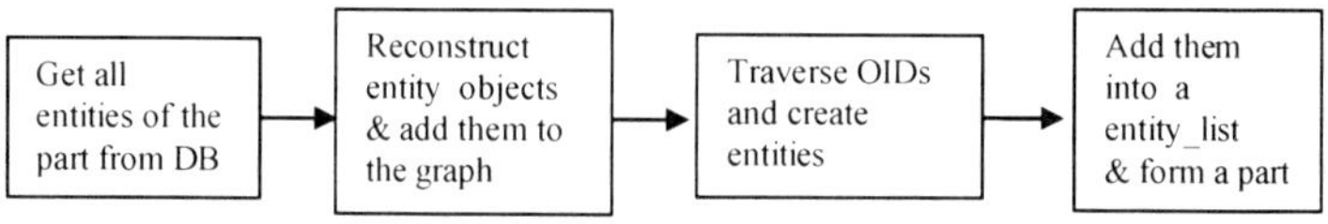

Figure 6.10. Restore algorithm

- All the entities of a part are retrieved from the database by searching their linked Object Identifiers (OIDs);
- Reconstruct new objects, *e.g.*, features, geometrical entities, topological entities. Upon reconstruction, all the objects will be validated;
- Add all the entities to a newly generated object graph map;
- Convert these OIDs to genuine pointers;
- Create an entity list and add all the entities to the list to form the part. Validation, *e.g.*, geometry and feature validation will be carried out during this procedure.

6.5 Feature Model Re-evaluation

6.5.1 Problems of Historical-dependent System

For most parametric and history-based modeling systems, feature model is re-evaluated by re-executing whole or part of the model history. The disadvantages of this method are the high computational cost and the considerable amount of storage space [16]. Moreover, history-based model re-evaluation causes ambiguous feature semantics due to the static chronological feature creation order in the model history. This is illustrated in the example shown in Figure 6.11(a). The simple part consists of a base block and a through hole. Later on, the designer wants to modify the part by adding another block and extending the depth of hole so that he can get the expected part model as shown in Figure 6.11(b). However, sometimes unexpected modeling results as shown in Figure 6.11(c) can be generated by the history-based reevaluation, because the feature creation order is *baseblock->hole->block*. In order to get the expected part model, the precedence order, in this example, should be changed to *baseblock->block->hole*. This semantic problem is caused by the static precedence order in the model history on which model re-evaluation is based. From this example, it is clear that the precedence relation among features should be dynamically maintained and updated after each modeling operation.

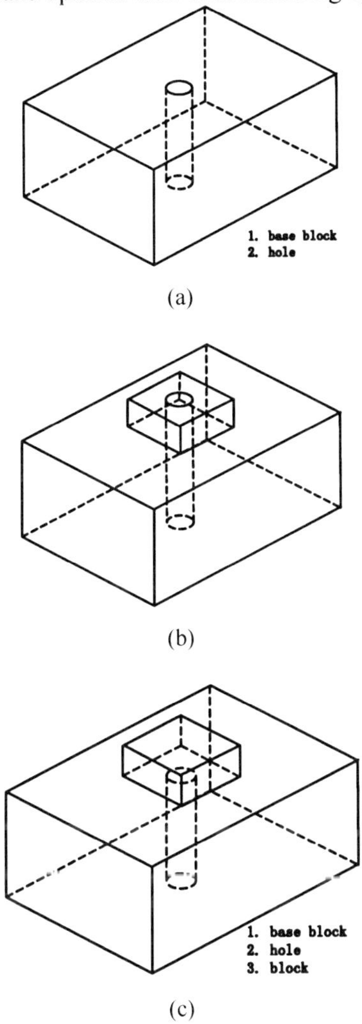

Figure 6.11. Semantic problem for historical-dependent system (a) example part at the initial state; (b) expected result after modification; (c) result of history-based re-evaluation after modification

6.5.2 Dynamically Maintaining Feature Precedence Order

In this work, feature precedence order is maintained dynamically based on a feature dependency graph. Relations between independent features can be determined by feature overlapping detection. Feature dependency relations are explicitly defined in the feature definition as explained in Section 6.2. The following rules are proposed for feature precedence determination. Note that, explicit rules always overrule implicit rules during dynamic maintenance of the global precedence order of all features. Stated differently, the explicit rules will be first used to determine the precedence relation; while if the global precedence order cannot be uniquely generated, implicit rules will be then considered to get a unique one.

Rule 1 (explicit rule)

For two dependent features, if feature f_2 depends on feature f_1, then f_1 precedes f_2 [16].

It is easy for us to derive from rule 1 that:

For n dependent features, if:

$$f_1 \rightarrow f_2 \rightarrow f_3 \rightarrow \ldots \rightarrow f_n$$

Then, there exist:

$$O_1 < O_2 < O_3 < \ldots < O_n$$

where:

$f_i \rightarrow f_j$: represents feature dependency relation(e.g. $f_1 \rightarrow f_2$ means f_2 depends on f_1);

O_i : represents the precedence order of feature f_i .

$O_i < O_j$: represents the jth feature is ordered after the ith feature.

Rule 2 (explicit rule)

For a feature in the feature dependency graph, if it depends on two or more features, the precedence order of this feature comes after the latest feature it depends on (we call it latest depended feature or LDF).

Note that in the feature dependency graph, LDF is always the feature that has the longest length of path (LLP) from the root node of the graph among all depended features of a particular feature.

Path: a path in a graph is a walk whose nodes are all distinct;

Walk: a walk in a graph is a finite alternating sequence of nodes and edges between its starting node and ending node;

Length of path: the length of a path is the number of edges that form the path.

Rule 3 (implicit rule)

For a group of features that have random precedence order, the feature creation sequence will be used to determine their precedence relations.

The feature creation sequence is defined as an attribute attached to the feature instance to record the sequence of the feature among all features in the part.

Rule 4 (implicit rule)

For two independent features, if they do not overlap with each other, the precedence relation between them is determined by LLP of these two features. There exists:

$O_1 < O_2$ if $LLP_1 < LLP_2$

In the case of $LLP_1 = LLP_2$, the precedence order can be determined by rule 3.

Rule 5 (implicit rule)

For two independent features with same natures (both negative or both additive), if they overlap with each other, the precedence relation between them is random and should be determined by LLP of these two features. There exists:

$O_1 < O_2$ *if* $LLP_1 < LLP_2$

In the case of $LLP_1 = LLP_2$, *the precedence order can be determined by rule 3.*

Rule 6 (explicit rule)

For two independent features (f_1 *and* f_2*) with different natures, if the overlap of these two features is caused by some modeling operation of* f_2*, then feature* f_1 *precedes feature* f_2 *[16].*

Based on the above rules for feature precedence determination, after each modeling operation, the following algorithm shown below is used to dynamically maintain feature precedence relations.

- Find all the features of the part and add them to a graph map (unsorted).
- Partially sort the graph map according to the existing feature dependency graph. This is done by using the algorithm shown in Figure 6.12 on the basis of rules 1 ~ 3.
- Sort the partially sorted graph with reference to the overlapping detection result based on rules 4 ~ 6.

In this way, a global feature precedence order can be updated dynamically.

```
(For i=1; i<n-1; i++)
{ (for j=i+1, j<n; j++)
  {if (Px_j>Px_i)
    {X_m= X_i;
     X_i=X_j;
     X_j=X_m;
    }
  }
}
```

Here:

X_i represents any feature in the feature set;

X_j represents depended feature of X_i;

Px_i represents the position of feature X_i in the feature map;

Figure 6.12. Algorithm for precedence order generation [40]

6.5.3 History-independent Feature Model Re-evaluation

First of all, re-evaluating the feature model requires that feature elements (cells, faces, edges and vertices) are correctly identified in the cellular model. This can be achieved by cellular entity owner list control.

6.5.3.1 Adding a New Feature Instance

This is carried out as follows:

- Create the shape of the new feature (one cell shape);
- Attach labels of the feature to each face of the feature instance; and
- Carry out Boolean operation (with the 'non-regular' option).
- During non-regular Boolean Union, intersection detection will be carried out for each cell (C_i) in the cellular model and the newly added feature cell (C). Upon cellular decomposition, the owner list of each cell and cell face should be controlled by the following rules [41]:
- The new cells that are in the intersection of C and C_i are assigned with an owner list that is the union of the owner lists of C and C_i;
- Other non-intersecting cells resulting from the decomposition get their owner lists which are the same as the original cells (either C or C_i);
- The new cell faces lying on the boundary of both C and C_i get the owner list that is the union of the owner lists of the overlapping cell faces from which it originates;
- The new cell faces lying on the boundary of either C or C_i inherit the owner list from their respective original cell faces;
- The remaining new cell faces get an empty owner list.

Figure 6.13(a) illustrates the creation of a *slot* feature on the *base_block*. The shape of the *slot* is first created as a one-cell shape. Then *non-regular-Boolean Union* is carried out to create the cellular model of the part. During the operation, upon intersection analysis, cell decomposition is performed. On the basis of above rules for cell and cell face owner list control, the result of the modeling operation is shown in Figure 6.13(b). Note that there are two cells in the cellular model. One is the original *base_block* cell (has *block* feature in its owner list). The other is a new cell generated by cell decomposition, namely the *slot* cell (which has *block* and *slot* in its owner list). Three double-side faces separate these two cells. Each double-side face has two corresponding cell faces (*e.g.*, CF_8 and CF_9); one (CF_8) is for the *block* cell boundary, the other (CF_9) is for the *slot* cell boundary.

Note that CF_i represents i^{th} cell face; *S* represents *slot* feature; *B* indicates *block* feature; and *( )* indicates the labeled entity's owner list.

6.5.3.2 Deleting a Feature Instance

This is carried out as follows (assume no other feature depends on the feature to be deleted) [16]:

- Traverse through all the cells and cell faces to remove from their owner list the feature to be deleted;
- Remove all the cells which has empty owner list. This can be realized by removing all one-side faces bounding the cell;
- Merge adjacent cells which have the same owner list. This can be realized by removing all double-side faces that separate the two cells;

- Clean up the model by merging the adjacent faces that have the same geometry and whose cell faces have the same owner list.

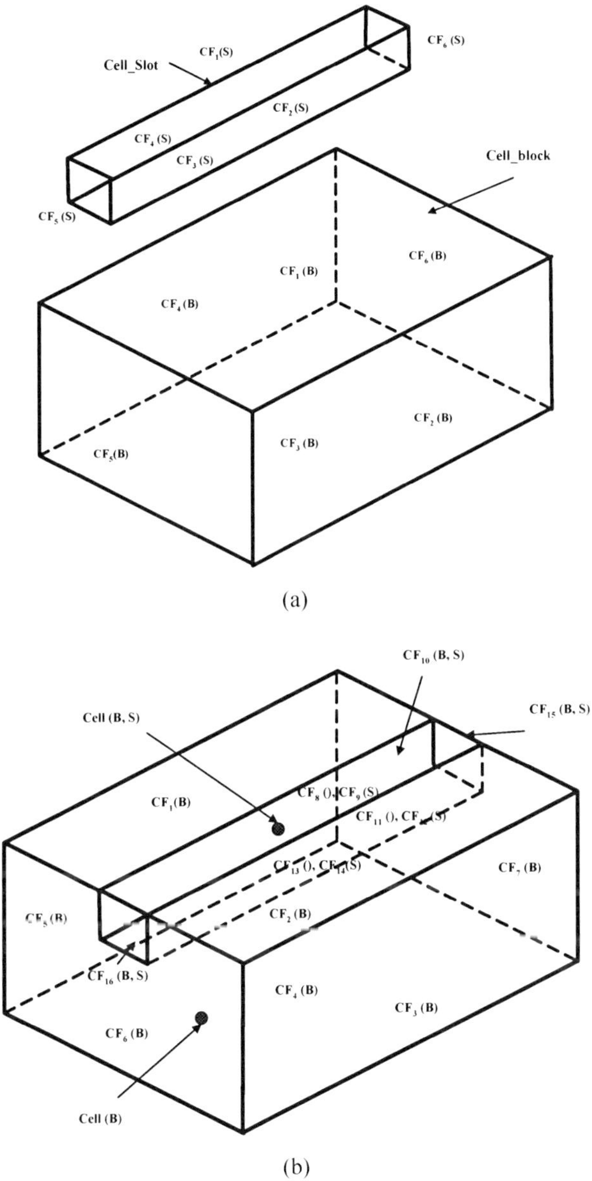

Figure 6.13. Creation of slot feature on the base block (a) base block and slot shape; (b) result of modeling operation

As shown in Figure 6.14, to delete the slot feature from the cellular model, all cells and cell faces in the cellular model are traversed through to remove from their owner list the slot feature.

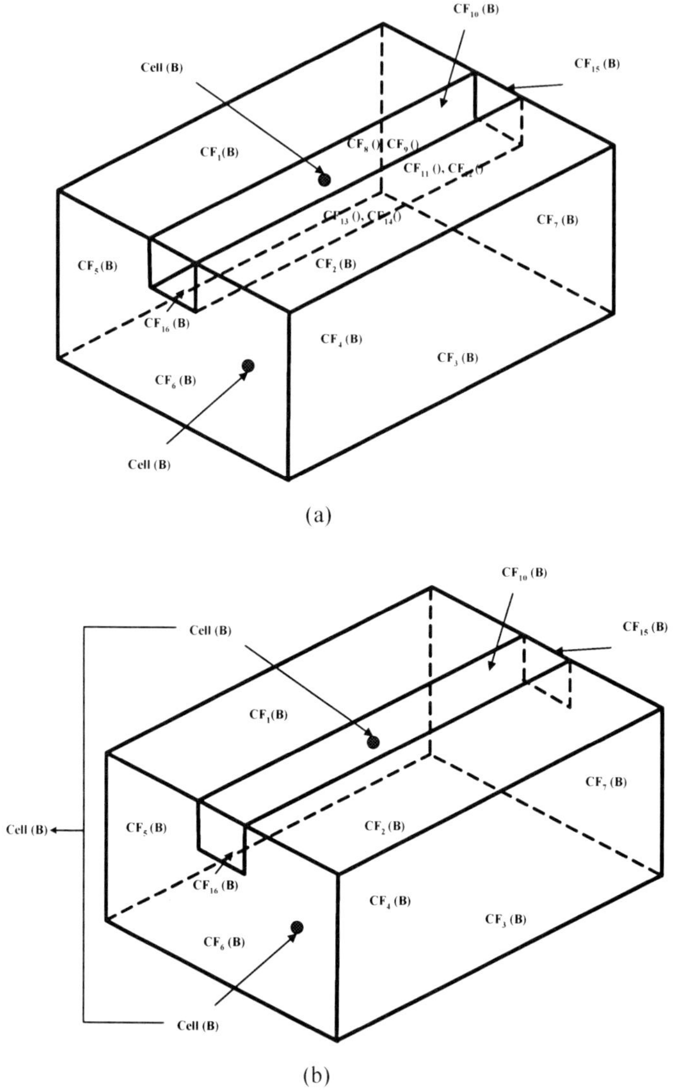

Figure 6.14. Feature deletion (a) remove slot from owner list of cell and cell face; (b) merge two cells

The result is shown in Figure 6.14(a). Then as two cells have the same owner list, the block feature, these two cells are merged by removing three double-side faces (the underlying faces of CF8 and CF9, CF11 and CF12, CF13 and CF14) that separate them. The result is shown in Figure 6.14(b). Finally, adjacent faces that have the same geometry and same owner list are merged. This option carries out the merging of the underlying faces of CF5 and CF16, CF7 and CF15, as well as CF1, CF2 and CF10. The result is the same as block feature shown in Figure 6.14(a) before creating the slot.

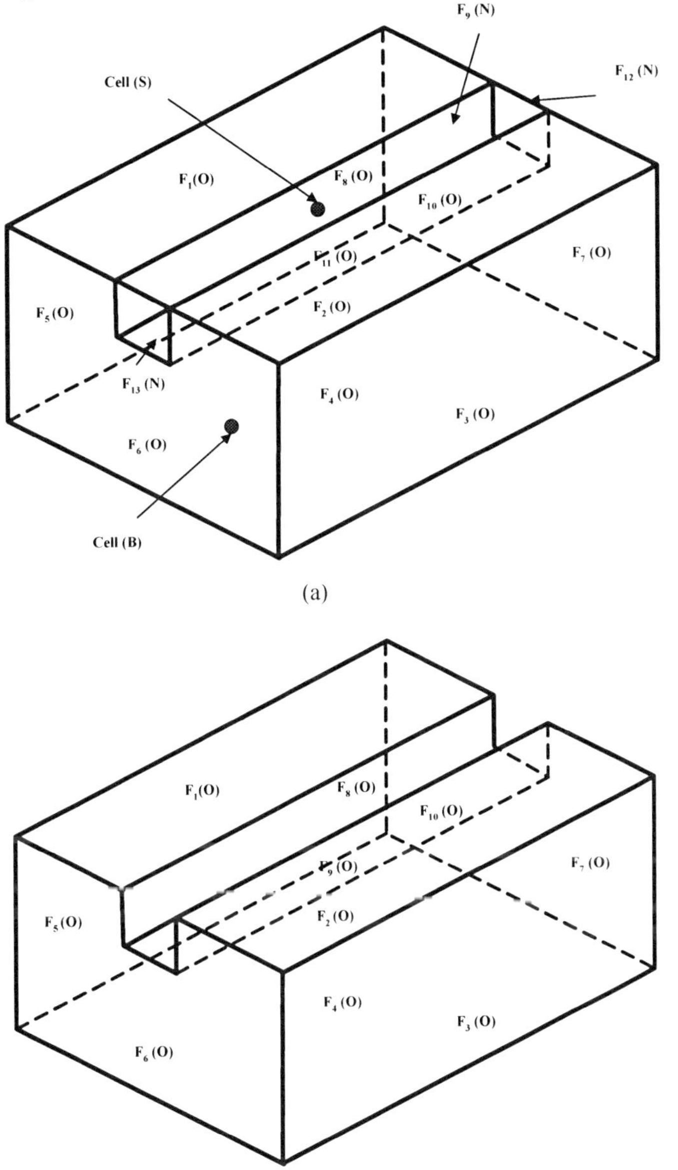

Figure 6.15. B-Rep evaluation (a) boundary detection; (b) boundary evaluation

6.5.3.3 Modifying a Feature Instance

To modify a feature instance, we shall first check which features are really involved in the modeling operation. This checking is based on the feature dependency graph and the global feature precedence order. Next, features involved will be removed from the part model. Finally, features with modified properties will be added to the model.

6.5.3.4 B-rep Evaluation

As the cellular model of a part contains much more information than only the boundary of the final part, which means it also contains faces that are not on the boundary. Therefore, the B-rep evaluation of the cellular model requires boundary detection of every face in the cellular model. The following rules are used to carry out the boundary detection of faces in the cellular model:

- For each single-side face of a cell, the nature of the cell determines whether it is on-boundary or not. Additive nature of the cell means the face is *on-boundary*; subtractive nature of the cell means the face is *not-on-boundary*.
- For each double-side face of a cell, if the cell has a different nature with the partner cell that shares the same face, this double-side face is *on-boundary*; otherwise, it is *not-on-boundary*.

Note that the nature of a cell is determined by the nature of its last owner in the cell owner list. The sequence of cell owner list is dynamically maintained according to the unique feature precedence order, see Section 6.5.2. On the basis of boundary detection of each face in the cellular model, the B-rep evaluation of the cellular model can be carried out in steps as follows:

- Walk through all the faces and find all the faces that are *not-on-boundary*;
- Remove all the cells and the faces that are *not-on-boundary*;
- Merge adjacent faces that have the same geometry.

Also taking the part shown in Figure 6.13(b) as an example, on the basis of rules for boundary detection, the result of boundary detection is shown in Figure 6.15(a). Then, the B-Rep evaluation of the cellular model can easily be realized by removing three faces (F_9, F_{12} and F_{13}) that are *not-on-boundary*. The result is shown in Figure 6.15(b). Note that in Figure 6.15, O represents *on-boundary*; N represents *not-on-boundary*.

6.6 A Case Study

The proposed feature-oriented database has been implemented coupled with a geometrical modeling kernel, ACIS. Design features and constraints have been defined and some example parts have been tested. Figure 6.16 illustrates the creation of an example part which is made up of a *base_block*, a *vertical_support*, a *rib*, a *cylinder and* two *through_hole* features.

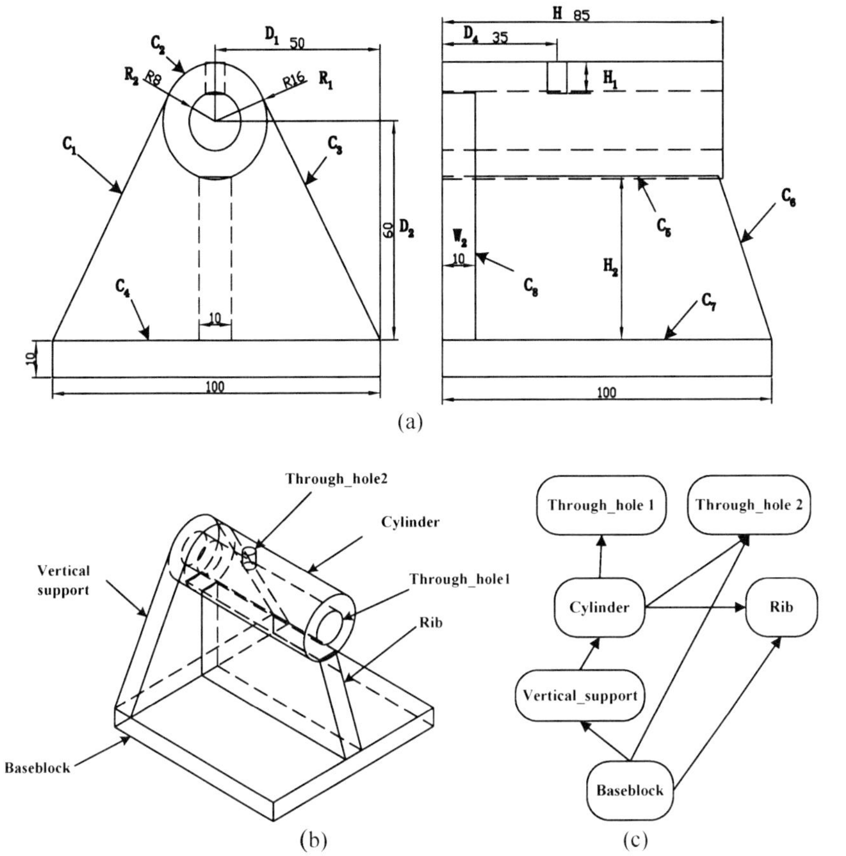

Figure 6.16. A case part (a) creation of the case part; (b) modeling result; (c) feature dependency graph

All the parameters and constraints are listed in Table 6.1. The precedence order of the part can be generated according to the rules described in Section 6.5.2 as follows: *Base_block* → *vertical_support* → *cylinder* → *rib* → *through_hole 1* → *through_hole 2*.

If the designer wishes to add a *cylinder_boss* feature on the top of *cylinder* overlapping with the *hole2* as shown in Figure 6.15, feature overlapping detection and semantic constraint checking (semantic constraints here refer to *through_hole2* must have both top and bottom face *not_on_boundary*) will be carried out. In this case, constraint conflict happens because the semantic constraint of the *through_hole2* cannot be satisfied if the current precedence relation is kept. Therefore, a message will be generated by the system to prompt the user on how to solve such problems (via changing the precedence order of those two features as shown in Table 6.2).

After modification, the feature precedence order of the part will be changed from: base_block→ vertical_support → cylinder → rib → through_hole1→

through_hole2→ boss to: base_block → vertical_support → cylinder → rib → through_hole1→ boss → through_hole2.

Table 6.1. Features, constraints and parameters in the example part

Feature	Constraints and parameters
Baseblock	Determined by two position points (0,0,0) and (100,100,10). length of baseblock = 100; width of baseblock = 100; height of baseblock = 10
Vertical_ support	Geometric constraints: verticalsupport_start coplane with baseblock_left; Radius of arc C_2 , R_1 =16; Center of arc C_2 determined by two distance constraints: D_1=50; D_2=60; C_1 tangent to C_2 ; C_3 tangent to C_2; Extrusion length W_2=10
Cylinder	Geometric constraint: Cylinder_top coplane with baseblock_left; Center of cylinder_top concentric to arc C_2 ; Height of cylinder H = 85
Rib	Distance constraint: distance between C5 and C7, $D_3 = D_2 - (R_1 - \sqrt{R_1^2 - (W_1/2)^2})$; Extrusion length W_1=10
Through_hole1	Geometric constraints: Through_hole1_top coplane with cylinder_top; Through_hole1_bottom coplane with cylinder_bottom; Center of through_hole1 concentric to the center of cylinder; Radius of through_hole1 = 8
Through_hole2	Radius of through_hole2 R_3 =3; Through_hole2_topcenter determined by three distance constraints: D_1, D_2+R_1, and D_4 ; Height of through_hole2 $H_2 = R_1 - \sqrt{R_2^2 - R_3^2}$

Table 6.2. Redefining two features

Feature	Constraints and parameters
Cylinder boss	Radius of cylinder boss R_4 = 6; The top center of cylinder boss determined by three distance constraints: D_1, D_4 and D_5 (distance between top center of cylinder boss and top of base block in Z axis; Height of cylinder boss $H_2 = D_5 - D_2 - \sqrt{R_1^2 - R_4^2}$
Through_hole2	Radius of through_hole2 R_3 =3; Top center of through_hole2 coplane with top center of cylinder boss; Top center of through_hole2 concentric with top center of cylinder boss; Height of through_hole2 $H_2 = D_5 - \sqrt{R_2^2 - R_3^2}$

Therefore, the result part model after modification can be generated as shown in Figure 6.17(a). The dependency graph of the modified part can be expressed as shown in Figure 6.17(b).

Subsequently, the designer wishes to remove the *boss* feature. According to the feature dependency graph shown in Figure 6.17(b) and the latest feature precedence order, the removal of the *boss* feature can be done by removing the boss as well as the *hole* feature that depends on the *boss* feature. The final part after removing the *boss* is shown in Figure 6.18.

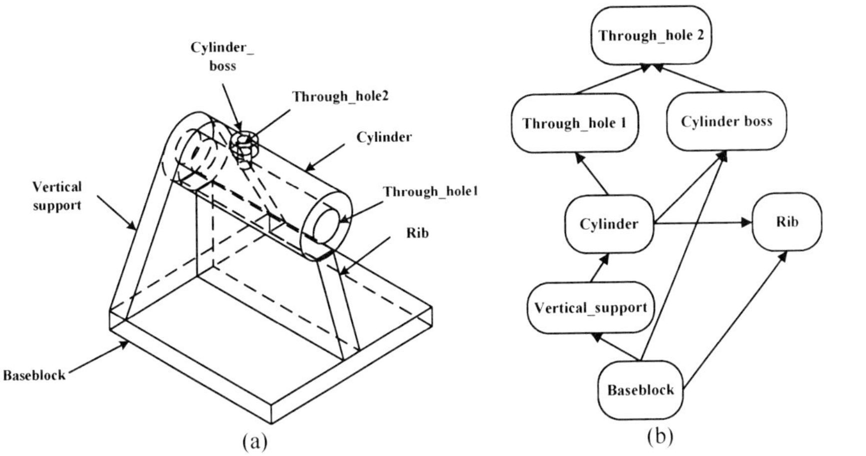

Figure 6.17. Feature model after adding (a) boss a. modeling result; (b) feature dependency graph after adding a boss

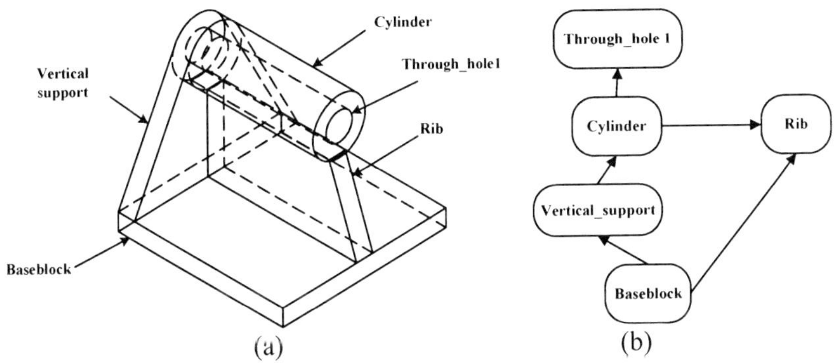

Figure 6.18. Feature model after removing the boss feature (a) modeling result after removing the boss feature; (b) feature dependency graph

6.7 Conclusions

In this chapter, the integration of a fine-grain, feature-oriented database and a solid modeler, is presented. The mapping mechanisms, from EXPRESS-defined generic feature model entities to ACIS workform format, and the integration with the repository database schema are described. Generic algorithms for feature manipulation with the solid molder and database methods are illustrated. Finally, a modeler-supported, history-independent feature model re-evaluation approach is described in detail. Based on the working prototype system, it can be concluded that solid modeler can be effectively integrated with the feature-oriented database to provide low-level geometrical modeling services. This kind of integration can further enable information sharing among different applications and Web enabled engineering collaboration.

6.8 Acknowledgements

The authors gratefully acknowledge the support of technical staff in Design Research Center and CAD/CAM lab of Nanyang Technological University.

6.9 References

[1] Nwagboso, C., Georgakis, P. and Dyke, D., 2004, "Time compression design with decision support for intelligent transport systems deployment," *Computers in Industry,* 54, pp. 291–306.

[2] Terwiesch, C. and Loch, C. H., 1999, "Measuring the effectiveness of overlapping development activities," *Management Science,* 26, pp. 44–59.

[3] Prasad, B., 1996, *Concurrent Engineering Fundamentals: Integrated Product and Process Organization*, Prentice Hall.

[4] Mittra, S. S., 1991, *Principles of Relational Database Systems*, Prentice Hall.

[5] Raflik, M., 1990, *CAD*I Database-An Approach to an Engineering Database,* ECSC-EEC-EAEC.

[6] Regli, W. C. and Gaines, D. M., 1997, "A repository for design, process planning and assembly," *Computer Aided Design*, 29, pp. 895–905.

[7] Kang, S. H., Kim, N., Kim, C. Y., Kim, Y. and O'Grady, P., 1997, "Collaborative design using the world wide Web," *Technical Report*, Dept. Industrial Engineering, Seoul National University, Korea.

[8] Loffredo, D., 1998, *Efficient Database Implementation of EXPRESS Information Models*, PhD Thesis, Rensselaer Polytechnic Institute, New York.

[9] Hoffmann, C. M. and Arinyo, R. J., 1998, "CAD and the product master model," *Computer Aided Design*, 30(11), pp. 905–918.

[10] Wang, H. F. and Zhang, Y. L., 2002, "CAD/CAM integrated system in collaborative development environment," *Robotics and Computer Integrated Manufacturing*, 18, pp. 135–145.

[11] Wang, H. F., Zhang, Y. L., Cao, J., Lee, S. K. and Kwong, W. C., 2003, "Feature-based collaborative design," *Journal of Material Processing Technology*, 139, pp. 613–618.

[12] Tang S. H., Ma Y. S. and Chen, G., 2004, "A feature-oriented database framework for web-based CAx applications," *Computer-Aided Design & Applications*, 1(1–4), pp. 117–125.

[13] Tang, S. H., Ma, Y. S. and Chen, G., 2004, "A Web-based collaborative feature modeling system framework," *Proceedings of the 34th International MATADOR conference*, pp. 31–36.

[14] Chen, G., Ma, Y. S., Thimm, G. and Tang, S. H., 2004, "Unified feature modeling scheme for the integration of CAD and CAx," *Computer-Aided Design & Applications*, 1(1–4), pp. 595–602.

[15] Martino, T. D., Falcidieno, B. and Habinger, S., 1998, Design and engineering process integration through a multiple view intermediate modeler in a distributed object-oriented system environment," *Computer Aided Design*, 30(6), pp. 437–452.

[16] Bidarra, R. and Bronsvoort, W. F., 2000, "Semantic feature modeling," *Computer Aided Design,* 32, pp. 201–225.

[17] Bidarra, R., van den Berg, E. and Bronsvoort, W. F., 2001, "Collaborative modeling with features," *Proceedings of DETC'01 ASME Design Engineering Technical Conferences*, Pittsburgh, Pennsylvania.

[18] Bronsvoort, W. F., Bidarra, R., Dohmen, M., van Holland, W. and de Kraker, K. J., 1997, "Multiple-view feature modeling and conversion," In: Strasser, W., Klein, R., and Rau, R. (Eds.), *Geometric Modeling: Theory and Practice - The State of the Art*, Springer, Berlin, 159–174.

[19] Bronsvoort, W. F., Bidarra, R. and Noort, A., 2001, "Semantic and multiple-view feature modeling: towards more meaningful product modeling," In: Kimura F, (Eds.) *Geometric Modeling - Theoretical and Computational Basis towards Advanced CAD Applications,* Kluwer Academic Publishers, 69–84.

[20] Kim, J. and Han, S., 2003, "Encapsulation of geometric functions for ship structural CAD using a STEP database as native storage," *Computer Aided Design*, 35, pp. 1161–1170.

[21] Bhandarkar, M. P., 2000, "STEP-based feature extraction from STEP geometry for agile manufacturing," *Computers in Industry*, 41, pp. 3–24.

[22] Dereli, T. and Filiz, H., 2002, "A note on the use of STEP for interfacing design to process planning," *Computer Aided Design*, 34, pp. 1075–1085.

[23] Fu, M. W., Ong, S. K., Lu, W. F., Lee, I. B. H. and Nee, A. Y. C., 2003, "An approach to identify design and manufacturing features from a data exchanged part model," *Computer Aided Design*, 35, pp. 979–993

[24] Holland, P., Standring, P. M, Long, H. and Mynors, D. J., 2002, "Feature extraction from STEP (ISO10303) CAD drawing files for metal-forming process selection in an integrated design system," *Journal of Materials Processing Technology*, 125–126, pp. 446–455.

[25] Suh, Y. S. and Wozny, M. J., (1997) "Interactive feature extraction for a form feature conversion system," In: Hoffmann, C. M. and Bronsvoort, W. F. (Eds.), *Solid Modeling '97, Fourth Symposium on Solid Modeling and Applications*, 11–122, New York, ACM Press.

[26] Li, W. D., Ong, S. K. and Nee, A. Y. C., 2002, "Recognizing manufacturing features from a design-by-feature model," *Computer Aided Design*, 34, pp. 849–868.

[27] Gao, J., Zheng, D. T., and Gindy, N., 2004, "Extraction of machining features for CAD/CAM integration," *International Journal of Advanced Manufacturing Technology*, 24, pp. 573–581.

[28] Noort, A., Hoek, G. F. M. and Bronsvoort, W. F., 2002, "Integrating part and assembly modeling," *Computer Aided Design*, 34, pp. 899–912.

[29] Bronsvoort, W. F. and Noort, A., 2004, "Multiple-view feature modeling for integral product development," *Computer Aided Design*, 36, pp. 929–946.

[30] Ma, Y. S. and Tong, T., 2003, "Associative feature modeling for concurrent engineering integration," *Computers in Industry*, 51, pp. 51–71.

[31] Ma, Y. S., Britton, G. A., Tor, S. B., Jin, L. Y., Chen, G. and Tang, S. H., 2004, "Design of an feature-object-based mechanical assembly library," *Computer-Aided Design & Applications*, 1(1–4), pp. 379–403.

[32] Geelink, R., Salomons, O. W., Van Slooten, F., Van Houten, F. J. A. M. and Kals, H. J. J., 1995, "Unified feature definition for feature based design and feature based manufacturing," *Proceedings of the 15th Annual International Computers in Engineering Conference and the 9th Annual ASME Engineering Database Symposium*, pp. 517–533.

[33] Chen, G., Ma, Y. S., Thimm, G. and Tang, S. H., 2005, "Knowledge-based reasoning in a unified feature modeling scheme," *Computer-Aided Design & Applications*, 2(1-4), pp. 173–182.

[34] Chen, G., Ma, Y. S., Ming, X. G., Thimm, G., Lee, S. S. G., Khoo, L. P., Tang, S. H. and Lu, W. F., 2005, "A unified feature modeling scheme for multi-applications in PLM," *Proceedings of The 12th ISPE International Conference on Concurrent Engineering (CE2005): Research and Applications -Next Generation Concurrent Engineering*, Sobolewski M and Ghodous P (Eds.), ISPE, Dallas, pp. 343–348.

[35] ISO, 1994, *Industrial Automation Systems and Integration — Product Data Representation and Exchange — Part 11: Description Methods: The EXPRESS Language Reference Manual*, ISO 10303-11, Geneva.

[36] *ACIS Online Help User's Guide*. Available at http://www.spatial.com.

[37] *ORACLE online documentation*. Available at http://www.oracle.com.

[38] Sannella, M., 1993, "The SkyBlue constraint solver and its applications," *First Workshop on Principles and Practice of Constraint Programming*.

[39] Kramer, G. A., 1992, *Solving Geometric Constraints Systems: A Case Study in Kinematics*, MIT Press, USA.

[40] Wirth, N., 1976, *Algorithms + Data Structure = Programs*, Prentice-Hall.

[41] Bidarra, R., Madeira, J., Neels, W. J. and Bronsvoort, W. F., 2005, "Efficiency of boundary evaluation for a cellular model," *Computer Aided Design*, 37, pp. 1266–1284.

7

A Web-based Framework for Distributed and Collaborative Manufacturing

M. Mahesh, S. K. Ong and A. Y. C. Nee

Department of Mechanical Engineering
National University of Singapore, Singapore

To develop a distributed and collaborative manufacturing system at an enterprise level, different domains like manufacturability evaluation, resource coordination, process planning, scheduling, fabrication, and logistics, have to be seamlessly integrated for product and process development. This integration necessitates a need to formalize, encode and share manufacturing related knowledge. Collaborative manufacturing provides a mechanism for information sharing and decision making between the various domains. Further, with the introduction of software agents, individual manufacturing elements are able to cooperate to promote collaborative manufacturing. This chapter addresses the development of a web-based framework for distributed and collaborative manufacturing.

7.1 Introduction

Decisions in integrated product development involve a number of independent elements like part design, evaluation, process planning, scheduling, production, *etc*., to final delivery of the manufactured part. While considering the development of distributive manufacturing at an enterprise level, different domains like manufacturability evaluation, resource coordination, process planning, scheduling, fabrication and logistics play important roles in integrated product and process development. This necessitates a need to formalize, encode and share manufacturing related knowledge between various domains. With the advent of collaborative manufacturing, the ease of information sharing for decision making between the participating elements has become very useful. Furthermore, with the introduction of software agents, individual manufacturing elements are able to cooperate to promote collaborative manufacturing. This research addresses the development of a web-based framework for distributed and collaborative manufacturing of engineering parts. In essence, a Multi-Agent System (MAS) is proposed where different domains in manufacturing are represented as functional modules. The implementation of the agent system is accomplished using the Java programming language. A case study is presented considering an engineering company with distributed facilities.

Generally, designers, process engineers and machine operators are all capable of individually handling complexities in decision making involved in their respective domains of manufacturing. However, in a distributed scenario, it is crucial for the different domains to cooperate to handle complexities. For example, a designer must evaluate the manufacturability of his designed part, a process engineer must evaluate and plan the job scheduling depending on the availability of machine-tools, and the machine operator must evaluate whether the part could be fabricated meeting deadlines, machine breakdown and so on. To conduct such evaluations it is essential for the distributed domains to cooperate, advertise, interact and advise each other to complete a job task. Such cooperation and collaboration ensures a successful job completion.

In this chapter we aim to address a framework for distributed and collaborative manufacturing of engineering parts. Apart from addressing a framework we further encapsulate the various manufacturing related elements into functional agents [1] for implementing a distributed MAS for manufacturing. Numerous researchers have applied agent technologies to perform tasks like part production control on either the shop floor level or in a distributed manner. Notably among them, Lin and Solberg [2] proposed a framework to realize integrated shop floor manufacturing, Tan, *et al.*, [3] proposed to integrate design, manufacturing and shop-floor control, Francisco and Douglas [4] developed a framework for distributed task planning and manufacturing, Sikora and Shaw [5] presented a coordination mechanism in a multi-agent scheduling system. Howley, *et al.*, [6] presented a compromising model in an agent-based environment, Klein and Lu [7] proposed a model for cooperative design, Lander, *et al.*, [8] also proposed a cooperating expert framework to support cooperative problem-solving, Werkman, *et al.*, [9] developed a Design Fabricator Interpreter system and so on. More recently, Odrey and Mejia [10] proposed an approach addressing the issue of combining the discipline of hierarchical systems with the agility of MASs. Blecker and Graf [11] discussed a coordinated application for mass customization using multi-agent systems in internet based production environments for production planning and control. Ong and Sun [12] have proposed a Web-based distributed architecture for developing a platform-independent real-time monitoring system through mobile agents. Shin and Jung [13] proposed a negotiation mechanism, called a Mobile Agent-based Negotiation Process (MANPro), which applies a mobile agent system to the process of information exchange. Boonserm, *et al.*, [14] described a framework to facilitate the collaboration of engineering tasks, particularly process planning and analysis for globalized manufacturing activities. Liu and Young [15] presented an approach which utilizes a combination of information and knowledge models to support global manufacturing coordination decision-making. Jiao, *et al.*, [16] applied the multi-agent system paradigm for collaborative negotiation in a global manufacturing supply chain network. Nahm and Ishikawa [17] discussed a MAS framework for integrated product design in a computer network-oriented Concurrent Engineering (CE) environment.

All the earlier proposed systems have in different ways addressed specific issues like using multiple-way and multiple-step negotiation, using black boards, sub-contracts, conflict resolution, *etc*. However, in this chapter we propose a framework to incorporate a flexible system for multiple coordinated tasks starting from part design to final scheduling.

7.2 Distributed and Collaborative Manufacturing

Figure 7.1 presents the various manufacturing elements involved in decision making in a distributed and collaborative manufacturing environment. However, considering its implementation as a distributed agent-based manufacturing system, there exist certain constraints while encoding and sharing of manufacturing related knowledge and information, between the participating agents. In this research we address the implementation of the manufacturing elements for decision making in a distributed fashion while taking advantage of the Internet and Web-based computing facilities. The goal of implementing a distributed system is to aid designers to send design data, perform evaluation and obtain reliable results pertaining to a manufacturing job, a mold fabrication for example.

In such a collaborative environment, it is advantageous for designers, process-engineers or machinists with insufficient experience to follow certain structured approaches to reason out critical design parameters affecting manufacturability. Hence during implementation, a proper rule-based knowledge repository for manufacturability evaluation and planning plays a crucial role in integrated product and process development.

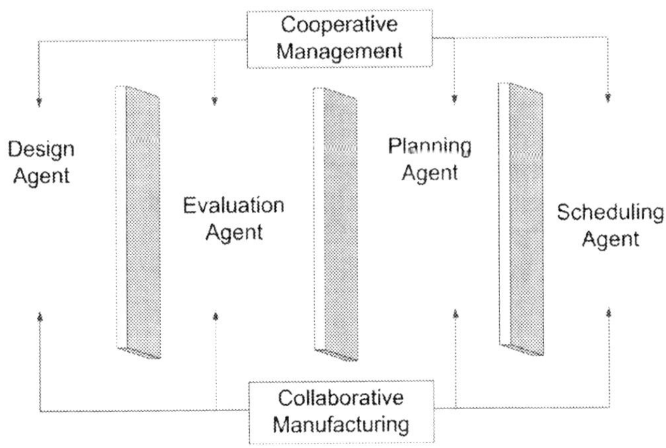

Figure 7.1. Elements involving a decision in manufacturing

Manufacturability evaluation focuses on product design and verification, to suggest modifications or redesign alternatives that are functionally acceptable and compatible with the selected manufacturing processes. Such evaluations are certainly skill-intensive activities demanding a wide variety of design expertise and knowledge of the manufacturing processes that are available. Decision-making processes in manufacturability evaluation are generally based on product geometry. In the proposed system such expertise information sharing is achieved through interaction between the various agents.

The rest of the chapter is organized as follows: the next section presents the proposed framework with implementation aspects for a distributed manufacturing system. The participating functional manufacturing agents are also briefly discussed. The subsequent section discusses a case study considering an engineering company with distributed facilities. Finally, the chapter is concluded with suggestions and scope for future related research work.

7.3 Proposed Framework and Implementation

Figure 7.2 presents an integrated framework comprising of a Designer Computer-Aided Design (CAD) interface communicating with a MAS. The MAS interface may comprise and include any number of manufacturing related functional agents, like Design Mediator Agent (DMA), Manufacturability Evaluation Agent (MEA), Manufacturing Capability Agent (MCA), Process Planning Agent (PPA), Manufacturing Scheduling Agent (MSA), etc. Figure 7.3 presents a more detailed framework of the agent interaction and communication as part of the multi-agent system interface. All manufacturing agents in the framework, although distributed physically, can connect, communicate and share information with each other through the Web.

In the proposed framework all manufacturing agents communicate over the Internet via a bundle of Knowledge Query and Manipulation Language (KQML) messages to transfer data and information among each other. KQML [18] is a common communication protocol used for negotiation and interaction between agents. Developed on JATLite [19] (Stanford University) each agent understands the message it receives and executes specific tasks. A performative header at the beginning of a message defines the implied message for the recipient agent to understand and act. Performatives developed as part of the Agent Language include instructions for manufacturing like: 'tell', 'evaluation', 're-evaluation', 'find_model', 'need_model', 'process_planning', 'scheduling', 'job_schedule', *etc.*

There exists a Router [18] in JATLite, which is a specialized application that receives messages from the registered agents and routes the messages to the appropriate receivers. To coordinate the activities of the functional agents, a central coordination agent exists, namely, the Facilitator or Manufacturing Managing Agent (MMA). Unlike most agents in previously reported systems that were intentionally designed with a compact all-in-one structure to obtain beneficial characteristics, such as prompt responsiveness, integrated control, *etc.*, a modularized structure is used in this research to implement the agents to acquire adaptiveness and upgradeability in these agents. Though these agents are dispersed, this modularized structure is wrapped and connected using the JATLite template. Thus the internal structure of each agent [20, 21] is composed of components that include the JATLite template, I/O modifier, work engine, and knowledge-based pool. Thus, all the separate components form a close entity and still maintain the desirable characteristics of agents with integrated structures. With limited types of manufacturing agents with knowledge of their specific stages/functions in a product development, a centralized control mechanism is adopted. The MMA assumes the role of central control and is responsible for solving conflicts relating to the coordination among agents, while the other agents by themselves are responsible for solving specific problems, such as manufacturability, process planning, *etc*.

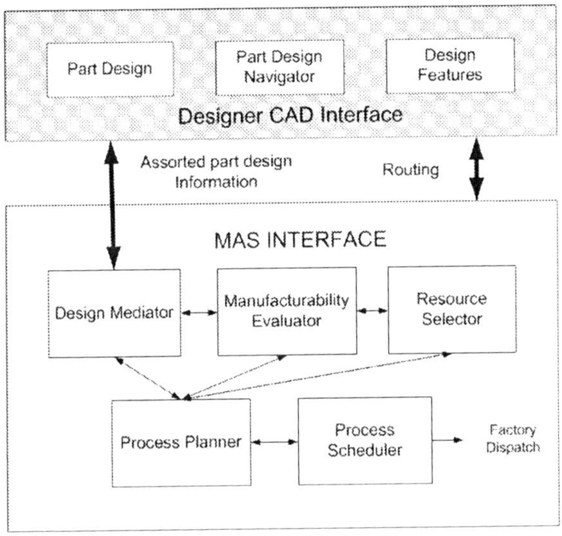

Figure 7.2. An integrated framework (the Designer CAD interface communicating with the multi-agent system interface)

In a distributed framework, each participating functional agent receives data input from the MMA, carries out its operation, and sends the data output back to the MMA. When a message is transmitted among the agents, it is wrapped by the KQML in a standard format. The destination agent can de-compose the KQML and retrieve the embedded message. Such architecture allows each functional agent to be independent in its task execution and the breakdown or malfunction of any functional agent will not affect the operation of other agents as long as the MMA is functioning. Table 7.1 presents a brief description of the work engine of each functional agent currently constituting the developed distributed network. The JATLite template and Java-based programming are very useful when developing newer agents to be part of the web based system. It is sufficient only to concentrate on developing the functional part of an agent since the basic structure already exists.

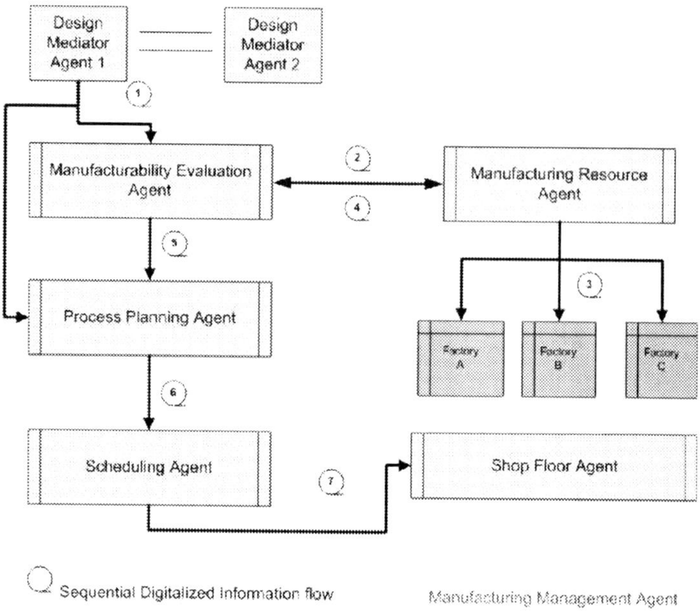

Figure 7.3. Distributed MAS interface

Table 7.1. Agents' work engine

Agent	**Description**
MEA	This work engine is an inference-reasoning engine connected with two rule-based knowledge pools: rules on the manufacturing principles and machining processes for individual features.
MCA	This work engine is a forward searching engine that compares the available factories with the required resources to identify eligible factories.
PPA	The PPA employs a genetic algorithm-based engine to perform process planning for a feature-based part.
MSA	The MSA work engine uses an integrated genetic algorithm and Gantt chart approach to determine an optimal production schedule in a selected factory.

7.4 A Case Study

A case study is given here considers an engineering company A with distributed facilities for fabricating engineering parts through machining. Generally information/data flow in a typical engineering company from design through manufacturing is in line with Figure 7.4. The designer does the part design followed by design evaluation after which the process engineer is responsible for resource coordination, planning and the final job scheduling.

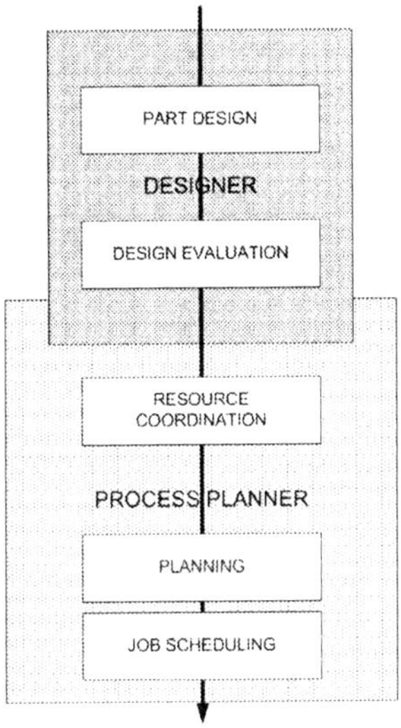

Figure 7.4. Information Flow in Manufacturing

In this example of a distributed scenario, the engineering and manufacturing facilities are assumed to be distributed over Asia, where all the distributed facilities are represented as different functional manufacturing agents responsible to process the tasks required of the engineering company A. Figure 7.5 illustrates such distributed engineering facilities. The distributed facilities may include an engineering design agent, manufacturability evaluation agent, process planning agent, scheduling agent and factories, all controlled by a manufacturing managing agent. A designed part as shown in Figure 7.6 is used to demonstrate the developed system. The designed part information in terms of individual geometric features, dimensions and location are presented in Table 7.2. The precedence information is presented in Table 7.3.

The Designer designs the engineering part using the Designer CAD interface (Figure 7.2), after which he submits the part information (geometric features, design dimensions, operation precedence) to the Design Mediator Agent (DMA as a message under the performative 'part_features'. The DMA in turn forwards this message to the MMA, which routes the message to the MEA and PPA under a performative header 'evaluation'. On receiving the message, MEA interprets the message from the performative header and performs the manufacturability evaluation [22]. Figure 7.7 presents a snap shot of the manufacturability evaluation and its inference process by MEA. During the manufacturability evaluation process, if a discrepancy arises, for example, an improperly defined geometric feature or design, an inappropriate error (conflict resolution) is triggered for a corrective action.

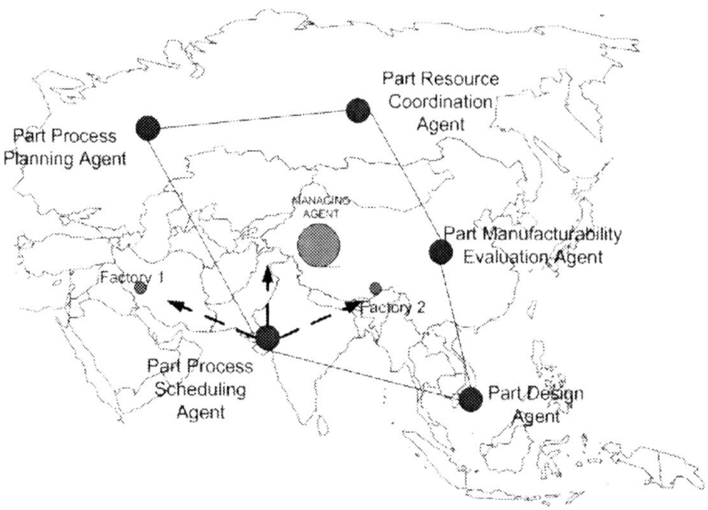

Figure 7.5. Web-based distributed engineering facilities

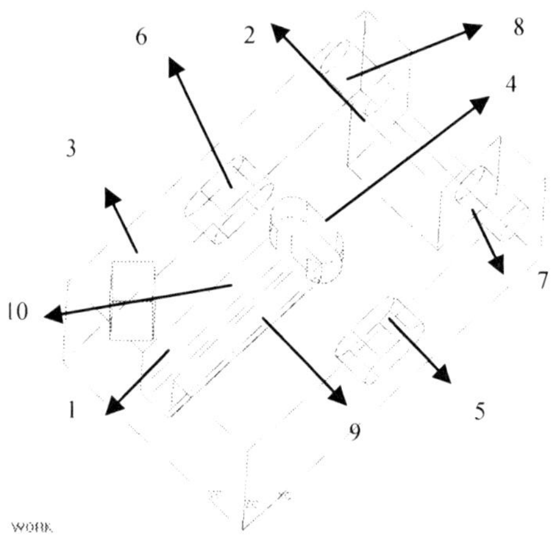

Figure 7.6. An engineering part

On successful evaluation however, the results with a performative header, 'need_model' is automatically sent to the MMA. Once the MMA receives the message, it routes the message to MCA with a performative 'find_model'. Figure 7.8 presents a snap shot of the MCA agent for resource selection. After the inference process, the results are sent back once again to the MEA by the MMA under the performance header 'evaluation_again' for re-evaluation of the inferred results. MEA re-evaluates the capabilities of the manufacturing resources selected for processing the individual features. On successful re-evaluation, the results are forwarded to MMA for the next step of process planning. If a conflict occurs and an unsuitable resource is encountered even after re-evaluation, an appropriate message is triggered to indicate unsuitable resources or alternatively requesting a correct model.

The successful re-evaluation results sent to MMA will be dispatched to the PPA agent (Figure 7.9) under the performance header 'process_planning'. The PPA agent having received the required messages performs the process planning task [23].

Table 7.2. Design data of the engineering part

Geometric Features	Location Parameters	Dimensions Parameters
Rect-Pocket (1)	0.0,25.0,20.0	50.0,10.0,20.0
Rect-Pocket (2)	80.0,10.0,10.0	20.0,30.0,30.0
Rect-Pocket (3)	5.0,50.0,20.0	20.0,10.0,10.0
Rect-Slot (4)	50.0,20.0,30.0	10.0,20.0,10.0
Rect-Slot (5)	40.0,0.0,20.0	20.0,10.0,10.0
Rect-Slot (6)	40.0,60.0,20.0	20.0,10.0,10.0
Single-Hole (7)	90.0,0.0,30.0	15.0,10.0D
Single-Hole (8)	90.0,45.0,30.0	15.0,10.0D
Fillet (9)	0.0,25.0,20.0	50.0,2.0D
Fillet (10)	0.0,35.0,20.0	50.0,2.0D

Table 7.3. Precedence information of the engineering part

Operation ID	Predecessor	Successor
Op1	--	Op4, Op9, Op10
Op2	--	Op7, Op8
Op3	--	--
Op4	Op1	--
Op5	--	--
Op6	--	--
Op7	Op2	--
Op8	Op2	--
Op9	Op1	--
Op10	Op1	--

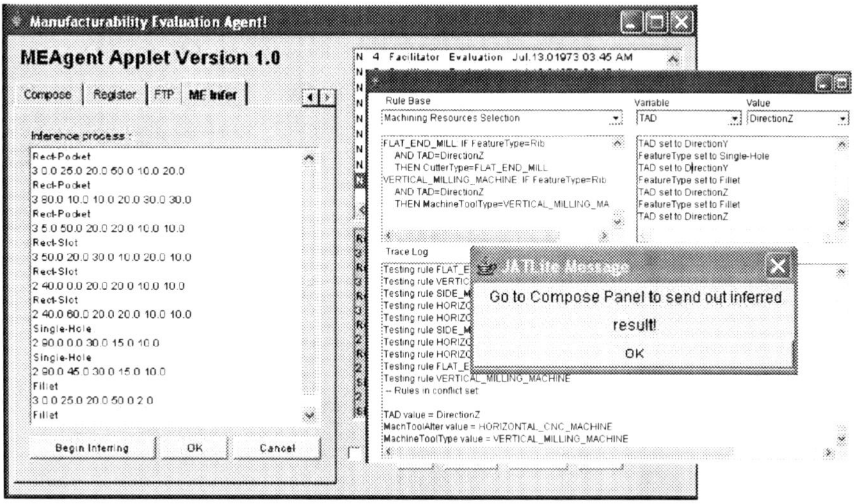

Figure 7.7. Manufacturing evaluation agent

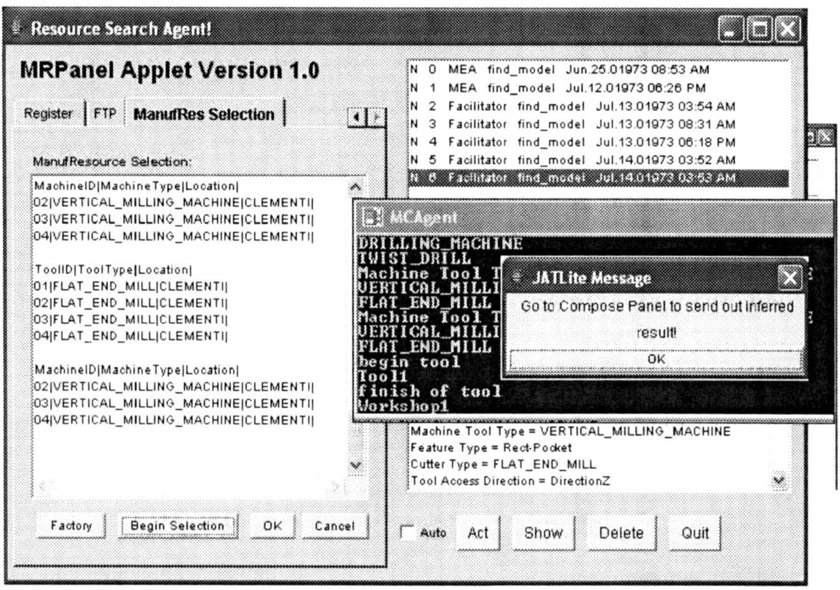

Figure 7.8. Manufacturing capability agent

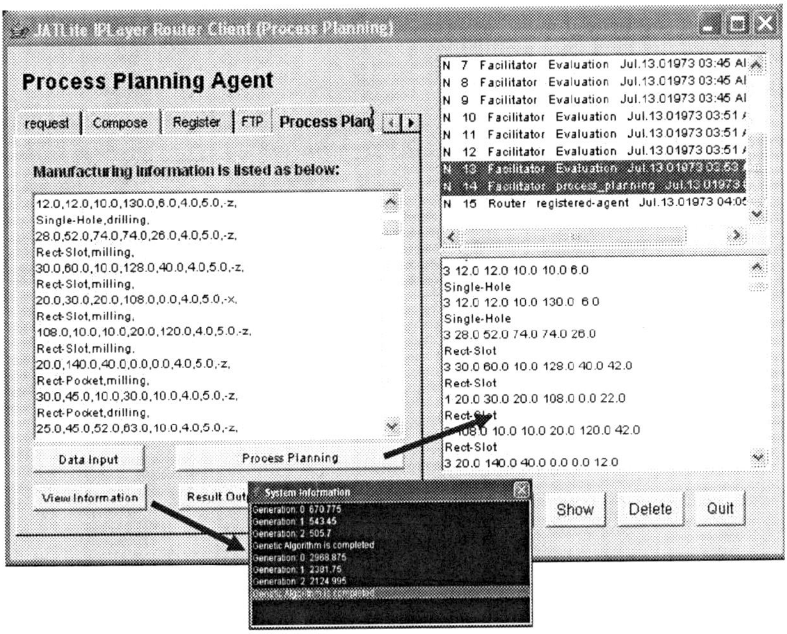

Figure 7.9. Process planning agent

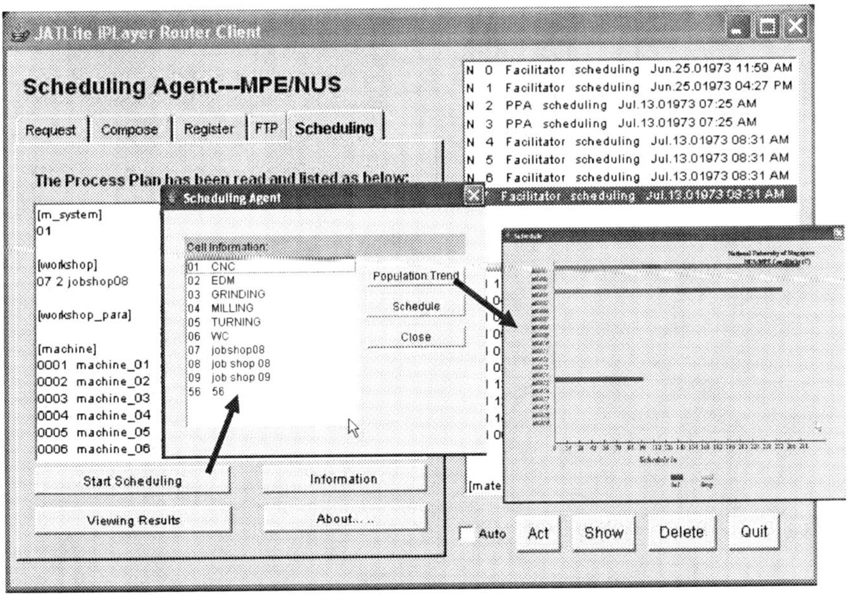

Figure 7.10. Manufacturing scheduling agent

On successful completion of the process planning task, the results are dispatched for manufacturing scheduling to the MSA (Figure 7.10) agent by the MMA agent [20, 21]. The scheduling results could be later dispatched to the shop floor for actual fabrication. In such engineering manufacturing applications the information/data flow follows a logical sequence to ensure the completeness of the job task as discussed above. The developed system offers the flexibility of adding newer functional agents. For example a new fault diagnosis agent can become a part of the system by registering its functionalities with the MMA. Thus the system allows for further expansion, improvement and customization according to the functional requirements.

7.5 Conclusions

Considering the development of distributive manufacturing at the enterprise level, this chapter presents a generic web-based framework for collaborative manufacturing for engineering parts. The proposed framework is implemented as an agent-based distributed manufacturing system based on JATLite and Java programming. The proposed framework and system offer the flexibility to be used for distributed and collaborative manufacturing of engineering parts. To envisage the implementation a case study is also presented in this chapter. Future works include real-time industrial implementation and testing besides exploring management, organizational limitations, and standards. Possible extension of the framework to other areas like integrated virtual manufacturing, rapid prototyping, *etc.*, can also be explored.

7.6 References

[1] Caglayan, A. and Harrison, C., 1997, *Agent Sourcebook*, John Wiley & Sons Inc., New York, NY, USA.

[2] Lin, G. and Solberg, J., 1992, "Integrated shop floor control using autonomous agents," *IIE Transactions*, 24(3), pp. 57–71.

[3] Tan, G. W., Hayes, C. C. and Shaw, M., 1996, "An intelligent-agent framework for concurrent product design and planning," *IEEE Transactions on Engineering Manufacturing Management*, 43(3), pp. 297–306.

[4] Francisco, P. M. and Douglas, H. N., 1996, "Multi-agent mediator architecture for distributed manufacturing," *Journal of Intelligent Manufacturing*, 7(4), pp. 257–270.

[5] Sikora, R. and Shaw, M. J., 1998, "A multi-agent framework for the coordination and integration of information systems," *Management Science*, 44(11), pp. 65–78.

[6] Howley, B., Cutkosky, M. and Biswas, G., 1999, "Compromising and sharing dynamic models in an agent-based concurrent engineering environment," *Proceedings of the American Control Conference*, California, USA, pp. 3147–3153.

[7] Klein, M. and Lu, S. C. Y., 1990, "Conflict resolution in cooperative design," *Artificial Intelligence in Engineering*, 4(4), pp. 168–180.

[8] Lander, S., Lesser, V. R. and Connell, M. E., 1991, "Conflict resolution strategies for cooperating expert agents," In Deen S. M. (Eds.), *CKBS-90 - Proceedings of the International Working Conference on Cooperating Knowledge Based Systems*, Springer-Verlag: Heidelberg, Germany, pp. 183–200.

[9] Werkman, K. J., Wagaman, S. J., Hillman, D. J., Barone, M. and Wilson, J. L., 1990, "Design and fabrication problem solving through cooperative agents: designer fabricator interpreter system," *NSF-ERC-ATLSS Technical Report No. 90-05*, Lehigh University Bethlehem.

[10] Odrey, N. G. and Mejia, G., 2003, "A re-configurable multi-agent system architecture for error recovery in production systems," *International Journal of Robotics and Computer Integrated Manufacturing*, 19(1-2), pp. 35–43.

[11] Blecker, T. and Graf, G., 2003, "Multi agent systems in internet based production environments- an enabling infrastructure for mass customization," *Proceedings of the Second Interdisciplinary World Congress on Mass Customization and Personalization*, Munich, Germany, pp. 1–27.

[12] Ong, S. K. and Sun, W. W., 2003, "Application of mobile agents in a Web-based real-time monitoring system," *International Journal of Advanced Manufacturing Technology*, 22(1–2), pp. 33–40.

[13] Shin, M. and Jung, M., 2004, "MANPro: mobile agent-based negotiation process for distributed intelligent manufacturing," *International Journal of Production Research*, 42(2), pp. 303–320.

[14] Boonserm, K., Richard, A. W., Hyunbo, C. and Albert, J., 2004, "Integration framework of process planning based on resource independent operation summary to support collaborative manufacturing," *International Journal of Computer Integrated Manufacturing*, 17(5), pp. 377–393.

[15] Liu, S. and Young, R. I. M., 2004, "Utilizing information and knowledge models to support global manufacturing co-ordination decisions," *International Journal of Computer Integrated Manufacturing*, 17(6), pp. 479–492.

[16] Jiao, J. R., You, X. and Kumar, A., 2006, "An agent-based framework for collaborative negotiation in the global manufacturing supply chain network," *Robotics and Computer Integrated Manufacturing*, 22(3), pp. 239–255.

[17] Nahm, Y. E. and Ishikawa, H., 2005, "A hybrid multi-agent system architecture for enterprise integration using computer networks," *International Journal of Robotics and Computer-Integrated Manufacturing*, 21(3), pp. 217–234.

[18] Finin, T., Fritzon, R., Mckay, D. and McEntire, R., 1994, "KQML as an agent communication language," *Proceedings of the 3rd International Conference on Information and Knowledge Management*, Gaithersburg, MD, USA, ACM Press, pp. 456–463.

[19] JATLite Home Page at Stanford: http://java.stanford.edu. Last Access 16 October 2005.

[20] Jia, H. Z., Ong, S. K., Fuh, J. Y. H., Zhang, Y. F. and Nee, A. Y. C., 2004, "An adaptive upgradable agent- based system for collaborative product design and manufacture," *Robotics and Computer-Integrated Manufacturing*, 20(2), pp. 79–90.

[21] Jia, H. Z., Fuh, J. Y. H., Nee, A. Y. C. and Zhang, Y. F., 2002, "Web-based multi-functional scheduling system for a distributed manufacturing environment," *International Journal of Concurrent Engineering: Research and Application*, 10(1), pp. 27–39.

[22] Zhang, Y. F., Fuh, J. Y. H. and Wang, G. J., 2002, "Agent-based manufacturing resource planning," *Proceedings of the International Manufacturing Leaders Forum*, Adelaide, Australia, pp. 136–141.

[23] Lin, G. and Solberg, J., 1992, "Integrated shop floor control using autonomous agents," *IIE Transactions*, 24(3), pp. 57–71.

8

Wise-ShopFloor: A Portal toward Collaborative Manufacturing

Lihui Wang

Integrated Manufacturing Technologies Institute
National Research Council of Canada, Canada

This chapter presents the principles of a Web portal system named Wise-ShopFloor, including system architecture, information flow, and a proof-of-concept prototype enabled by Web and Java technologies. It is designed to use the popular client-server architecture, VCM (View-Control-Model) and publish-subscribe design patterns for effective data sharing during collaboration. A case study of Distributed Process Planning (DPP) linking to Web-based rapid machining is carried out to demonstrate the effectiveness of this approach toward Web-based collaboration.

8.1 Introduction

Recently, collaborative manufacturing has emerged as the norm of manufacturing in a distributed environment. This is largely due to the global business decentralization and manufacturing outsourcing. To stay competitive in the dynamic global market, companies with distributed factories or divisions are demanding a new way of effective collaborations among themselves and even between their suppliers and outsourced service providers. Among many other factors, flexibility, timeliness and adaptability are identified in this research as the major characteristics to bring dynamism to collaborative manufacturing. Distributed manufacturing processes are complex, especially at machining shop floors where a large variety of products, usually in small batch sizes, are handled dynamically. The dynamic environment requires an adaptive system architecture that enables distributed planning, dynamic scheduling, real-time monitoring, and remote control. It should be responsive to both varying collaboration needs and unpredictable changes of distributed production capacity and functionality. An ideal shop floor should be the one that uses real-time manufacturing intelligence to achieve the best overall performance with the least unscheduled downtime. However, traditional methods are based on off-line advance processing and thus are impractical if applied directly to this dynamic collaborative environment. In response to the requirements and to coordinate the dynamic activities in collaborative manufacturing, a sensor-driven and Web-based planning and control approach is needed to achieve the dynamism in the distributed manufacturing environment.

The objective of this research is to develop methodologies and a *Wise-ShopFloor* (Web-based integrated sensor-driven e-ShopFloor) framework for distributed planning, dynamic scheduling, real-time monitoring, and remote control supported by sensors, Java technologies and the Web infrastructure. The Wise-ShopFloor is designed to use the popular client-server architecture, VCM (View-Control-Model) and publish-subscribe design patterns for effective information sharing during collaborative planning and control.

This chapter is organized as follows. In Section 8.2, enabling technologies including Web, Internet, Java 3D and Java servlets are introduced. It is followed by a brief description of the Wise-ShopFloor framework in Section 8.3. Details on adaptive and distributed process planning are presented in Section 8.4, which leads to a Web-based real-time monitoring and control documented in Section 8.5. A case study using planning results for Web-based remote machining are described in Section 8.6. Finally, our contributions are summarized in Section 8.7.

8.2 Enabling Technologies

With the growing manufacturing decentralization, products and services might be distributed everywhere and sourced anywhere along supply chains. Product design and fabrication have shifted rapidly from intra-corporation to global networks. How to coordinate manufacturing activities and keep them under control is a challenging issue. Flexibility, timeliness and adaptability of manufacturing operations are the essential requirements for collaborative manufacturing in such a dynamic environment. Fortunately, the Web infrastructure today is mature enough to form a distributed manufacturing network through client-server interconnections. During the past decade, the Web has been widely used for development of collaborative applications to support dispersed working groups and organizations because of its platform, network and operating system transparency, and its easy-to-use user interface – the Web browser. In addition to the Web technology, Java has brought about a fundamental change in the way that applications are designed and deployed. Java's "*write once, run anywhere*" model has reduced the complexity and cost traditionally associated with producing software solutions on multiple distinct hardware platforms. With Java, the browser paradigm has emerged as a compelling way to produce collaborative applications over the Web. Examples include *WebCADET* [1] for collaborative design and *CyberCut* [2] for rapid machining. In terms of technologies used in the existing systems, HTML, Java applets, ActiveX, and VRML are widely adopted for developing client-side user interfaces. At the server side, technologies including JSP (JavaServer Pages), Java Servlets, and XML are quickly obtaining attentions for new system development. To facilitate a viable collaborative system, its application server must engage users in a 3D graphical interaction in addition to the dialog-like data sharing, because remote users need active and visual aids to coordinate their efforts in a distributed environment. Web and Java technologies are adopted in our research as the enabling technologies for collaborative manufacturing realization. In the current implementation, a thin-client user interface has been developed as a Java applet that runs inside a Web browser. Java 3D has been used to model a physical device that can replace or supplement cameras in providing visual help during remote planning, monitoring and control. A set of decision-making logics have been designed as server-side components for multi-client collaborations. For example, a Java 3D model can communicate with server-side Java servlets for real-time monitoring. Details on how the different technologies can work together are explained below in the Wise-ShopFloor Framework.

8.3 Wise-ShopFloor Framework

The *Wise-ShopFloor* framework [3] has been designed to provide users with a Web-based and sensor-driven intuitive environment where distributed process planning, dynamic scheduling, real-time monitoring and remote control are undertaken. Within the framework, each machine should become an information node and be a valuable resource in the information network. A direct connection to sensors and machine controllers is used to continuously monitor, track, compare, and analyze production parameters. Instead of camera images (usually large in data size), a physical device of interest (*e.g.*, a milling machine) can be represented by a Java 3D scene graph model with behavioral control nodes embedded. Once downloaded from its application server, the 3D model is rendered by the local CPU and can work on behalf of its remote counterpart showing real behavior for visualization at a client side. It remains alive by connecting with the physical device (via servlets) through low-volume message passing (sensor data). As the 3D model is entirely driven by the sensor data and rendered locally for visualization, there is no need of transmitting camera images over the Internet. The largely reduced network traffic makes real-time monitoring and remote control practical for dispersed users connected through the Web. It also enables engineers to make accurate decisions in a timely manner, and to ensure that machines are operating within the defined expectations. Being able to plan and control dynamic shop floor operations from anywhere at any time collaboratively is what this research is aiming at. Figure 8.1 illustrates the scope of the Wise-ShopFloor.

As a constituent component in manufacturing supply chain, the Wise-ShopFloor links physical shop floors with the upper manufacturing systems. Similar to the e-manufacturing and e-business, the four major Wise-ShopFloor activities shown in Figure 8.1 are conducted in a collaborative cyber workspace.

In more detail, the interactions among the modules are illustrated in Figure 8.2, where the framework has been designed into a client-server architecture using VCM design pattern with built-in secure session control. The mid-tier application server handles major security concerns, such as session control, session registration, sensor data collection/distribution, planning and scheduling, as well as real device manipulation. A central Session Manager has been designed to look after the issues of user authentication, session synchronization, and sensitive data logging. All initial transactions need to go through the Session Manager for access authorization. In a multi-client environment, different users may require different sets of data or logic for different tasks. For example, in the case of monitoring, it is not efficient to have multiple users who share the same model talking with the same device at the same time. Publish-subscribe design pattern is adopted to collect and distribute sensor data at the right time to the right user, efficiently. As a server-side module, the Signal Collector is responsible for sensor data collection from networked physical devices. The collected data are then passed to another server-side module Signal Publisher who in turn multicasts the sensor data to the registered subscribers (clients) through applet-servlet communication. A Registrar has been designed to maintain a list of subscribers with the requested sensor data. A Java 3D model thus can communicate indirectly with sensors no matter where the client is, inside a firewall or outside. HTTP streaming is chosen as the communication protocol between server and clients.

Although the global behaviors of a Java 3D model are controlled by the server based on real-time sensor signals, users still have the flexibility of viewing the model from different perspectives (zooming, orbiting, panning and tilting, *etc.*) at a client side. In order to control a device, an authorized user can send control commands to the application server which in turn manipulates the physical device. Although the Wise-ShopFloor framework provides an alternative of camera-based monitoring through Java 3D models, an off-the-shelf Web-ready camera can easily be switched on remotely to capture unpredictable (un-modeled) scenes for diagnostic purposes.

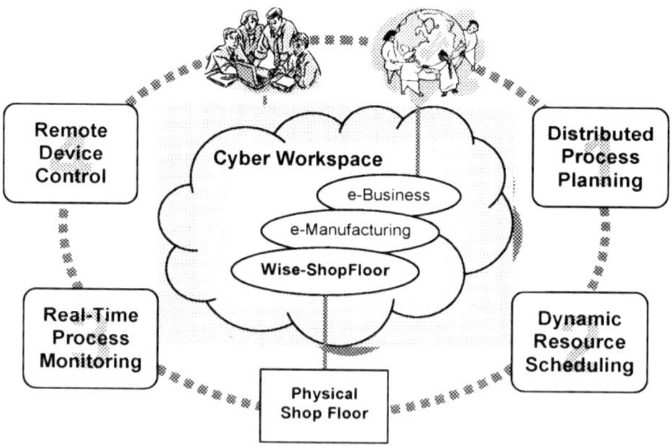

Figure 8.1. Scope of Wise-ShopFloor

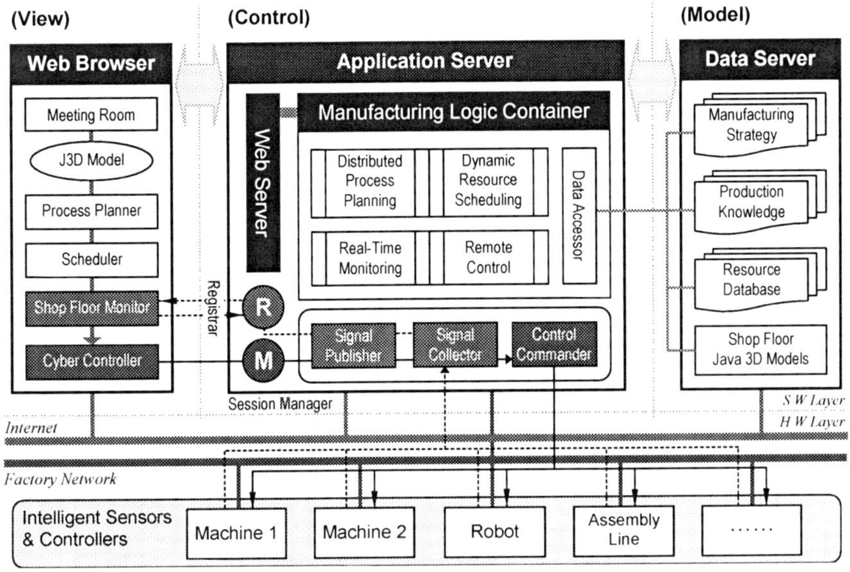

Figure 8.2. Wise-ShopFloor framework

8.4 Adaptive Process Planning and Scheduling

The four business modules shown in Figure 8.1 are interrelated, that is, the output of one module may be the input of another. For dynamic scheduling, real-time information from the monitoring module plays an important role. For the sake of page limitation, only the process planning is presented, leaving an interface to the scheduling open.

8.4.1 Architecture Design

Figure 8.3 shows the detailed architecture of our adaptive process planning.

Within the Wise-ShopFloor, our approach to adaptive process planning is realized by a two-layer structure of shop-level *Supervisory Planning* and machine-level *Operation Planning*. A process plan generally consists of two parts: generic data (machining method, machining sequence, and machining strategy) and machine-specific data (tool data, cutting parameters, and tool paths). Such a two-layer structure is, therefore, considered suitable to separate the generic data from those machine-specific ones. Since the resources, knowledge/database, and decision-making are logically and geographically distributed, such an adaptive process planning approach is also named *Distributed Process Planning* (DPP) [4]. The Supervisory Planning focuses on product data analysis, machining feature (m-feature) parsing, setup planning, machining process sequencing, and machine selection, while the Operation Planning considers jig/fixture selection and the detailed working steps for each machining operations, including cutting tool selection, cutting parameters assignment, tool path planning, and control code generation. Optimization is only performed at the latter stage, when specific resources (machine, tool and fixture) are known, and within a relatively small search space.

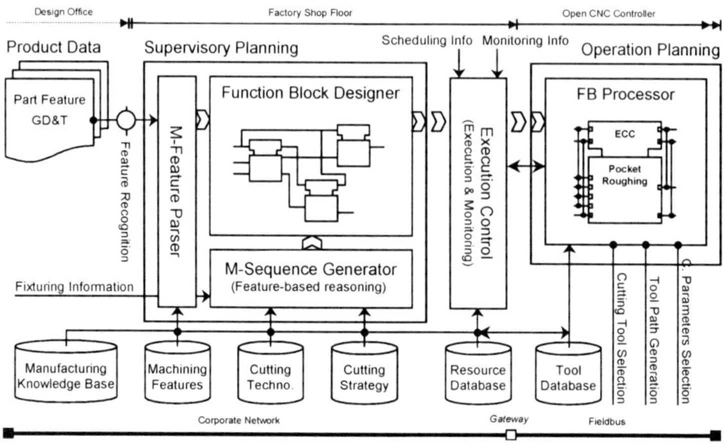

Figure 8.3. Architecture for adaptive process planning

8.4.2 Machining Process Sequencing

One critical task in process planning is machining sequence generation. Since a part design can be decomposed into basic m-features (such as hole, slot, pocket, *etc.*) either through feature-based design or via a third-party feature recognition solution, the task of machining process sequencing is literally treated as the task of putting m-features into proper setups and in proper sequence, which is called m-sequencing in DPP. A high-level process plan as a result of m-sequencing only consists of machine-neutral information in the form of generic machining sequences, including both critical and non-critical machining operations. Some of the non-critical ones are presented in a parallel order, whose sequence will be determined by a CNC controller during low-level operation planning. Before an m-feature can be machined, it must be grouped into a setup for the ease of fixturing. The basic idea of feature grouping is to determine a primary locating direction of a setup, and group the appropriate m-features into the setup according to their pre-defined tool access directions. This process is repeated for a secondary locating direction and so on until all the m-features are properly grouped.

Here, a primary locating direction is the surface normal $\vec{V}$ of the primary locating surface (*LS*). It can be determined by the following equations:

$$LS = \left\{ f\left(A^*, T^*\right) \middle| W_A \times \frac{A^*}{A_{\max}} + W_T \times \frac{T^*}{T_{\max}} \right.$$

$$= \max\left(W_A \times \frac{A}{A_{\max}} + W_T \times \frac{T}{T_{\max}}\right)\Bigg\} \quad (8.1)$$

$$\vec{V} = \left[\frac{\partial f}{\partial x}, \frac{\partial f}{\partial y}, \frac{\partial f}{\partial z}\right] \quad (8.2)$$

where, A^* and T^* are the surface area and the generalized accuracy grade of an *LS*; W_A is the weight factor of A^*; W_T is the weight factor of T^*; $A_{\max}$ and $T_{\max}$ are the maximum values of A^* and T^* of all candidate locating surfaces. A generalized accuracy grade *T* can be obtained by applying the algorithms described in [5-7]. Based on the primary locating direction $\vec{V}$, those m-features whose tool access directions $\vec{T}_{EMF}$ are opposite to $\vec{V}$ are grouped into setup $ST_{\vec{V}}$, as denoted below.

$$ST_{\vec{V}} = \left\{EMF \mid \vec{T}_{EMF} = -\vec{V}\right\} \quad (8.3)$$

To be generic, the setups at this stage are planned for 3-axis machines only. A setup merging is handled by the *Execution Control* module for 4-axis or 5-axis machines, if needed, after a specific CNC machine is selected.

In order to further sequence m-features in each setup, we proposed a geometry reasoning algorithm using IMV (intermediate machining volume) [8]. An IMV of an m-feature is the intersection of its maximum machining volume (MMV) and the current workpiece. Figure 8.4 schematically shows the concept of IMV through a hole, where the IMV of the hole varies between its MMV and its actual machining volume (AMV) during the machining.

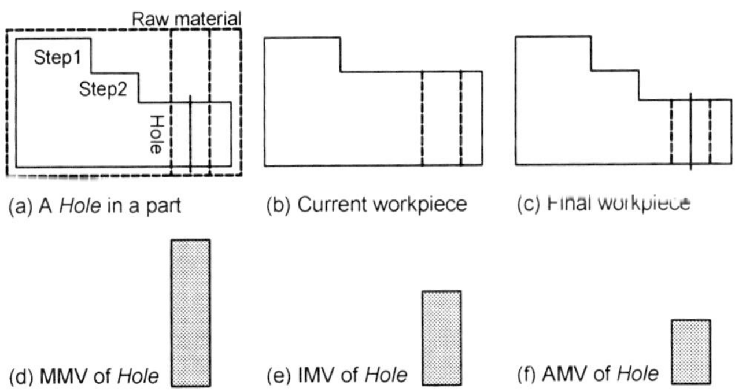

Figure 8.4. Intermediate machining volume of a hole feature

Based on the concept of IMV, five reasoning rules are defined below for m-sequencing.

Rule-1: *If the IMV of an m-feature equals to the AMV of the m-feature, or IMV=AMV, it is the time to machine the m-feature.*

Rule-2: *If the IMV of m-feature A is to be divided into more than one piece as a result of the machining operation of m-feature B, m-feature A should be cut first.*

Rule-3: *If an m-feature is to be changed to another m-feature type as a result of its own machining operation, this m-feature should be cut later.*

Rule-4: *A bigger machining volume is to be cut first.*

Rule-5: *In a setup, the m-features sharing the same tool types are grouped into clusters.*

The above five reasoning rules are used effectively for m-sequencing as demonstrated in the case study in Section 8.6. The sequenced m-features are then embedded in a set of function blocks with built-in decision-making functions of cutting parameters selection, tool path generation, and G-code generation at the individual m-feature level. The function blocks can be dispatched to a selected machine where detailed operation planning is accomplished before part fabrication. The built-in functions of each function block are resource- and event-driven, and can be called at runtime upon request so as to adapt to any environmental changes. Details on function block design and its utilization are explained in the next section.

8.4.3 Function Block Design and Utilization

"Function blocks" (or IEC 61499-1) [9] is an IEC standard for distributed process measurement and control, particularly for PLC control. A function block is a reusable functional module based on an explicit event-driven model, and provides for data flow and finite state automata based control. It is relevant to CNC control in machining data encapsulation and process plan execution. In the DPP, we use function blocks to address the manufacturing uncertainty through resource-driven algorithms embedded in each function block. The event-driven model (or resource-driven algorithms) of a function block gives a CNC machine more intelligence and autonomy to make decisions on how to adapt a generic process plan to match the actual machine capacity and dynamics. It also enables dynamic task scheduling, execution control, and process monitoring.

Three basic function block types are defined in the DPP: (1) *machining feature function block* (MF-FB), (2) *event switch function block* (ES-FB), and (3) *service interface function block* (SI-FB). Figure 8.5(a) depicts a typical *4-Side Pocket* MF-FB. A basic function block like this can have multiple outputs and can maintain its unique internal state, meaning that it can generate different outputs even if the same inputs are applied. The fact is of vital importance for adaptive cutting condition modification, after the function block has been dispatched to a machine, by changing the internal hidden state of the function block. For example, the same *4-Side Pocket* MF-FB can be used for *roughing* and/or *finishing* at the same machine (or at a different machine) with different cutting parameters and tool paths, by adjusting the internal state of the function block to fine-tune the algorithms in use. Such a behavior is controlled by a finite state machine, whose operation is represented by an ECC (execution control chart) as shown in Figure 8.5(b).

The START state is an initial idle state ready for receiving event inputs. EI_INI (an incoming event requesting initialization) triggers the state transition from START to INI for function block initialization, and when the state INI is active, the algorithm ALG_INI is being executed for the initialization. Upon its completion, ALG_INI will trigger an event output EO_INI indicating the success of the initialization.

Similarly, for other state transitions to RUN, UPDATE and MON (execution monitoring), different algorithms ALG_RUN (MF-FB execution), ALG_UPDATE (cutting condition update), and ALG_MON (MF-FB monitoring) are triggered, correspondingly. An event "1" means a state transition is always true. That is to say, the state will transit back to the START state and be ready for receiving the next event input.

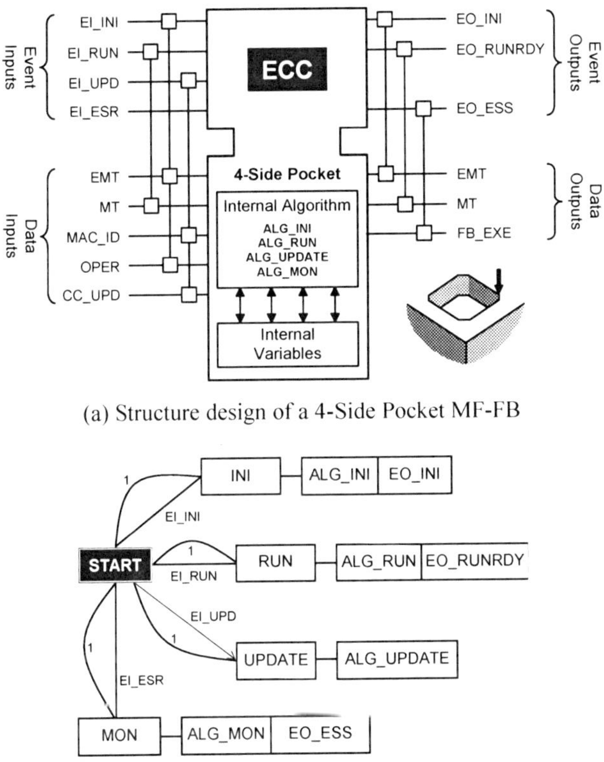

(a) Structure design of a 4-Side Pocket MF-FB

(b) Execution control of embedded algorithms

Figure 8.5. A basic machining feature function block

While basic MF-FBs define the functional relationships of events, data and algorithms for individual machining features fabrication, their combination can form a composite function block representing a setup. A composite function block may consist of several basic and/or composite function blocks with partially sequenced connections via events and data. The event flow among MF-FBs determines their machining sequence. Figure 8.6(a) shows a composite function block, where the event flow (or sequence) among three MF-FBs is facilitated at run-time by an Event Switch Function Block (ES-FB). For instance, if a sequence of "342" is given, the ES-FB will fire events accordingly to appropriate MF-FBs for feature fabrications in the order of 3→4→2. It thus adds flexibility to the composite function block. Figure 8.6(b) illustrates the graphical definition of the ES-FB, where ROUTE is the only data input to the function block. It is used as a reserved port for controller-level operation planning to do the local optimization of machining sequence.

Once the final sequence becomes explicit for those parallel MF-FBs, a string of integer numbers indicating the sequence is applied to the port. Event switching is realized by the internal algorithm ALG_SWITCH, which parses the data string and triggers one execution event at a time until the entire string is exhausted.

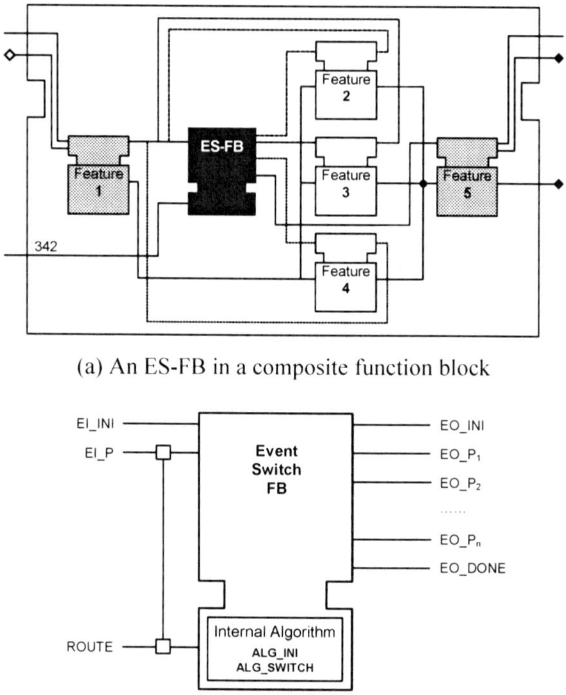

(a) An ES-FB in a composite function block

(b) Structure design of an ES-FB

Figure 8.6. Event switch function block for parallel m-features sequencing

In addition to MF-FBs and ES-FB, a Service Interface Function Block (SI-FB) is defined, as shown in Figure 8.7(a), to facilitate the execution control of MF-FBs. It also enables machining process monitoring during function block execution. In DPP, all MF-FBs are grouped in setups before being dispatched to appropriate machines. Each setup is a Composite Function Block (CFB). An SI-FB is plugged to each setup with the following assigned duties: (1) collects runtime *execution status* of an MF-FB including FB id, cutting parameters, and job completion rate; (2) collects *machining status* (cutting force, cutting heat, and vibration, *etc.*) if made available; and (3) reports any *unexpected situations* to DPP, *e.g.*, security alarm and tool breakage, *etc.*

Similar to other function block types, an SI-FB has been designed with five embedded algorithms for requesting and reporting execution status (ES), machining status (MS), and Unexpected Situation (US) from MF-FBs and to the Execution Control module (see Figure 8.3), respectively. Figure 8.7(b) is such an example. In order to monitor the machining process during execution, an SI-FB is plugged to the composite function block. Per the request from the Execution Control module, the SI-FB will pass the request (EI_ESR, execution status request) to the composite function block, which will then return an array of FB_EXE containing runtime execution status back to the SI-FB and finally to the Execution Control module. The SI-FB is of vital importance for machining process monitoring and dynamic re-scheduling in case of machine failure.

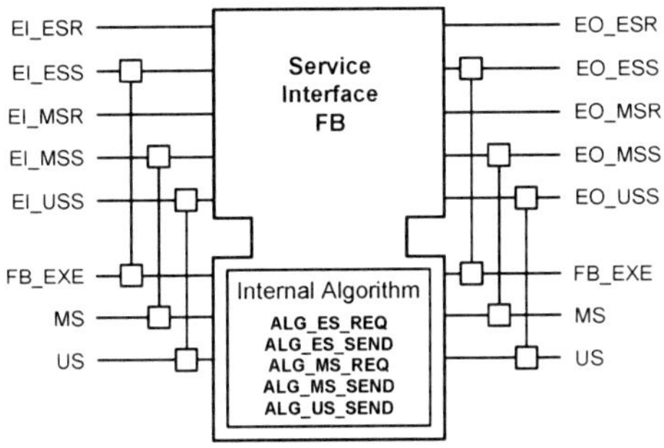

(a) Structure design of an SI-FB

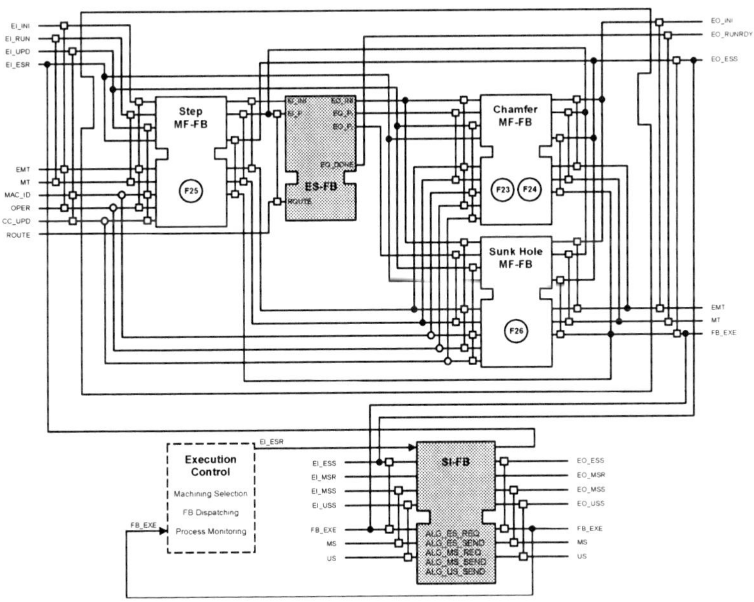

(b) SI-FB for process monitoring

Figure 8.7. Service interface function block for execution control and monitoring

The DPP prototype has been implemented in Java, including a module dedicated to function block design (see Figure 8.3). This function block designer consists of a Basic FB Designer, a Composite FB Designer, and an FB Network Designer. As the name suggests, each FB designer performs a specific function. Figure 8.8(a) illustrates a 4-side pocket MF-FB being designed using the Basic FB Designer, whereas Figure 8.8(b) depicts the result of a composite function block (a setup). In the DPP, a set of sequenced machining features can be mapped to a network of composite function blocks easily using this design tool. An example for a test part machining is explained in detail in Section 8.6.

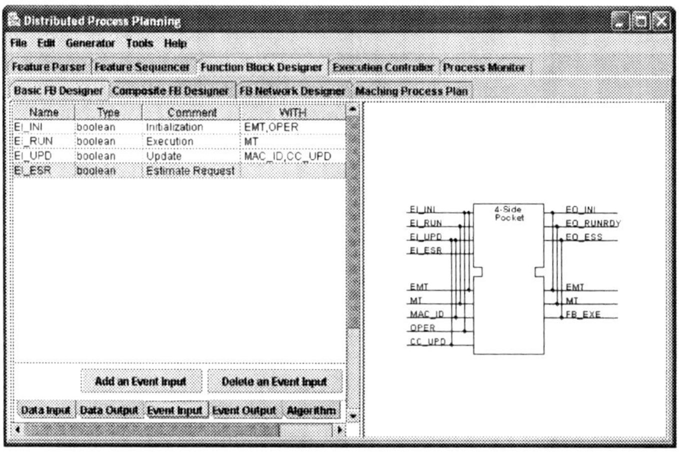

(a) Designing a basic function block

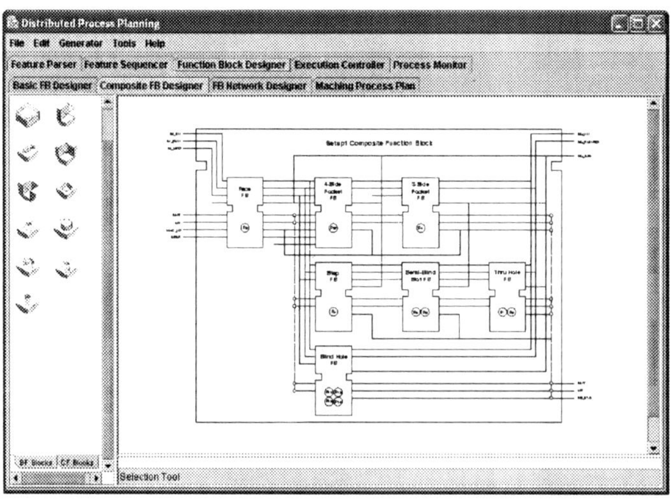

(b) Designing a composite function block

Figure 8.8. Function block design in DPP

8.4.4 Shop Floor Integration

Enabled by the Wise-ShopFloor, true shop floor integration can be realized by combining the following three systems: 1) the DPP system, 2) an agent-based scheduling system, and 3) a Web-based monitoring and control system, where DPP is treated as the main thread (Figure 8.9). The scheduling system is relatively standalone, to which the integration is loosely coupled. More details on dynamic scheduling can be found in [10].

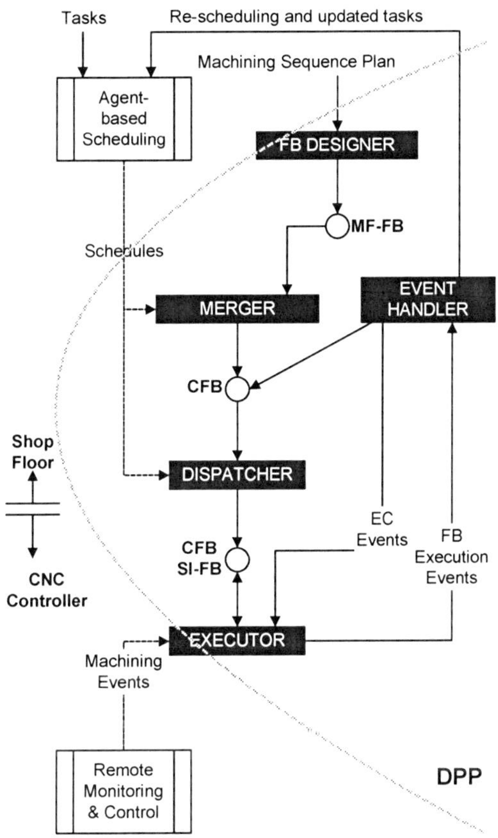

Figure 8.9. Shop floor integration enabled by Wise-ShopFloor

In DPP, a generic process plan is embedded in a set of function blocks that are portable to different machines. The machine-specific data, however, is determined at runtime by the function block embedded algorithms that are adaptive to unpredictable situations. For example, an alternative resource (cutter or machine tool) has to be used due to tool shortage or machine breakdown. In this case, the function blocks can apply appropriate algorithms to dynamically figure out the best cutter parameters and tool path for the alternative resource without re-doing the entire process planning. A snapshot of the adaptive process planning and its integration with Web-based remote machining is demonstrated in Section 8.6 through a simple case study.

8.5 Web-based Real-time Monitoring and Control

Obtaining real-time monitoring, control, and inspection data for a machine is limited by the available bandwidth for the data transfer. Broadcasting data about all machines to all clients would require sending more messages than necessary, slowing down the transfer of each message, and reducing the application's ability to display data and images in real time. Polling initiated by a client requires two-way communication, while only the information sent to the server from the client is of any use. The best solution to reducing network congestion and ensuring quick transfers is to have data multicast to only the clients requiring that data, with an open connection established for data streaming, and sending data whenever the data is changed. This section presents in detail the system configuration, sensor data collection and distribution, and Java 3D-based visualization.

8.5.1 System Configuration

Figure 8.10 illustrates a typical configuration for Web-based rapid machining, where a 5-axis horizontal milling machine is hooked up to the network for remote monitoring and CNC machining. The milling machine is equipped with a PC-based open architecture controller that serves as a gateway between itself and the application server. For security reason, TCP (Transmission Control Protocol) has been adopted for data communication between the machine and the application server, whereas HTTP streaming is used for data sharing from the server to the remote users. While the former is better for hardware protection with handshaking, the latter is firewall-transparent and suitable for Web-based application. Based on this configuration, it allows a remote user to monitor the absolute and relative motions of all axes as well as to control the spindle speed and feed rate for CNC machining.

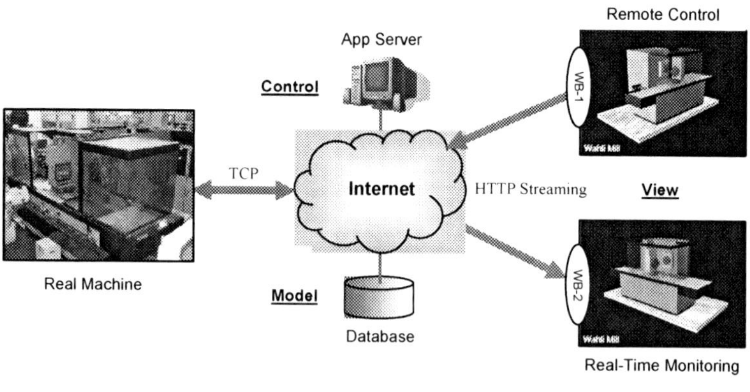

Figure 8.10. Configuration of Web-based rapid machining

8.5.2 Sensor Data Collection for Real-time Monitoring

To this end, the Wise-ShopFloor implements a *Publish-Subscribe* design pattern. A client (end user) subscribes to information pertaining to a specific machine, leaving an open connection to receive events. When a new event for that machine is posted, it is published only to those clients who have subscribed to it. In the Wise-ShopFloor, this communication is handled by a modification of the *Pushlet* [11]. Figure 8.11 shows the communication pathway for events to and from clients and a real machine.

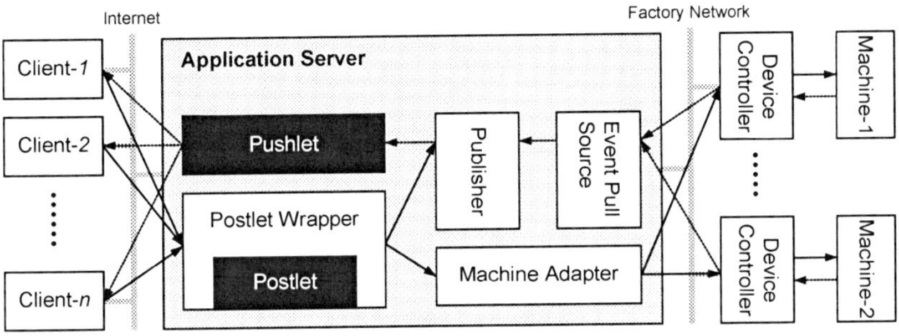

Figure 8.11. Event and data flow between clients and devices

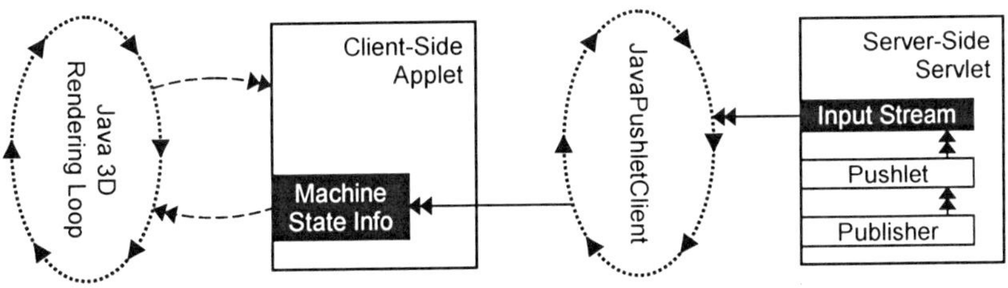

Figure 8.12. Streaming based applet-servlet communication

The client-side applet of the Wise-ShopFloor communicates with the Pushlet, an HTTP servlet. Invoking the Pushlet with an HTTP "Get" request with a "subject" parameter allows a client to subscribe to that subject. When receiving a subscription request, the Pushlet leaves the connection to the client open, allowing data to be streamed in without reopening a connection for each event. On the client-side of the Pushlet package is the *JavaPushletClient*. The JavaPushletClient sends the request for the Pushlet subscription, and opens an input stream from the socket. The Pushlet client then loops continuously and checks for data in the data stream. If the publisher has written new data to the stream, the Pushlet client overwrites the next most recent data with the new data, ensuring that only the most recent information is used to update a Java 3D image. The actual update of the image, however, comes from a different loop. Java 3D provides an interface, the *InputDevice*, which can be registered to the Java 3D Physical Environment. Once registered, a schedule is created to call a Polling and Processing method from the

InputDevice. In the Wise-ShopFloor, this schedule is designed such that the method is called each time a frame is rendered, so that each frame renders a machine with only the most recent information about the machine. Figure 8.12 shows the pathway of applet-servlet communication.

The Publisher sends information through the connection established by the Pushlet. This data is found by the *JavaPushletClient* loop, and is pushed into a client-side storage location. On a different thread, the Java3D rendering loop retrieves the data and updates the on-screen image for monitoring.

The Pushlet also provides a *Postlet* servlet, used by clients to "Post" events to the Publisher. When a client wishes to control a machine, he/she needs to seek permission from the application server and then enters into the control mode. At any given time, only one client can be granted the control authority for manipulating a given machine. The client-side applet then connects to the Postlet, sending an HTTP "Get" request with the desired instructions as a parameter. When the Postlet passes the data to the Publisher, the connection is closed, while the Publisher sends the data to all clients who subscribe to the indicated subject.

On the real machine side, data collection is slightly different. There are many different types of machines and robots that usually have different types of controllers. The Pushlet package provides an adapter, the *Event Pull Source* (see Figure 8.11), which can be extended to obtain data from a required source (real device). Events are "pulled" from an *Event Pull Source* at a regular interval, which can be set to a desired increment to approximately replicate real-time monitoring. A comprehensive data flow is shown in Figure 8.13, where the needed sensory data are directed to the right users using the HTTP streaming.

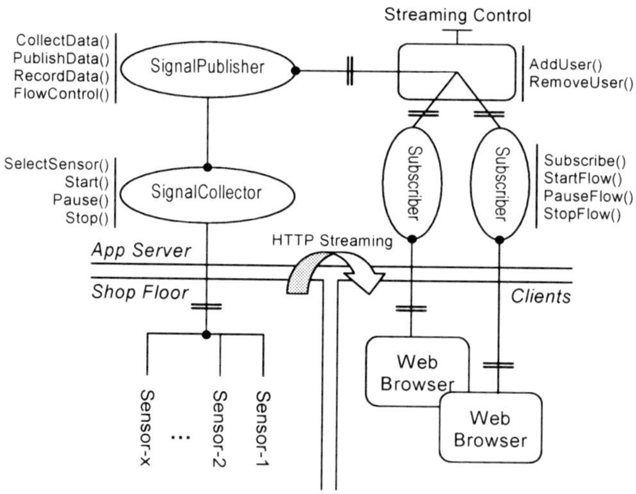

Figure 8.13. Streaming based applet-servlet communication

In the collection of sensor data from real machines, the server containing the Pushlet actually acts as a client of the machine controllers, establishing a socket connection and working with the provided interface of each machine controller. The concrete implementation of the *Event Pull Source* is one adapter between the interface of a machine controller and the interface required by the Pushlet. However, the communication to the real machine must be in both directions to achieve control, although the *Event Pull Source* communicates in only a single direction – from the machine to the application server. A machine controller is not able to interpret the Pushlet event, and thus will not be a client of the Pushlet. Another Pushlet adapter, the *Machine Adaptor*, is required to take information from the Postlet (*i.e.*, from the client), and send it to a machine controller in the required format. As the Pushlet does not provide this functionality, the Wise-ShopFloor uses a wrapper for the Postlet, which determines whether data is destined for the publisher or the machine, and thus directs it appropriately.

8.5.3 Data Packet Format

As shown in Figure 8.10, runtime sensory data collection from the milling machine is accomplished over the TCP connection using a series of 12 floating numbers and one long integer that form one data packet. In the current implementation, a typical data packet is defined as follows,

1	2	3	4	5	6	7	8	9	10	11	12	13
Relative position of 5 axes					Absolute position of 5 axes					FR	SS	CW

where, FR, SS and CW denote feed rate, spindle speed and NC control word, respectively. A control word is a reserved long integer indicating the status of the machine, including operation mode, such as manual (0x0001), auto (0x0002), or jogging (0x0040), coordinate system, axis status, *etc*. Same as the real machine controller, the data packet provides both relative and absolute positions of the five motion axes that are used for joints transformation and Java 3D model rendering for the ease of off-site monitoring and CNC control.

8.5.4 Java 3D Enabled Visualization

For the sake of network bandwidth conservation, Java 3D is chosen for geometric modeling of the CNC machine, as an alternative of camera-based solutions. Java 3D is designed to be a fourth-generation 3D API [12]. What sets a fourth-generation API apart from its predecessors is the use of scene-graph architecture for organizing 3D objects in the virtual world. Enabled by the scene-graph architecture, Java 3D provides an abstract, interactive imaging model for behavior control of 3D objects. Different from other scene graph-based systems, a Java 3D scene graph is a directed acyclic graph. The individual connections between Java 3D nodes are always forming a direct relationship: parent to child.

The 5-axis milling machine requires linear motion control of X, Y, and Z axes, as well as rotary motion control of B and C (around Y and Z axes, respectively). A combined rotary stage having two rotary motions is mounted on top of an X-table, whereas the spindle head of the machine provides the other two linear motions along Y and Z axes. Figure 8.14 illustrates the Java 3D scene graph model of the machine.

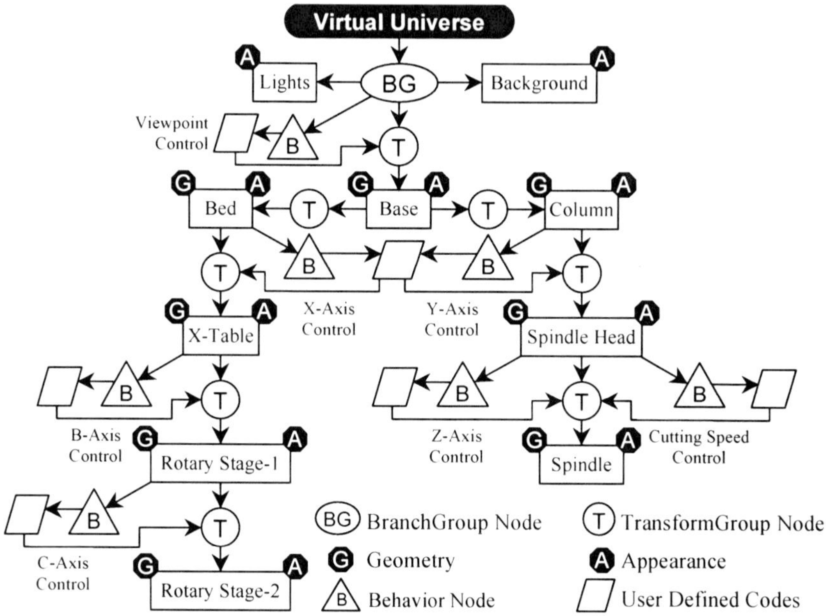

Figure 8.14. Java 3D scene graph model of a 5-axis milling machine

The scene graph contains a complete description of the entire scene. It includes the geometries, the attributes, and the viewing information needed to render the scene from a particular point of view. All Java 3D scene graphs must connect to a *Virtual Universe* object to be displayed. The *Virtual Universe* object provides grounding for the entire scene. A scene graph itself, however, starts with *BranchGroup* (BG) nodes (although only one BG node in this case). A *BranchGroup* node serves as the root of a sub-graph, or branch graph, of the scene graph. The *TransformGroup* nodes inside of a branch graph specify the position, the orientation, and the scale of the geometric objects in the virtual universe. Each geometric object consists of a *Geometry* object, an *Appearance* object, or both. The *Geometry* object describes the geometric shape of a 3D object. The *Appearance* object describes the appearance of the geometry (color, texture, material reflection characteristics, *etc.*). The behavior of the machine is controlled by *Behavior* nodes, which is subject to sensor data and is implementation-specific. The results of sensor data processing can be embedded into the codes for remote monitoring. Once applied to a *TransformGroup* node, the so-defined behavior control affects all the descending nodes. In our case, the 5-axis motions (X-Table, Rotary Stage-1, Rotary Stage-2, Spindle Head, and Spindle) are controlled by their corresponding behavior control nodes, for both on-line monitoring/control and off-line simulation. As the Java 3D model is connected with its physical counterpart through the control nodes by low-volume message passing (real-time sensor signals and control commands), it becomes possible to remotely machine a part on the real machine through the Wise-ShopFloor, where the physical security is addressed separately.

8.5.5 Web-based Remote CNC Control

Web-based CNC control is possible by sending proper NC commands through the applet-servlet (or *CyberController-ControlCommander-Machine*) communication as shown in Figure 8.2. In order to remotely machine a part, user authentication and authorization must be accomplished for the client who demands this operation. Control right authorization is done by setting a bit in the control word in a data packet that is sent to the client. If the client has requested the control right and the bit is set, a message will appear on the screen notifying the user that he/she is now in control of the machine. For the purpose of remote machining, a control word, similar to CW in the monitoring data packet, is sent back to the machine controller, augmented by a text string containing lines of an NC program. Thus not just manual control can be exercised off-site, but a complete NC program generated by the DPP can be remotely executed. For example, the following NC line tells the machine controller to proceed from the current position to the next, incrementally by (20, -30, 10) in linear rapid traverse mode. At the same time, the controller sets the spindle speed to 3,000 rpm and turns the flood coolant on.

```
G0 X+20 Y-30 Z+10 S3000 M8
```

Most existing Web-based systems rely on camera-based monitoring to guide remote operations. Compared with one 8-bit VGA camera image of 640×480 (307,200 bytes), our data packet size is only 52 bytes – a significant size reduction suitable for Web-based real-time applications.

8.6 A Case Study

A test part shown in Figure 8.15(a) is chosen for the case study. After applying the five feature-based reasoning rules defined in Section 8.4.2, the 14 m-features are grouped into two setups, each of which consists of two or more partially sequenced m-features as shown in Figure 8.15(b). While each m-feature can be mapped to a function block, a setup forms a composite function block. Figure 8.16 shows the composite function block for Setup-2.

In the Wise-ShopFloor, the adaptive process plan shown in Figure 8.16 can be dispatched to a milling machine for rapid fabrication utilizing the real-time monitoring and control functions discussed in Section 8.5. The motions of the five axes of this machine are driven by either sensor data for client-side monitoring or user commands for remote control. As the 3D model is connected with its physical counterpart through the message passing, it becomes possible to remotely manipulate the real machine through its Java 3D model. For example, the jogging control is with the use of the individual control buttons as labeled in Figure 8.17, whereas feature machining can be remotely achieved through *NC Control* mode.

The data packet format and the current implementation provide all information needed by the Java 3D model and its physical counterpart, the milling machine, for process plan execution. The 3D model ignores the first five numbers, while the machine controller ignores the second five numbers.

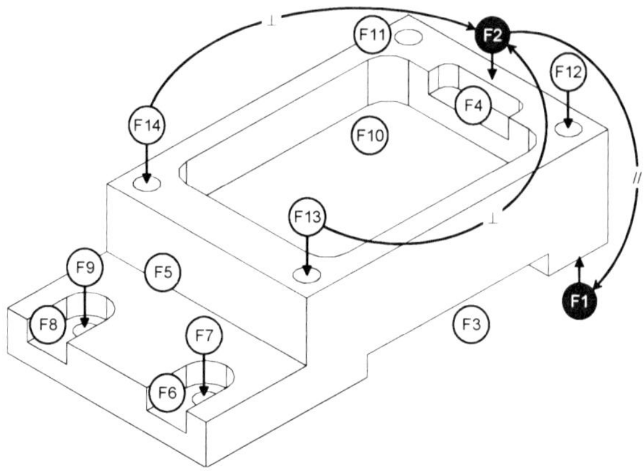

(a) Test part design

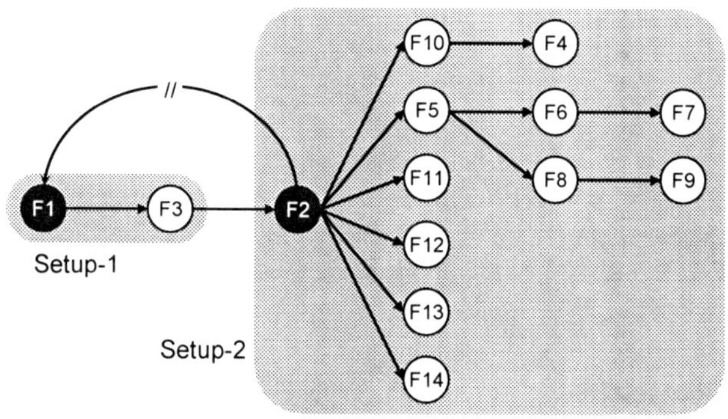

(b) Sequenced m-features

Figure 8.15. A test part and its machining sequence

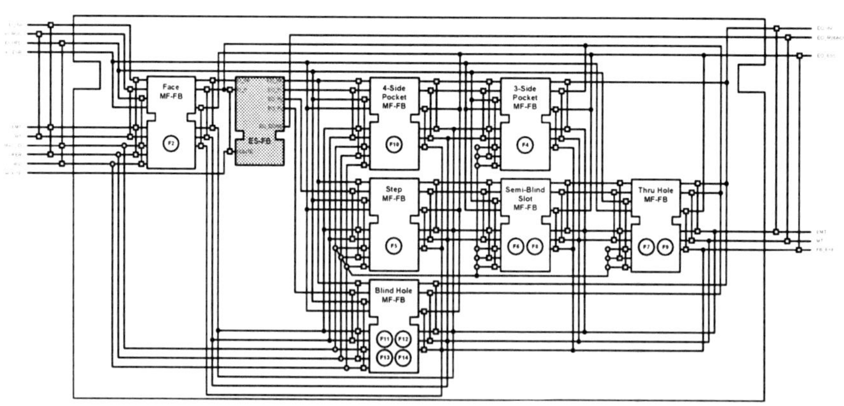

Figure 8.16. An adaptive process plan embedded in a composite function block

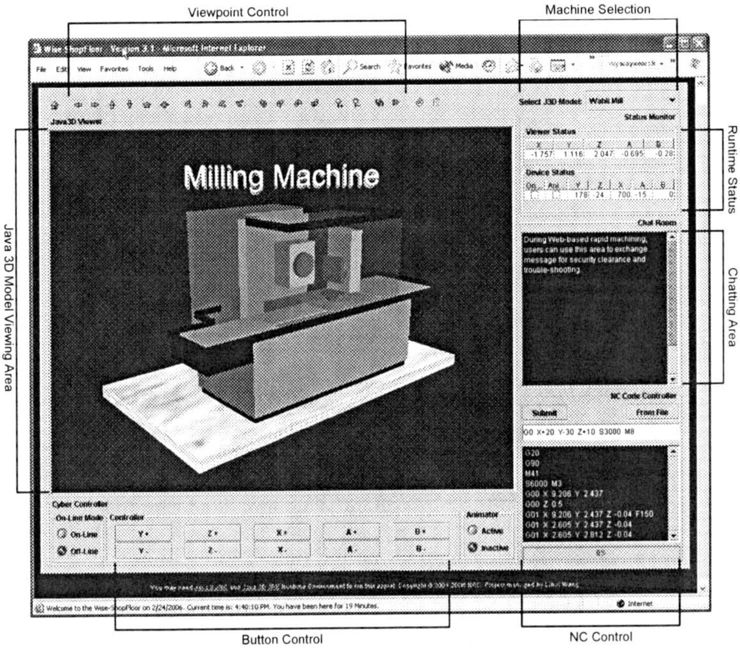

Figure 8.17. User interface for Web-based remote machining

As mentioned in Section 8.3, although the Wise-ShopFloor provides an alternative of camera-based monitoring, an off-the-shelf Web-ready camera can easily be switched on remotely to capture unmodeled scenes for trouble-shooting. Figure 8.18 illustrates one snapshot of a real scene of CNC machining during the case study.

Figure 8.18. A snapshot of Web-based remote machining

The Wise-ShopFloor prototype system provides users with a Web-based collaborative environment for real-time monitoring and control of manufacturing devices in the shop floor. It utilizes the latest technologies, including Java 3D and Servlets, for system design and implementation. Figure 8.19 shows the modular user interfaces that form the integrated system.

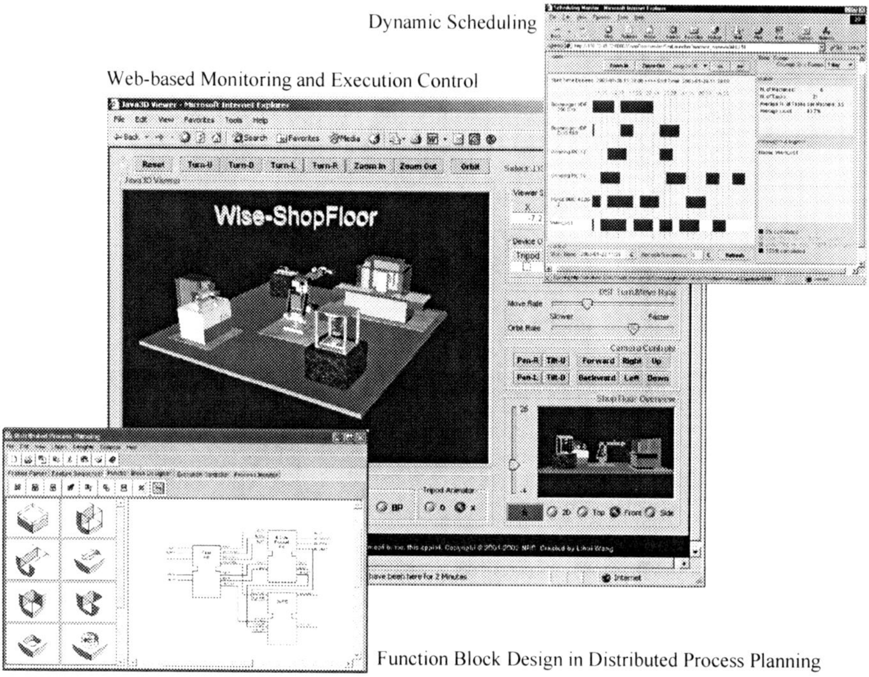

Figure 8.19. An integrated system for collaborative manufacturing

8.7 Conclusions

This chapter presents a novel approach toward Web-based collaborative manufacturing, including adaptive process planning, dynamic scheduling, real-time monitoring and remote control. On top of a *Wise-ShopFloor* framework, our prototype system has been designed into view-control-model architecture and developed using publish-subscribe design pattern for sensor data collection and distribution. In terms of adaptive process planning, our approach is to separate machine-specific data from generic ones using two-layer Supervisory Planning and Operation Planning. A generic process plan has been embedded into function blocks with built-in algorithms for machine level adaptive decision-making. A planning-machining case study demonstrates its feasibility and shows promise of this approach in a distributed manufacturing environment. As decentralization of business grows, a large application potential of this research is anticipated, such as control simulation, operator training, facility touring, off-site trouble-shooting and collaborative design verification, in addition to remote real-time monitoring and control.

8.8 Acronyms

AMV	actual machining volume
API	application programming interface
CC_UPD	updated cutting condition
CFB	composite function block
CNC	computer numerical control
CW	control word
DPP	distributed process planning
ECC	execution control chart
EI_x	event input x
EMT	estimated machining time
EO_y	event output y
ES-FB	event switch function block
FB	function block
FB_EXE	execution status of function block
FR	feed rate
HTTP	hypertext transfer protocol
IMV	intermediate machining volume
MAC_ID	machine ID
MF-FB	machining feature function block
MMV	maximum machining volume
MS	machining status
MT	machining time
OPER	operator's input
PLC	programmable logic controller
SI-FB	service interface function block
SS	spindle speed
TCP	transmission control protocol
US	unexpected situation
VCM	view-control-model
_ESR	execution status request
_ESS	execution status sent
_INI	initialization
_MSR	machining status request
_MSS	machining status sent
_RUN	function block execution
_RUNRDY	execution completed
_UPD	cutting parameter update
_USS	unexpected status sent

8.9 References

[42] Caldwell, N. H. M. and Rodgers, P. A., 1998, "WebCADET: Facilitating Distributed Design Support," *IEE Colloquium on Web-based Knowledge Servers*, U.K., pp. 9/1–9/4.

[43] Smith, C. S. and Wright, P. K., 1996, "CyberCut: a world wide Web based design-to-fabrication tool," *Journal of Manufacturing Systems*, 15(6), pp. 432–442.

[44] Wang, L., Shen, W. and Lang, S., 2004, "Wise-ShopFloor: A Web-based and sensor-driven e-shop floor," *ASME Journal of Computing and Information Science in Engineering*, 4(1), pp. 56–60.

[45] Wang, L., Feng, H.-Y. and Cai, N., 2003, "Architecture design for distributed process planning," *Journal of Manufacturing Systems*, 22(2), pp. 99–115.

[46] Boerma, J. R. and Kals, H. J. J., 1989, "Fixture design with FIXES: the automated selection of positioning, clamping and support features for prismatic parts," *Annals of CIRP*, 38, pp.399–402.

[47] Rong, Y., Liu, X., Zhou, J. and Wen, A., 1997, "Computer-aided setup planning and fixture design," *International Journal of Intelligent Automation and Soft Computing*, 3(3), pp.191–206.

[48] Ma, W., Li, J. and Rong, Y., 1999, "Development of automated fixture planning systems," *International Journal of Advanced Manufacturing Technology*, 15, pp. 171–181.

[49] Wang, L., Cai, N. and Feng, H.-Y., 2004, "Generic machining sequence generation using enriched machining features," *Transactions of NAMRI/SME*, 32, pp. 55–62.

[50] IEC 61499-1, 2005, *Function Blocks – Part 1: Architecture, International Electrotechnical Commission*, Switzerland.

[51] Shen, W., Lang, S. and Wang, L., 2005, "iShopFloor: An Internet-enabled agent-based intelligent shop floor," *IEEE Transactions on Systems, Man, and Cybernetics*, Part C, 35(3), pp. 371–381.

[52] van den Broecke, J., 2000, "Pushlets, part 1: send events from servlets to dhtml client browsers," *JavaWorld*, March.

[53] Barrilleaux, J., 2001, *3D User Interfaces with Java 3D*, Manning Publications Co., Greenwich, CT, USA.

9

Real Time Distributed Shop Floor Scheduling: An Agent-based Service-oriented Framework

Chun Wang[1], Kewei Li[1], Hamada Ghenniwa[1], Weiming Shen[1,2], Ying Wang[1]

[1]*Dept. of Electrical & Computer Engineering*
University of Western Ontario, Canada
[2]*Integrated Manufacturing Technologies Institute*
National Research Council, Canada

This chapter presents a distributed manufacturing scheduling framework at the shop floor level. The shop floor is modeled as a collection of multiple workcells. Each workcell is modeled as a flexible manufacturing system. The framework consists of a distributed shop floor control structure, dynamic distributed scheduling algorithms, multi-agent system modeling of workcells, and service-oriented integration of the shop floor. At the workcell level, a designated scheduler allocates jobs to resources and deals with any dynamic events locally, if possible. Otherwise, it collaborates with other workcells' schedulers. Workcells are modeled as multi-agent systems. Local dynamic scheduling is achieved by the cooperation among the scheduler agent, the real time control agent and resource agents. Distributed scheduling is conducted through Web services facilitated by the service-oriented shop floor integration. The proposed distributed control structure, dynamic distributed scheduling algorithms, and system integration have been designed and implemented using an agent-based service-oriented approach. Our experiments have shown promising results in enhancing shop floor flexibility and agility. The possible application of the proposed framework to other levels of manufacturing scheduling is discussed as well.

9.1 Introduction

Globalization of markets has driven manufacturing enterprises to shed the security of mass production and shift to a new paradigm, mass customization. An essential goal of this transformation is to respond to market changes in a timely and cost effective manner. As an integral component of manufacturing manangement, scheduling needs to be effectively integrated with other components of manufacturing systems such as supply chain management, ERP (Enterprise Resource Planning), and shop floor control. Dynamic changes can derive from either outside parties in the market, such as the supply side (representing suppliers), demand side (representing customers) or within the enterprise, such as real-time events from the shop floor. In a real world shop floor environment, it is rarely the case to execute exactly as planned. Operation durations tend to vary, machines break down, raw materials fail to arrive on time, new customer orders appear, others get cancelled, *etc*. Such disrupted execution incurs higher costs due to missed customer delivery dates, higher work-in-process inventory, and lower resource utilization. To deal with these issues, practical scheduling systems need to be able to effectively reorganize the shop floor production plan and repair or redo the production schedule accordingly. Scheduling systems with the capability of revising or re-optimizing a schedule in response to unexpected events become a key for companies to sustain their productivity.

This research is concerned with developing real time distributed scheduling systems at the shop floor level. The shop floor is modeled as a collection of workcells. The workcells, in turn, are modeled as Flexible Manufacturing Systems (FMS). The scheduling is performed cooperatively and collectively by the group of schedulers, each delegated to a specific workcell. Dynamic scheduling in this environment requires real time scheduling algorithms and their effective integration with the distributed shop floor control structure. Dynamic scheduling has been extensively studied in the literature [1, 2]. There have been research efforts focusing on distributed scheduling as well [3-5]. In this chapter, we study dynamic and distributed scheduling algorithms in the multi-workcell shop floor setting. In addition to the individual algorithms, we investigate how to integrate them in a way that the overall shop floor scheduling agility and solution quality are enhanced.

9.2 Scheduling Problems in Multiple Workcell Shop Floor

The shop floor considered here consists of a collection of workcells (as illustrated in Figure 9.1). Each of them is modeled as a flexible manufacturing system. Within a workcell, jobs need to be scheduled on resources. The scheduling problem at this level is a dynamic FMS scheduling problem, which is handled by a designated scheduler of the workcell. At the shop floor level, due to workcell capability limitation or unexpected events, some workcells may have jobs to be assigned to other workcells. From this point of view, it is a dynamic distributed scheduling problem, which needs to be solved collectively by a group of schedulers through cooperation. In this chapter, we only consider the type of reactive dynamic scheduling algorithms [6]. Thus it is not necessary to capture the randomness caused by dynamic events in the models of the scheduling problems.

9.2.1 Workcell Scheduling Problem

In our workcell scheduling problem, workcells are implemented by FMS systems. Among many FMS scheduling models, we focus on a class of problems in which machines have Partially Overlapping capabilities [7]. As illustrated in Figure 9.2, a workcell consists of various types of resources, such as computer numerical controlled machines, Automated Guided Vehicles (AGV), workpiece storage system. These resources are controlled by resource controllers and the whole workcell is controlled by a real time controller. An operator can program the processing of the workcell by interacting with the real time controller. At the workcell level, we are interested in the impacts of partially overlapped characteristics of resources. Therefore, we have simplified the workcell scheduling model by assuming that the transportation times of jobs between machines are equal and have been modeled in machine processing times. Therefore, there is no need to explicitly model the AGV. At the same time, storage and port are treated as independent resources, like machines. This allows us to differentiate the resources only by their capability sets, not by the relationships among them. For some resources, *e.g.*, machines, their capability sets may be partially overlapped.

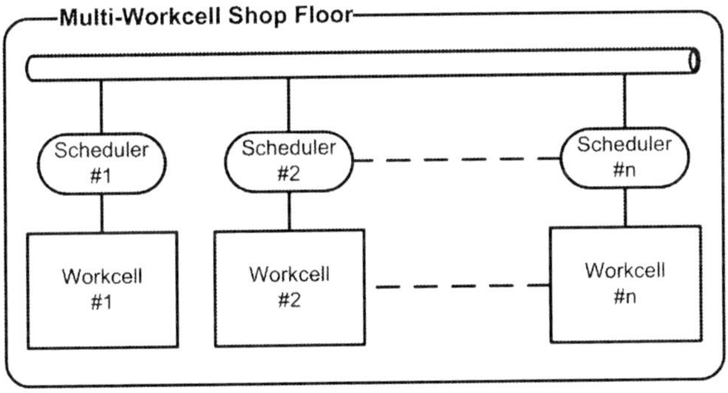

Figure 9.1. Multiple Workcell Shop Floor environment

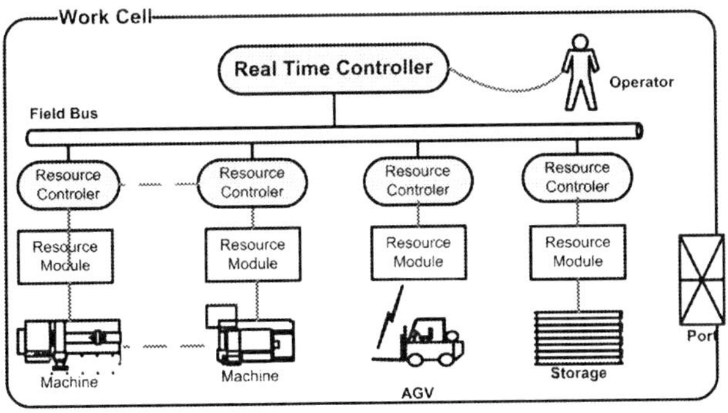

Figure 9.2. Workcell resources and the control structure

Formally, an instance of the class of scheduling problems in partially overlapping systems consists of a set of n jobs, denoted by $J = \{J_1, J_2, ..., J_n\}$, to be processed by a set of m resources, denoted by $M = \{M_1, ..., M_m\}$. Each job J_j ($j = 1, ..., n$) requires the processing of a sequence operations $o_{j,k}$ $(k = 1, ..., n_j)$, where n_j is the number of operations belong to J_j. An operation $o_{j,k}$ corresponds to an uninterrupted physical process which has to be performed on a resource. Each resource $M_i (i = 1, ..., m)$ is defined by a set of operations, which represent its capability. If $o_{j,k} \in M_i$, $o_{j,k}$ can be processed by resource M_i. For any two resources $M_i, M_l \in M$, $M_i \cap M_l$ may not be empty, which means that resources have overlapping capabilities. If a resource M_i $(1 \le i \le m)$ is capable of processing an operation $o_{j,k}$ $(1 \le k \le n_j; 1 \le j \le n)$, a processing time $p_{i,j,k} \in R^+$ is given. $p_{i,j,k}$ may not be equal to $p_{l,j,k}$, for $i \ne l$, $i, l \in \{1, ..., m\}$, which means the same operation may have different processing times on different resources. In the workcell scheduling problem, we do not model the resource eligibility constraints. Instead, we assign processing time $p_{i,j,k} = +\infty$, if M_i cannot process $o_{j,k}$. In addition, with each job $J_j (j = 1, ..., n)$ we associate two values: release time of a job $J_j - r_j$, due date for the completion of a job J_j - d_j. There are precedence constraints among operations of each job. The objective is to minimize makespan of the solution schedule. Using the following variables,

$S_{j,k}$, the starting time of the operation k of job j,

$$X_{i,j,k} = \begin{cases} 1 & \text{if machine } i \text{ is chosen to perform operation } k \text{ of job } j \\ 0 & \text{otherwise.} \end{cases}$$

$$Y_{j,k,\hat{j},\hat{k}} = \begin{cases} 1 & \text{if operation } k \text{ of job } j \text{ is performed before operation} \\ & \hat{k} \text{ of job } \hat{j} \text{ on the same machine;} \\ 0 & \text{if operation } k \text{ of job } j \text{ is performed after operation} \\ & \hat{k} \text{ of job } \hat{j} \text{ on the same machine; } j \ne \hat{j}. \end{cases}$$

the partially overlapping scheduling problem can be formulated as a mixed integer programming as follows.

$$\min \left\{ \max_{J_j \in J} \left\{ S_{j,n_j} + \sum_{i=1}^{m} p_{i,j,n_j} X_{i,j,n_j} \right\} \right\} \tag{9.1}$$

Subject to

$$S_{j,1} \ge r_j \quad , \forall J_j \in J \tag{9.2}$$

$$S_{j,k-1} + \sum_{i=1}^{m} p_{i,j,k-1} - S_{j,k} \le 0, \forall j, 1 < k \le n_j \tag{9.3}$$

$$S_{j,k} + \sum_{i=1}^{m} p_{i,j,k} X_{i,j,k} - S_{\hat{j},\hat{k}} + HX_{i,j,k} + HX_{i,\hat{j},\hat{k}} + HY_{j,k,\hat{j},\hat{k}} \le 3H, \\ \forall j, \hat{j}, j \ne \hat{j}, 1 \le k \le n_j, 1 \le \hat{k} \le n_{\hat{j}} \tag{9.4}$$

$$Y_{j,k,\hat{j},\hat{k}} + Y_{\hat{j},\hat{k},j,k} = 1, \quad \forall j, \hat{j}, j \ne \hat{j}, 1 \le k \le n_j, 1 \le \hat{k} \le n_{\hat{j}} \tag{9.5}$$

$$\sum_{i=1}^{m} X_{i,j,k} = 1, \forall j, 1 \le k \le n_j \tag{9.6}$$

$$X_{i,j,k} \in \{0,1\}, \forall i, j, 1 \le k \le n_j \tag{9.7}$$

$$Y_{j,k,\hat{j},\hat{k}} \in \{0,1\}, \forall j, \hat{j}, j \ne \hat{j}, 1 \le k \le n_j, 1 \le \hat{k} \le n_{\hat{j}} \tag{9.8}$$

$$S_{j,k} \ge 0, \forall j, 1 \le k \le n_j \tag{9.9}$$

The objective function (9.1) is to minimize the makespan of the solution schedule. The set of constraints (9.2) ensure that a job does not start before its release time. The set of constraints (9.3) ensure that an operation does not start before the previous operation of the same job has completed. The set of constraints (9.4) and (9.5) ensure that at most one job can be processed by a resource at a time. In (9.4) H is a large finite positive number. Constraints (9.6) say one operation can and only can be processed by one resource. Constraints (9.7), (9.8), and (9.9) are non-negative and integer constraints.

9.2.2 Dynamic Scheduling Problem

Manufacturing is a process often fraught with contingencies. It is rarely the case to execute exactly as planned. Small disruptions such as minor deviations in operation durations often do not warrant major modifications to the schedule. However, as the impact of small disruptions accumulate or as more sever disruptions occur, such as long machine breakdowns, it is sometimes desirable to re-optimize the schedule from a more global perspective [6]. In many cases, this re-optimization means re-schedule all operations that have not been processed by the time of disruption. We distinguish two types of dynamic scheduling situations, namely minor disruption and severe disruption. In the case of minor disruption, a schedule repair procedure which minimizes the perturbation to the original schedule is appropriate. On the other hand, if the disruption is severe (caused either by the accumulation of small disruptions or a major resource malfunction), a re-optimization from a more global perspective is usually desirable. An obvious issue is how to decide the severity of a dynamic disruption. The threshold of distinguishing minor and severe disruptions should be set by the workcell operator as it involves the trade-off, depending on the conditions within which a breakdown occurs, between the overall solution quality and the perturbation to the original schedule. Frequent schedule re-optimization can result in instability and lack of continuity in detailed shop floor plans, resulting in increased costs attributable to what has been termed "shop floor nervousness" [8].

9.2.3 Distributed Scheduling Problem

At the shop floor level, scheduling is to coordinate the local schedules of workcells in a way that the good solution quality of the shop floor schedule is achieved. In this context, individual workcell scheduling problems are tied together by two elements: shop floor level objective and inter-workcell constraints. Because each workcell tries to minimize/maximize its own objective function, at the shop floor level, the scheduling problem can be modeled as a multi-objective optimization problem. The overall solution quality can be measured by Pareto efficiency or some forms of aggregation of the individual objectives. In the workcell scheduling problem formulated in the previous section, the objective of each workcell scheduler is to minimize the makespan. Accordingly, we define the objective of the shop floor scheduling problem as weighted sum of makespans over all workcells.

The inter-workcell constraints in the distributed shop floor scheduling problems are derived from the machine capacity dependency among workcells. Solving the constraints is to achieve coordination among schedulers of workcells. The process of solving the inter-workcell constraints is to find a value assignment to some shared variables that satisfies all local constraints of workcells involved. Specifically, when solving the local scheduling problem, each scheduler has some variables and tries to determine their values. However, because of the capacity dependency there exist inter-workcell constraints, and the value assignment must satisfy these constraints. Formally, there exist l workcells $1,2,...,l$. X_h ($h = 1,...,l$) is the set that contains all variables the scheduler h needs to assign values to in order to determine a schedule for workcell h. Because of the inter-workcell constraints, some schedulers need to share a subset of their variables. However, such a case can be formalized as these schedulers having different variables, and there exist constraints that these variables must have the same value. We say that the constraints of a distributed shop floor scheduling problem are satisfied, if and only if

1. for any scheduler h and $\forall x \in X_h$, the value of x is assigned to d, any constraints in L_h is satisfied under the assignment $x = d$, where L_h is the set of local constraints of workcell h;

2. if $\bar{x}$ is a shared variable between workcell h and workcell q, $\bar{x} = d_h$ in workcell h, $\bar{x} = d_q$ in workcell l, then $d_h = d_q$.

Let $S_{h,q}^I$ denote the set of schedules which satisfy inter-workcell constraints between workcell h and q; let S_h^L denote the set of schedules which satisfy local constraints of workcell h. The distributed shop floor scheduling problem can be formulated as follows:

$$min \sum_{h=1}^{l} w_h M_h(S)$$

s.t. $S \in S_h^L$, $h = 1,\ldots,l$,

$S \in S_{h,q}^I$, $h = 1,\ldots,l$, $q = 1,\ldots,l$, $h \neq q$.

where w_h is the weight of workcell h, $M_h(S)$ is the latest completion time of all jobs belong to workcell h, the allocations of jobs of all workcells form an overall shop floor schedule S.

9.3 Scheduling Algorithms for Multiple Workcell Shop Floor

This section presents three algorithms for the scheduling problems formulated. Workcell scheduling algorithm is the fundamental job allocation procedure at the workcell level. This algorithm is used as a component of dynamic scheduling algorithm and distributed scheduling algorithm. The relationship among these three algorithms in the context of real time distributed shop floor scheduling can be depicted in terms of UML use case diagram as shown in Figure 9.3.

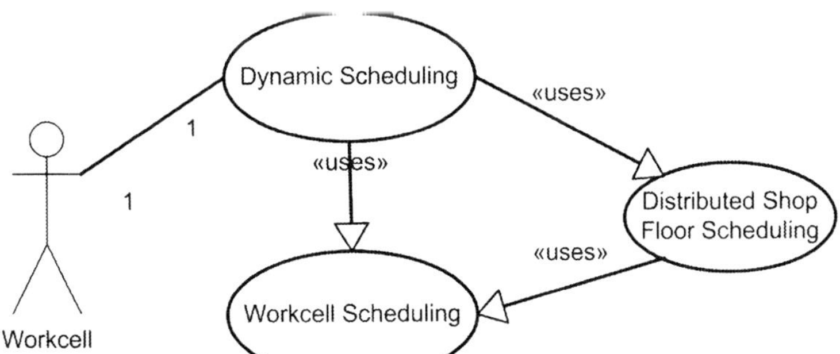

Figure 9.3. Relationship of the three scheduling algorithms

9.3.1 Workcell Scheduling Algorithm

The workcell scheduling problem modeled in Section 9.2.1 is a class of FMS scheduling problems. The majority of the approaches developed for these problems are heuristic oriented [*e.g.*, 9-11] and artificial intelligence-based [*e.g.*, 6, 12, 13]. Some Genetic Algorithm-based approaches [*e.g.*, 14, 15] can be considered as meta-heuristic methods. Most of these methods use simulation to generate or evaluate schedules. A comprehensive survey on simulation approaches in FMS scheduling can be found in [16]. Basnet and Mize [17] reviewed the literature concerning the operational aspects of FMSs. Zweben and Fox [18] provided a comprehensive reference for artificial intelligence based scheduling approaches. The problem we are focusing on is a class of FMS scheduling problems with partially overlapping system structure, which has been proved to be NP-hard [19]. Exact methods which find optimal schedules are not practical because of the prohibitive computation demanded. While many approaches proposed in the literature focus on other aspects of the FMS scheduling problems, we propose a dispatching rule based heuristic algorithm leveraging the partially overlapping characteristics of the problem.

The heuristic algorithm combines a set of dispatching rules. The basic ideas are to: (1) effectively utilize the flexibility provided by the partially overlapping characteristics of the system to balance the work loads, and at the same time, (2) assign operations to resources which can finish them faster. Based on the above heuristics, we propose two dispatching rules, Flexible Operation Last (FOL) and Earliest Finishing Time First (EFT). FOL is a composite of two elementary dispatching rules, Longest Processing Time first (LPT) [20] and Least Flexible Job first (LFJ). As a composite dispatching rule, FOL is modeled as a ranking expression that combines LPT and LFJ. This combination can be implemented as the following function:

$$f_{j,k} = \frac{\exp\left(\dfrac{n_{j,k}}{Q}\right)}{l_{j,k}}$$

where $f_{j,k}$ is the ranking index of FOL, defined as the Flexibility of $o_{j,k}$; $n_{j,k}$ is the number of resources in the workcell that can perform $o_{j,k}$; $l_{j,k}$ is the average processing time of the operation, which is calculated based on historical data. Q is the scaling parameter that can be determined empirically. If Q is very large, the FOL rule reduces to the LPT rule. If Q is very small, the rule reduces to the LFJ rule. FOL selects operations to be scheduled according to their Flexibilities. Operations with higher Flexibilities (short average processing time and more eligible resources) are placed towards the end of the schedule, where they can be used to balance loads more effectively. Once the operation to be scheduled has been selected by FOL, EFT finds a resource for the operation based on its completion time. The resource with earliest completion time is chosen.

The algorithm consists of two steps. Before scheduling an operation to a resource, the FOL rule is used to select an operation from Eligible Operation Set (EOS, which contains operations for which the job release time and operation preceding constraints are satisfied). Then the EFT rule is used to designate the selected operation to a resource based on the current partial schedule. The algorithm implements the EFT rule by considering two factors, the workload of a resource in the current partial schedule which has been established by previous operations and the processing speed of this resource for the selected operation. The resource which can finish the operation first is chosen. Briefly, the algorithm can be described as shown in Figure 9.4.

```
1. Set EOS= ϕ;
2. If (J_j has unscheduled operations, ∀J_j ∈ J) then
   2.1. Move eligible operations from J_j to EOS;
   2.2. Select an operation to be scheduled from EOS based on FOL rule;
   2.3. Allocate selected operation to a capable machine based on EFT rule.
   2.4. Remove the allocated operation from EOS;
3. Go to 2.
4. If EOS is not empty
   4.1. Go to 2.2
5. Terminate
```

Figure 9.4. The workcell scheduling algorithm

9.3.2 Dynamic Scheduling Algorithm

In Section 9.2.2 dynamic scheduling problems are classified based on the severity of the disruption happened in workcells. In the cases of minor disruption, schedule repair procedures are appropriate to provide a fast response and small permutation. In the cases of severe disruption, re-optimization algorithms need to be considered. While the workcell scheduling algorithm proposed in the previous section can be applied directly to the severe disruptions as a re-optimization algorithm, this section presents a scheduling repair procedure.

In repairing a schedule, the schedule repair algorithm first identifies a number of operations that are affected by the disruption and need to be unscheduled, then, allocates them to available resources using the workcell scheduling algorithm. Once a dynamic disruption happens in a workcell, some operations are affected by the event directly. At the same time, others may be affected indirectly by the conflict propagation caused by various constraints. For example, if resource m breaks down at time t_m, an operation $o_{j,k}$ scheduled on m which has not started the processing or has not been finished, $c_{j,k} > t_m$, can no longer be processed on m (where $c_{j,k}$ is the completion time of $o_{j,k}$ and assume m cannot be recovered before the end time of the schedule's execution). We call $o_{j,k}$ a directly affected operation. All directly affected operations form a set, denoted by O_{DA}, $o_{j,k} \in O_{DA}$. Operations belong to O_{DA} have to be rescheduled on other capable resources. In addition to operations in O_{DA}, an operation $o_{\hat{j},\hat{k}}$ scheduled on other resources which have precedence constraints with an operation in O_{DA}, say $o_{j,k}$, and has been scheduled after $o_{j,k}$, $s_{\hat{j},\hat{k}} > c_{j,k}$, may need to be rescheduled as well because $c_{j,k}$ may change too much after its rescheduling such that $s_{\hat{j},\hat{k}} > c_{j,k}$ is no longer true. We call $o_{\hat{j},\hat{k}}$ an indirectly affected operation. All indirectly affected operations form a set, denoted by O_{IA}, $o_{\hat{j},\hat{k}} \in O_{IA}$. Clearly, not all operations in O_{IA} need to be rescheduled in order to generate a valid schedule repair. To minimize the perturbation to the original schedule, we propose a two-step scheduling repair procedure: (1) a schedule repair first un-schedules operations in O_{DA}; (2) if these operations are not sufficient to enable a new solution to be generated, the unscheduled operations are expanded incrementally to operations in O_{IA} until a solution is found. Based on the modeling and analysis mentioned above, we propose a dynamic scheduling repair algorithm described in Figure 9.5. We assume that resource m breaks down at time t_m and cannot be recovered within the time period to be scheduled in a workcell.

1. Set Eligible Operation Set $EOS = \phi$;
2. Evaluate operations scheduled on m, find all $o_{j,k} \in O_{DA}$, such that $c_{j,k} > t_m$;
3. Find all indirectly affected operations $o_{\hat{j},\hat{k}} \in O_{IA}$ based on the operations in O_{DA};
4. Unschedule operations in O_{DA};
5. If (J_j has unscheduled operations, $\forall J_j \in J$) then
    5.1. Move eligible operations from J_j to EOS;
    5.2. Select an operation $o_{j,k}$ to be scheduled from EOS based on FOL rule;
    5.3. Allocate $o_{j,k}$ to a capable resource based on EFT rule;
    5.4. If no allocation is feasible (violate constraints),
        5.4.1. Find the nearest successor of $o_{j,k}$ which is not in EOS based on precedence constraints;
        5.4.2. Unschedule the successor found;
        5.4.3. Go to 5;
    5.5. Remove the allocated operation from EOS;
    5.6. Go to 5;
6. If EOS is not empty, go to 5.2;
7. Terminate

Figure 9.5. The scheduling repair algorithm

The workcell states keep changing in dynamic scheduling situations. These changes can be modeled as a Finite State Machine as shown in Figure 9.6. The workcell scheduling system has six states. Among them, Monitoring, Scheduling, and Deploying are of importance. After Initialization, the schedule has been calculated and deployed to the workcell. The system will be in the state of Monitoring. Dynamic events, which represent the changes from a workcell can trigger the transition from Monitoring state to Scheduling state at which repairing or rescheduling algorithms respond to the events occurred and work out a new schedule. If dynamic events happened in the state of Scheduling, rescheduling procedure may be restarted to accommodate the newly happened events. Once new schedule is ready, the system changes to Deploying state where the new schedule is deployed to the workcell. If deploying failed because of unexpected changes in the workcell, system backs to Scheduling state, and rescheduling will be restarted again.

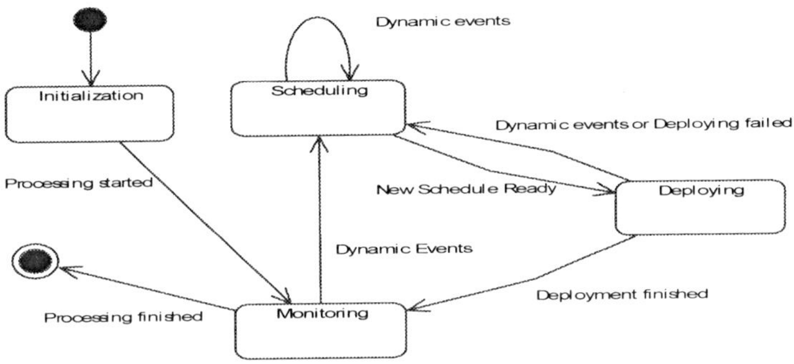

Figure 9.6. Finite State Machine model of the real time scheduling system

9.3.3 Distributed Scheduling Algorithm

The scheduling problem at the shop floor level contains multiple distributed workcells. Thus, it is actually a distributed scheduling problem. In a workcell, some dynamic events, such as a resource malfunction, may happen. If this disruption cannot be contained inside the workcell, in other words, some disrupted jobs can no longer be scheduled in the same workcell because of the lack of processing capabilities caused by the resource breakdown, the scheduler of this workcell needs to outsource the unscheduled operations to other workcells. Agent-based approach has been proposed as a new paradigm for developing distributed scheduling algorithms. Examples can be found in [21-25]. An extensive survey regarding multi-agent systems for manufacturing can be found in [26]. We design a distributed scheduling algorithm for the multi-workcell shop floor scheduling. The algorithm uses the Contract Net [13] as its interaction protocol. We will implement the algorithm in an agent-based service-oriented system framework in the next section. Briefly, the algorithm can be described as follows:

1. Dynamic events at a workcell (called an initiator in terms of the Contract Net protocol) cause some operations no longer being able to be processed in the workcell. The scheduler of the workcell finds eligible workcells

(called responders) which can process the unscheduled operations through a resource discovery mechanism (*e.g.*, UDDI in the case of Web service-based implementation). Note that, one responder does not have to be able to process all unscheduled operations of the initiator. The definition of an eligible workcell requires that the workcell can process at least one of the unscheduled operations.

2. The initiator sends out a call for proposal (CFP) to all responders. The CFP contains unscheduled operations and their associated constrains, such as precedence and release dates.
3. The responders try to accommodate the unscheduled operations from the initiator into their own local schedules based on their scheduling objectives respectively. Once the scheduling on the responders is finished, the responders send proposals back to the initiator. Each of the proposals contains a solution schedule for the unscheduled operations of the initiator. Note that, the schedule solution from a responder may not contain all unscheduled operations. Therefore, for some responders, they can just provide schedules for some of the unscheduled operations due to their capability constrains or the availability of the resources.
4. Upon receiving the proposals from the responders, the initiator selects one or a combination of them based on its scheduling objectives to form a final schedule for the unscheduled operations and inform the responders which are included in the final schedule by sending Award messages.

The coordination structure of the distributed scheduling algorithm is actually a negotiation process that can be depicted using a sequence diagram as shown in Figure 9.7.

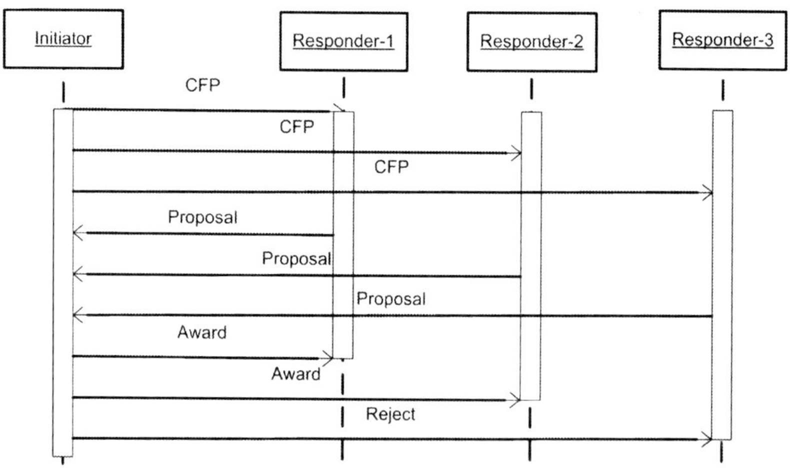

Figure 9.7. Negotiation process of the distributed scheduling algorithm

9.4 Agent-based Service-oriented System Integration

The distributed shop floor control, the scheduling algorithms and their integration have been implemented using an agent-based Web service integration framework (AWS) [27]. Agent-orientation is an appropriate design paradigm to enable automatic and dynamic collaborations. It is a natural system design and implementation choice in capturing the distributed and dynamic natures of the distributed real time shop floor scheduling. In addition, software agent paradigm has attained technical advantages in software modularization, legacy systems integration, distributed problem solving, and semantics-based interaction with complex and distributed transactions. Established technologies in these areas provide necessary foundation for the design and implementation of distributed real time scheduling systems. On the other hand, service-orientation is suitable in designing system integration at the shop floor level. Web Services paradigm is fast evolving and has been supported by several industrial leaders. This led to the development of various supporting technologies for Web Services that enable deploying, publishing, discovering, invoking and composing services in a standard and consistent way. This enables an open, flexible, standardized integration of manufacturing control at the shop floor, enterprise, and supply chain levels.

The merging of service-oriented and agent-based approaches has been a hot topic of research in recent years. Petrie, *et al.*, [28] discussed the shortcomings of Web services standards and how logical AI techniques like declarative commands, agents, and AI planning techniques can be used to address some of these shortcomings. They proposed an FX-Agent approach to address Web services discovery and composition of Web services. Matskin, *et al.*, [29] identified Web services composition as an important issue for efficient selection and integration of inter-organizational and heterogeneous services on the Web and they believed that software agents can help make Web services "pro-active". In their system, provider's Web services are wrapped into individual Providers' Agents on an agent-based marketplace providing services for Customers' Agents. Maamar, *et al.*, [30] presented an agent-based and context-oriented approach that supports the composition of Web services. During service composition process, software agents engage in conversations with their peers to agree on the Web services that participate in this process. Liu, *et al.*, [31] proposed a conceptual model of agent-mediated Web services for intelligent service matchmaking. In fact, most of research efforts in the literature like above mentioned approaches can be roughly categorized in to "agentification" of Web services into an agent community. We proposed a different approach for agent and Web services integration [32]. In our AWS framework, an agent core is built into each Web service, so that a Web service is itself an agent. No matter the agentification of Web services as agents in a multi-agent system [30] or encapsulation of agents as Web services over the Internet [32], both approaches share the common goal that, by taking the advantages of Web services and agents, the resultant integrated solution will produce a sophisticated paradigm for Internet computing.

9.4.1 System Overview

Figure 9.8 illustrates an agent-based Web service integration for the distributed real time shop floor scheduling system. The proposed system integration is composed of two levels. At the shop floor level, communication among schedulers is based on Web Services standards; at the workcell level, an agent-based scheduling system is implemented. The functionalities of agents and other software entities are described as follows:

1. Resource agents represent resources in a workcell. Each resource agent is on behalf of one resource. Resource agents receive job assignments from the Real Time Controller agent and report the working status of their resources to the Real Time Controller agent. The status information including routine data of processing and unexpected disruptions
2. Directory Facilitator (DF) has the registration service functionalities for other agents in a multi-agent system, keeps up-to-date agent registration, informs all registered agents with updated registry, and provides lookup and matchmaking services to the multi-agent system.
3. Real Time Controller is an agent that represents the overall control of a workcell. It accepts production schedules from the workcell scheduler and distributes them to resources in the workcell. At the same time it monitors the processing status of resources, analyzes and aggregates the raw resource processing data. If unexpected changes in the workcell affect the execution of the schedule, it will report to the scheduler with high level scheduling related processing information.
4. Scheduler performs the scheduling functionality in the system. At the workcell level it works with the Real Time Controller to implement the dynamic scheduling within the workcell. At the shop floor level it cooperatively works with other peer schedulers in achieving distributed

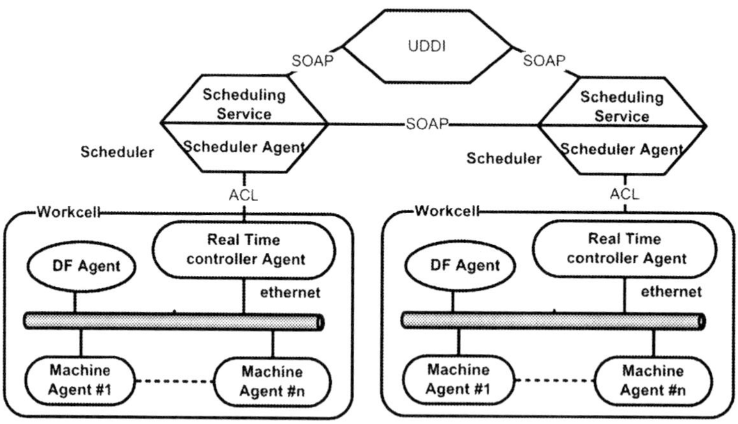

Figure 9.8. Agent-based Web service integration

shop floor scheduling. As shown in Figure 9.8, the scheduler has two identities, scheduling service and scheduling agent. When working with Real Time Controllers, it is exposed as an agent communicating using ACL. On the other hand, it is exposed as a Web service when working with its peer schedulers at the shop floor level.

5. UDDI is a static repository that provides schedulers' information with standard terms that contains workcell's capabilities and constraints.

9.4.2 Agent Architecture

The architecture adopted for the agents in our distributed shop floor scheduling system is Coordinated, Intelligent Rational Agent (CIR-Agent) architecture [33]. Detailed and logical architectures of CIR agent are shown in Figure 9.9.

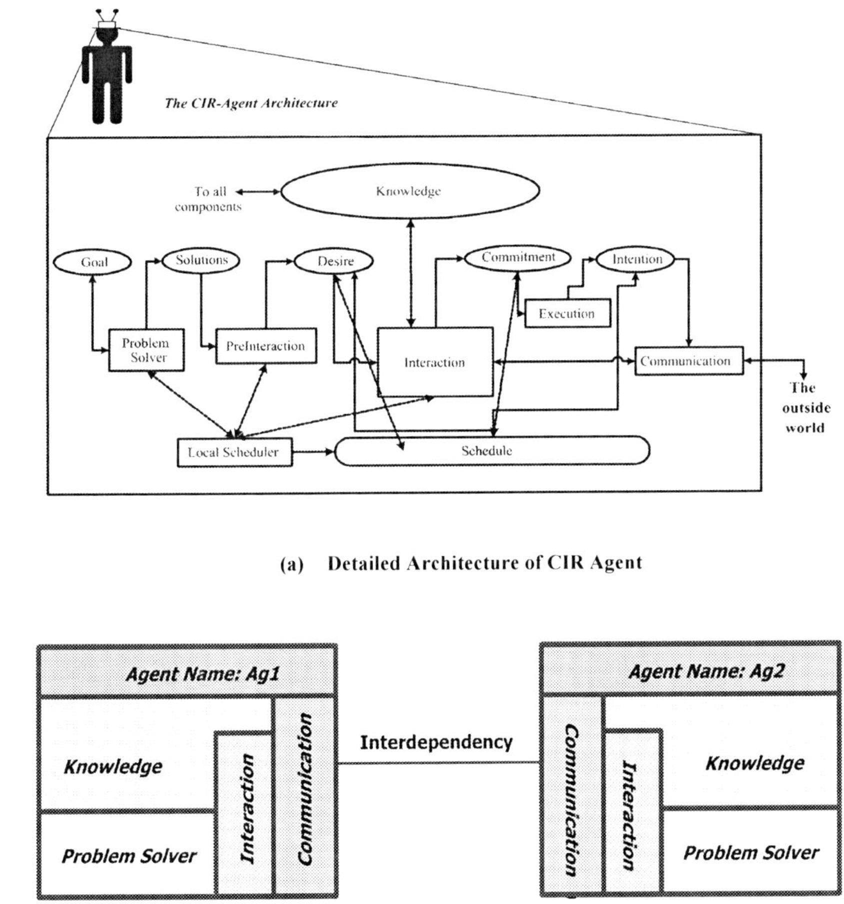

(a) Detailed Architecture of CIR Agent

(b) Logical Architecture of CIR Agent

Figure 9.9. Detailed and logical CIR-Agent architecture

In the CIR-Agent model, an agent is an individual collection of primitive components. Each component is associated with a particular functionality that supports a specific agent's mental state as related to its goal. The agent's mental state regarding the reasoning about achieving a goal, in the CIR model, can be in one of the following:

1. **Problem solving**: determines the possible solutions for achieving a goal.
2. **Pre-interaction**: determines the number and the type of interdependencies as well as the next appropriate domain action.
3. **Interaction:** resolves the problems associated with the corresponding type of interdependencies. The mechanisms used in the interaction are called interactive devices.
4. **Execution**: affects the world.

Based on these mental states, the CIR-Agent's architecture can be considered as a composition of four components: problem solver, pre-interaction, interaction, and execution. As an example, we describe the scheduler agent design based on the CIR-Agent architecture in the next section.

9.4.3 Scheduler Agent Design

In a distributed shop floor scheduling system, workcells are modeled as multi-agent systems. However, at the shop floor level, these multi-agent systems are integrated through Web services. The coexistence of these two different environments poses challenges in systems integration. Because all interactions between the two environments are facilitated by the scheduler, it is not really necessary to implement a general Web services agent gateway between agent and Web service environments. Our approach is to encapsulate the gateway functionality into the scheduler agent, such that it can communicate with both environments concurrently. Based on the CIR-Agent architecture, we design problem solver, interaction and communication components in the scheduler agent. However, in order to communicate to different environments, both the interaction and communication components in the scheduler agent are split into two parts. As shown in Figure 9.10, the workcell scheduling interaction and ACL communication are used by the local controller of the problem solver to interact with the real time controller agent in the workcell; the scheduling service interaction and SOAP communication are used by the remote controller of the problem solver to interact with scheduling services provided by other workcells. The problem solver component consists of the local controller, remote controller and scheduling algorithms designed for the scheduling at different levels of the shop floor.

Other agents in the system, such as resource agents and real time controller agents are also designed based on the CIR-Agent architecture. Because they only exist in agent environment, they are not equipped with SOAP communication and Service interaction components.

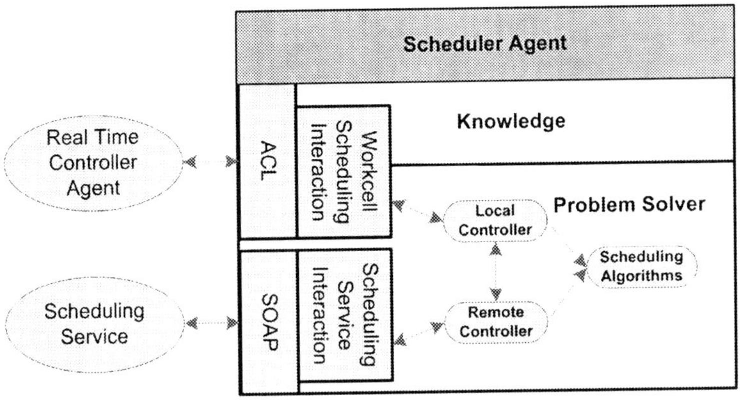

Figure 9.10. Design of scheduler agent

9.4.4 Coordination Between Scheduler Agent and Real Time Controller Agent

Coordination between the scheduler agent and real time controller agent implements the monitoring and control functionalities required by the dynamic scheduling. Through the control, the scheduler passes the generated schedules to the real time controller agent to be executed in the workcell. Workcell resource statuses can be reported to the scheduler through the monitoring. The protocol adopted for the coordination between the scheduler agent and the real time controller agent is FIPA Query Protocol (http://www.fipa.org). The protocol has been implemented in different ways to fulfill different functional requirements of the coordination.

Figure 9.11 depicts the schedule deployment protocol between the scheduler agent and the real time control agent. Once a new schedule is calculated, scheduler agent deploys the schedule to the workcell by sending a Deploy message. Real time control agent receives the updated schedule and passes it to the workcell resources to be deployed. Depending on different deploying results it may reply with Inform-done (deployment finished), Failure (deployment failed) or Not-understood if part of the schedule is not understandable to it.

Figure 9.12 shows the disturbance reporting protocol between the scheduler agent and the real time control agent. Once a disturbance happens in the workcell, the real time controller agent reports it to the scheduler agent by sending a Request message. The scheduler agent receives the disturbance report and replies with an Inform-done message. If the reported message is not understandable, it will ask the real time control agent to re-send the request by sending a Not-understood message.

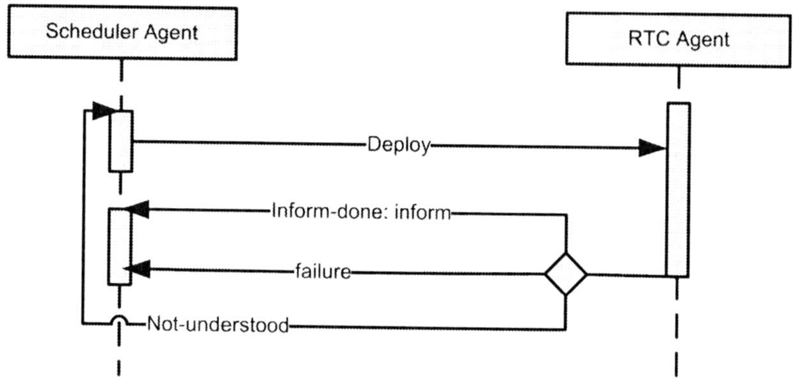

Figure 9.11. Schedule deployment protocol

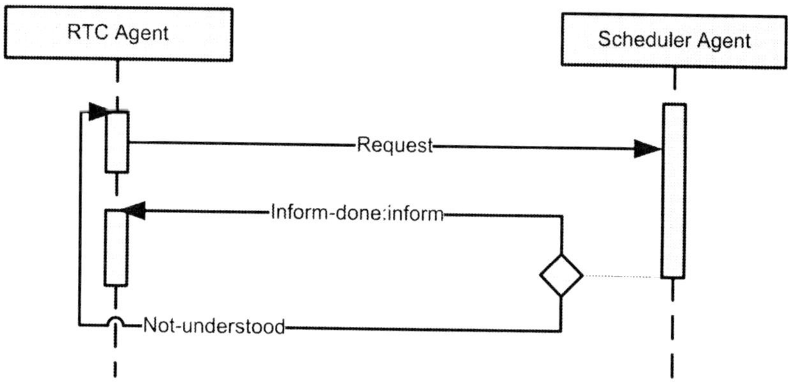

Figure 9.12. Disturbance reporting protocol

9.4.5 Coordination Between Scheduling Services

The coordination between scheduling services is required by the distributed scheduling algorithm in assigning jobs among workcells. We have implemented the coordination mechanism using FIPA Contract Net Protocol. Note that the messages between the initiator and responders defined in FIPA Contract Net are in the format of Agent Communication Language (ACL) which cannot be used directly in our Web services integration of multiple workcells. In this particular integration, ACL messages are encoded to their XML representation based on FIPA00071 specification (http://www.fipa.org) before they are sent to other scheduling services through SOAP. Upon receiving the SOAP message, a scheduling service decodes the XML and recovers the ACL message. Through this mechanism, the ACL based coordination protocol can be used in the Web services integration. Figure 9.13 depicts the Contract Net protocol between two schedulers, one as the initiator and the other one as the responder. The initiator has a set of incompletely scheduled jobs (it could be the case that some operations of a job have been scheduled in the initiator's workcell. However, some operations remain unscheduled). It sends a CFP message to the responder. Upon receiving the CFP, the responder tries to accommodate the jobs assigned to it into its local schedule. If it is not feasible for the responder to schedule the assigned jobs or the responder is not interested in the job assignment, it will send back a Refuse message. Otherwise, it sends back a proposal to the initiator with the contingent schedule. If the initiator is satisfied with the contingent schedule, it will send the responder an Award message: Accept-Proposal. Otherwise, it will reject the proposal. If the proposal is accepted by the initiator, the responder deploys the contingent schedule and sends the initiator a message indicating the result of deployment. If no dynamic events happen during the negotiation process, the responder should be able to successfully deploy the schedule and send the initiator an Inform-done message. If some dynamic events happened during the negotiation process make the contingent schedule impossible to be deployed, a Failure or Inform-Result message will be sent to the responder indicating how the contingent schedule is impacted.

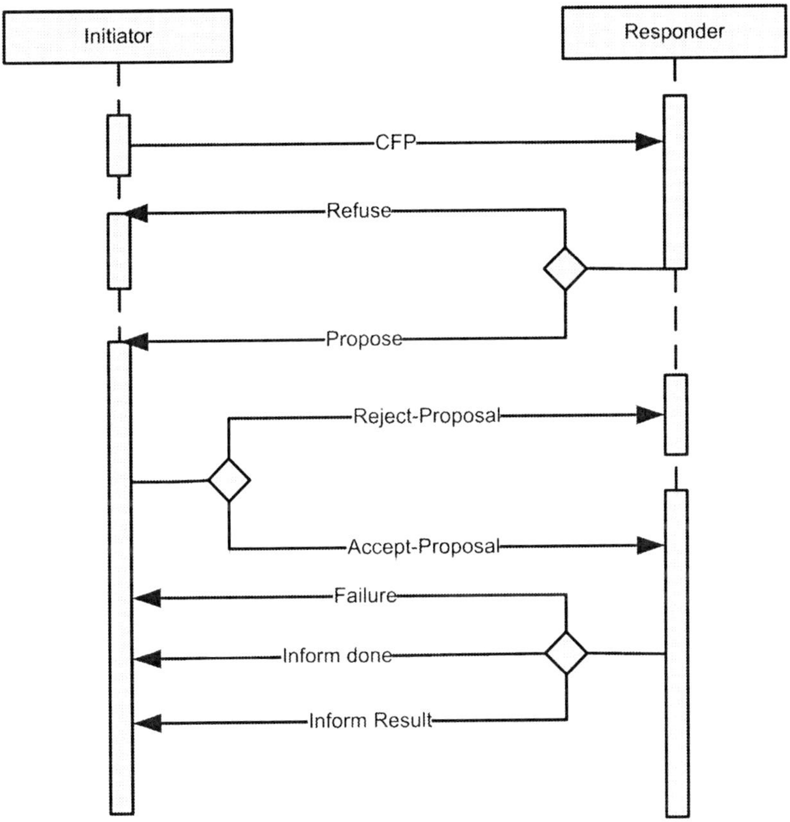

Figure 9.13. Contract Net Protocol for distributed scheduling

9.4.6 System Implementation

The Real Time Distributed Shop Floor Scheduling system has been implemented in Java on the JADE agent development platform (http://jade.tilab.com) and Java Web Services tools (http://java.sun.com/webservices). Figure 9.14 illustrates a two-workcell deployment of the system. The scheduling system for a workcell contains a scheduler agent, a real time controller agent, several resource agents and a scheduling service. All agents sit on a distributed JADE platform across several hosts. Java Web Service environment is installed on the same host that the scheduler agent sits on, which allows the scheduling service of a workcell to be connected with the scheduling services of other workcells. Together, the scheduling service and the scheduler agent fulfill the functionality of the scheduler of a workcell.

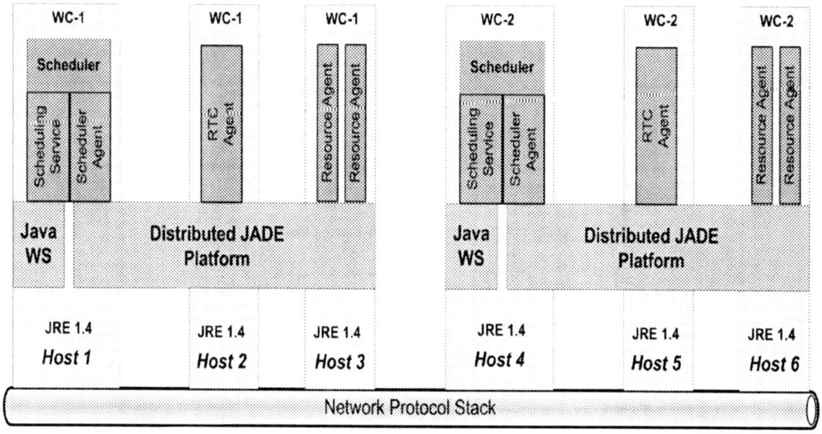

Figure 9.14. Deployment of the real time distributed scheduling system

9.5 A Case Study

This section presents a case study which demonstrates how the proposed scheduling algorithms at different levels of the multiple workcell shop floor are integrated in providing robust scheduling under the proposed Agent-based Web service integration framework.

Consider a shop floor with two workcells (Workcell A and Workcell B). Each of them has a set of jobs to be scheduled. The experimental scenario goes as follows.

Workcell Scheduling. The schedulers of Workcells A and B perform scheduling using the workcell scheduling algorithm described in Section 9.3.1. At this stage we assume that all jobs can be scheduled in local workcells. The generated schedules are passed to the Real Time Controllers of Workcells A and B respectively. Figure 9.15(a) shows the assigned schedule for Workcell A and

Figure 9.15(b) shows the assigned schedule for Workcell B in the form of Gantt chart. Workcell A and Workcell B have the same workcell configuration (only include three machines). The two job sets that need to be allocated have the same configuration as well (however, different job names are used). In the chart, the horizontal bars indicate the length of time allocated to each operation. The x-axis of the chart is subdivided into equal units of time (say hours in our case). The y-axis, on the other hand, lists all the resources in the workcell.

Dynamic Scheduling. For dynamic scheduling, we demonstrate how a machine down event is accommodated by the dynamic scheduling algorithm proposed in Section 9.3.2. Say Machine 3 of Workcell A breaks down at hour 25. This disruption is passed to the Scheduler through the Real Time Controller. The schedule repair algorithm first identifies operations (job13-op6, job13-op7, and job11-op8 in this case) are affected by the disruption and need to be re-scheduled, then, allocates them to available machines using the workcell scheduling algorithm. The repaired schedule is passed to the Real Time Controller and executed in Workcell A. As illustrated in Figure 9.15(c), job13-op6 is rescheduled on machine 2. To accommodate job13-op6, job13-op7 is shuffled two hours towards the end of the schedule. However, job11-op8 can no longer be processed by Workcell A because Machine 3 is the only one eligible in Workcell A. It needs to be assigned to other workcells on the shop floor by the distributed scheduling algorithm

Distributed scheduling. To assign job11-op8 to other workcells, the scheduler of Workcell A first tries to find all eligible workcells on the shop floor that can process the operation through the lookup service provided by the UDDI (Workcell B turns out to be the only eligible one). The scheduler A sends out a service request which contains a call for proposal to Scheduler B including the operation name (job11-op8) and the operation release time (at hour 18 because its precedent operation job11-op5 ends at hour 18). Upon receiving the request from Scheduler A, Scheduler B calculates a solution for job11-op8 and sends back a bid indicating when the operation will be processed. Scheduler A awards this operation to scheduler B. Scheduler B passes the modified schedule (including the assignment of job11-op8) to the Real Time Controller of Workcell B for execution. As shown in Figure 9.15(d), the operation is added to the end of Machine3's schedule in Workcell B.

To demonstrate the integration of scheduling algorithms more clearly and intuitively, we have intentionallly used a simple scenario in this case study. The performance of the algorithms has been tested using more complicated problem sets. Interested readers may refer to [34].

9.6 Conclusions

Generally speaking, any manufacturing enterprise is distributed. Distribution can be geographical, logical, temporal, or spatial. In manufacturing domain, it is not uncommon for production to be distributed geographically, sometimes on a continental scale (the automobile industry is a prime example).

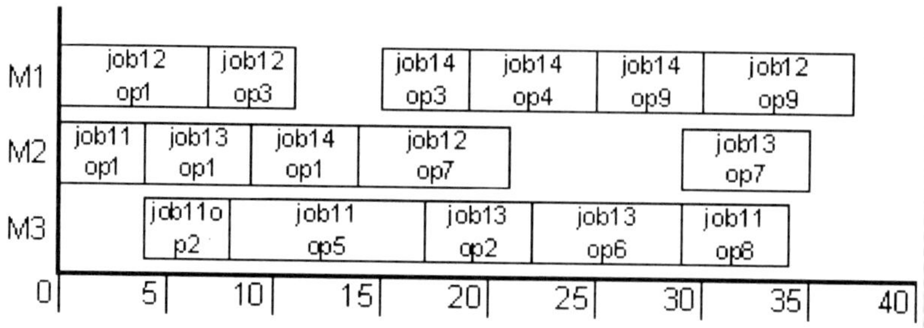

(a) Original schedule of workcell A.

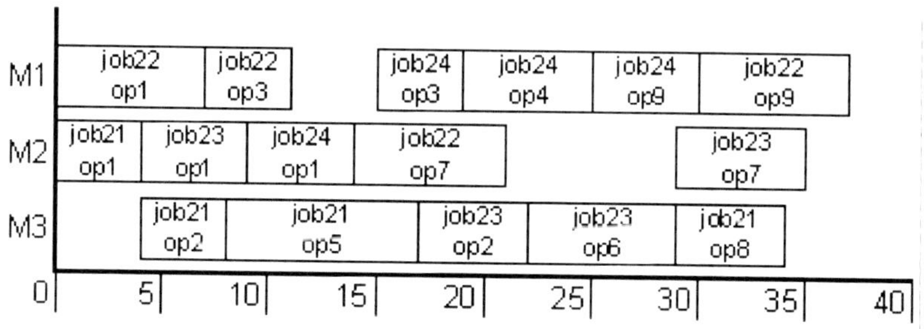

(b) Original schedule of workcell B.

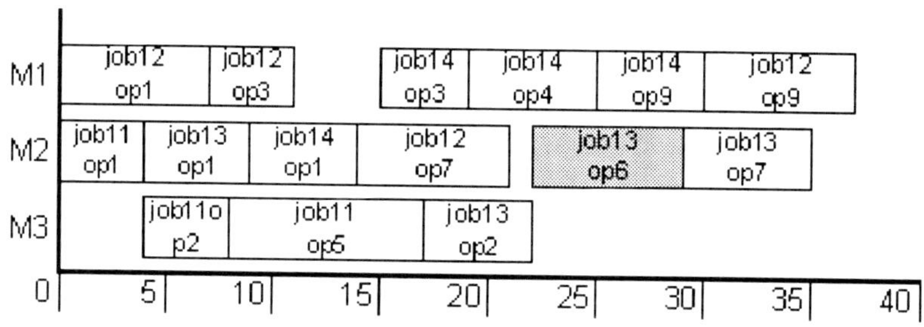

(c) Repaired schedule of workcell A.

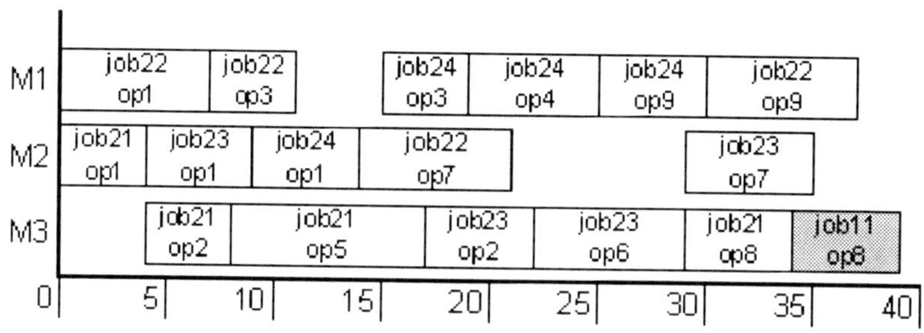

(d) Refined schedule of workcell B accomodating the op8 of job 11 from workcell A.

Figure 9.15. Gantt chart of the schedules

An enterprise can logically be distributed, reflecting its organizational structure. Organizational structuring can be a necessity in order to decompose the enterprise's problems into manageable chunks and to better exploit available expertise. Scheduling is an essential functionality required by manufacturing control and management at various levels of manufacturing. We have proposed a real time distributed scheduling framework for multi-workcell shop floors. Since distributed environments exist at other levels of manufacturing management, in many cases, it is justified to apply the proposed distributed control structure and even some algorithms (*e.g.*, the distributed scheduling algorithm) to inter-enterprise, enterprise and plant environments as well. For example, at the enterprise level, if a set of customer orders need the cooperation of several divisions of an enterprise, in a dynamic market environment, the scheduling problem involved is a real time distributed one. Currently, most of the enterprise planning and scheduling as in ERP/MRP systems are conducted in a centralized way. One of the criticisms on these systems is the fact that they are complex and inflexible. As a result, there has been interest in the development of decentralized strategies for enterprise systems. We see this as a potential application domain of real time distributed scheduling systems.

In many real world environments, scheduling exhibits decentralized nature and is conducted through negotiation processes. This observation triggers one of our important future research directions, which is the application of economic based resource allocation mechanisms, such as various auctions, to real time distributed manufacturing scheduling. In many business to business transactions, production scheduling parameters (*e.g.*, due dates) are set through a negotiation process between the customer and the service or product provider. In some cases, a firm may consider the possibility of "outsourcing" some time-sensitive orders through a negotiation mechanism if the system is highly congested and completing all the orders in-house would lead to very high tardiness penalties. As many manufacturing management applications require scheduling functionality in decentralized environments, we see that economic based scheduling mechanisms are good candidates in such environments.

9.7 References

[1] Kocjan, W., 2002, "Dynamic scheduling: state of the art report," *Technical Report*, T2002:28, SICS.

[2] Shanker, K., Tzen, Y. J., 1985, "A loading and dispatching problem in a random flexible manufacturing system," *International Journal of Production Research*, 23, pp. 579–559.

[3] Baker, A. D., (1998) A Survey of Factory Control Algorithms which Can be Implemented in a Multi-Agent Heterarchy: Dispatching, Scheduling, and Pull. Journal of Manufacturing Systems, 17(4), pp. 297–320.

[4] Neiman, D., Hildum, D., Lesser, V. and Sandholm, T., 1994, "Exploiting meta-level information in a distributed scheduling system," *Proceeding of Twelfth National Conference on Artificial Intelligence (AAAI-94).*

[5] Sycara, K., Roth, S., Sadeh, N. and Fox, M., 1991, "Distributed constrained heuristic search," *IEEE Transactions on Systems, Man, and Cybernetics*, 21(6), pp. 1446–1461.
[6] Sadeh, N., 1991, *Look-Ahead Techniques for Micro-Opportunistic Job Shop Scheduling*, PhD Thesis, School of Computer Science, Carnegie Mellon University, Pittsburgh, PA.
[7] Kamel, M. and Ghenniwa, H., 1995, "Partially-overlapped systems: the scheduling problem," In *Design and Implementation of Intelligent Manufacturing Systems*, Parsaei, H. and Jamshidi, M. (Eds.), Prentice-Hall, pp. 241–274.
[8] McKay, K. N., Safayeni, F. R. and Buzacott, J. A., 1998, "Job shop scheduling theory: what is relevant?" *Interfaces*, 18(4), pp. 84–90.
[9] Nof, S., Barash, M. and Solberg, J., 1979, "Operational control of item flow in versatile manufacturing systems," *International Journal of Production Research*, 17(5), pp. 479–489.
[10] Stecke, K. E. and Solberg, J., 1981, "Loading and control policies for flexible manufacturing systems," *International Journal of Production Research*, 19(5), pp. 481–490.
[11] Chang, Y. L., Sullivan, R. S., Bagchi, U. and Wilson, J. R., 1985, "Experimental investigation of real-time scheduling in flexible manufacturing systems," *Annals of Operations Research*, 3, pp. 355–377.
[12] Fox, M. S. and Smith, S. F., 1984, "ISIS: a knowledge-based system for factory scheduling," *Expert Systems*, 1(1), pp. 25–49.
[13] Smith, R. G., 1980, "The contract net protocol: high-level communication and control in a distributed problem solver," *IEEE Transactions on Computers*, C-29(12), pp. 1104–1113.
[14] Fang, J. and Xi, Y., 1997, "A rolling horizon job ship rescheduling strategy in the dynamic environment," *International Journal of Advanced Manufacturing Technology*, 13, pp. 227–232.
[15] Qi, J. G., Burns, G. R. and Harrison, D. K., 2000, "The application of parallel multipopulation genetic algorithms to dynamic job-shop scheduling," *International Journal of Advanced Manufacturing Technology*, 16, pp. 609–615.
[16] Chan, F. T. S., Chan, H. K., Lau, H. C. W., 2002, "The state of the art in simulation study on fms scheduling: a comprehensive survey," *International Journal on Advanced Manufacturing Technologies*, 19, pp. 830–849.
[17] Basnet, C. and Mize, J., 1994, "Scheduling and control of flexible manufacturing systems: a critical review," *International Journal of Computer Integrated Manufacturing*, 7 (6), pp. 340–355.
[18] Zweben, M. and Fox, M. S., (Eds.), 1994, *Intelligent Scheduling*, Morgan Kaufman Publishers, San Francisco, CA.
[19] Wang, C., Ghenniwa, H. and Shen, W., 2006, "Scheduling multi-operation jobs in partially overlapping systems," *International Journal of Computer Integrated Manufacturing*, 19(5), pp. 453–462.
[20] Panwalkar, S. and Iskander, W., 1977, "A survey of scheduling rules," *Operations Research*, 25(1), pp. 45–61.

[21] Shaw, M. J. and Whinston, A. B., 1983, "Distributed planning in cellular flexible manufacturing systems," *Management Information Research Center Technical Report*, Purdue University, West Lafayette, IN.

[22] Duffie, N. A., Piper, R., SHumphrey, B. J., *et al.*, 1986, "Hierarchical and non-hierarchical manufacturing cell control with dynamic part-oriented scheduling," *Proceedings of NAMRC, North American Manufacturing Research Conference,* Minneapolis, MN, pp. 504–507.

[23] Lin, G. Y–J. and Solberg, J. J., 1989, "Flexible routing control and scheduling," *Proceedings of the Third ORSA/TIMS Conference on Flexible Manufacturing Systems,* Stecke, K. E. and Suri, R. (*Eds.*), Elsevier, Amsterdam, pp. 155–160.

[24] Ramaswamy, S. E. and Joshi, S., 1995, "Distributed control of automated manufacturing systems," *Proceedings of 27th CIRP International Seminar on Manufacturing Systems,* Ann Arbor, MI.

[25] Baker, A. D. and Merchant, M. E., 1993, "Automatic factories: how will they be controlled," *IEEE Potentials*, 12(4), pp. 15–20.

[26] Shen, W. and Norrie, D. H., 1999, "Agent-based systems for intelligent manufacturing: a state-of-the-art survey," *Knowledge and Information Systems*, 1(2), pp. 129–156.

[27] Shen, W., Li, Y., Hao, Q., Wang, S. and Ghenniwa, H., 2006, "A service oriented integration framework for collaborative intelligent manufacturing," *Robotics and Computer-Integrated Manufacturing*, (in press).

[28] Petrie, C. and Bussler, C., 2003, "Service agents and virtual enterprises: A survey," *IEEE Internet Computing*, 7(4), pp. 68–78.

[29] Matskin, M., Küngas, P., Rao, J., Sampson, J. and Peterson, S. A., 2005, "Enabling Web services composition with software agents," *Proceedings of the Ninth IASTED International Conference on Internet and Multimedia Systems, and Applications (IMSA 2005 #477-122)*, Honolulu, Hawaii, USA.

[30] Maamar, Z., Mostéfaoui, S. K. and Yahyaoui, H., 2005, "Toward an agent-based and context-oriented approach for Web services composition," *IEEE Transactions on Knowledge and Data Engineering*, 17(5), pp. 686–697.

[31] Liu, F., Yao, L., Zhang, W., Liu, H. and Zhang, H., 2004, "A conceptual model of agent mediated Web service," *Proceedings of IEEE International Conference on Services Computing (SCC 2004)*, Shanghai, China, pp. 638–642.

[32] Li, Y., Ghenniwa, H. H., Shen, W., 2003, "Integrated description for Agent-based Web Services in eMarketplaces," *Proceedings of the Business Agents and the Semantic Web Workshop*, Halifax, Nova Scotia, Canada. pp. 11–17.

[33] Ghenniwa, H. and Kamel, M., 2000, "Interaction devices for coordinating cooperative distributed system," *Automation and Soft Computing*, 6(2), pp. 173–184.

[34] Wang, C., Ghenniwa, H. and Shen, W, 2005, "Heuristic scheduling algorithm for flexible manufacturing systems with partially overlapping machine capabilities," *Proceedings of IEEE ICMA 2005*, Niagara Falls, Canada, pp. 1139–1144.

10

Leveraging Design Process Related Intellectual Capital – A Key to Enhancing Enterprise Agility

Jitesh H. Panchal, Marco Gero Fernández, Christiaan J. J. Paredis, Janet K. Allen and Farrokh Mistree

Systems Realization Laboratory
G.W. Woodruff School of Mechanical Engineering
Georgia Institute of Technology, USA

The sustained improvement of Product Development Processes (PDPs) has long been the focus of research in manufacturing and more recently that of research in design as well. This is due in part to the key realization that a PDP constitutes not only a central component of the engineering effort but also a core business process. During the last decade, a strategic business approach for the effective management and use of corporate intellectual capital has emerged. This approach has come to be known as Product Lifecycle Management (PLM) and promises to further a holistic consideration of product design, emphasizing integration, interoperability, and sustainability throughout a product's lifecycle in order for an engineering enterprise to remain agile with respect to the constantly evolving demands of a global market. Intellectual capital, thus far, has been comprised mainly of product related knowledge and exploited mostly via the reusability and scalability of existing products through product platform and product family design. However, we strongly believe that focusing solely on product knowledge is not sufficient and limits agility to variant design (and adaptive design, to a limited extent). In order to effectively support the generation of entire portfolios of products (via derivative and original design), we believe that the design process should also be considered to constitute a crucial component of an engineering enterprise's intellectual capital. Hence, we propose a paradigm shift that is centered on *leveraging design process knowledge* derived from previous designs towards the design of entirely new products.

Rather than proposing new technologies or standards under the 'PLM umbrella', in this chapter our objectives are: (1) to highlight design processes as key elements of an engineering enterprise's intellectual capital, and (2) to motivate fundamental research directions. In this chapter, an overview of the requirements and research challenges inherent in leveraging previously expended resources and *designing design processes* is provided. These challenges are illustrated in the context of designing Linear Cellular Alloys (LCAs). Finally, we assert the importance of including the lifecycle considerations of design processes in PLM, thereby motivating Design Process Lifecycle Management (DPLM).

10.1 Design Processes – An Enterprise's Fundamental Intellectual Capital

The sustained improvement of Product Development Processes (PDPs) has long been the focus of manufacturing and more recently that of design as well. This is due in part to the key realization that a PDP constitutes not only a central component of the engineering effort but also a core business process [1]. As pointed out by Wheelwright and Clark [61], it is those firms that are able to develop and bring their products to market the fastest that are able to create a significant competitive advantage for themselves. Efforts aimed at reducing product development times, however, are faced with several challenges, identified by Lu [24] as pertaining to (1) increases in product complexity, (2) increases in time-to-market (TTM) pressure, (3) globalization and segmentation, and (4) increasing customer demands. While a number of recent research activities focus on addressing the needs, underlying these challenges, a majority are aimed at meeting the intensive information requirements posed. One of the most notable recent efforts along these lines is Product Lifecycle Management (PLM). PLM is taken to be a strategic business approach for the effective management and use of corporate intellectual capital [9, 16, 21]. PLM involves activities from the initial conception to retirement of a product and is aimed at improving the product development process. The goal in PLM is to integrate all the product realization activities including market planning, concept development, design, production, sales, marketing, *etc.* Considering the field's extensive scope there are numerous interpretations, each highlighting different facets of import. Examples include *a)* interoperability issues and standardization in CAD/CAM/CAE (Computer-Aided Design/Computer-Aided Manufacturing/Computer-Aided Engineering), *b)* overarching management considerations, *c)* collaboration, *d)* product information management and sharing, and *e)* integration of tools. In Figure 10.1, we present three key components of an enterprise's intellectual capital – process information (top-left corner), product information (top-right corner) and the supporting PLM infrastructure (bottom) that consists of various software tools. Arrows between tools are used to represent flow of information among them. Dashed and solid lines are implemented to illustrate the fact that some of the links are more developed than others. As indicated in Figure 10.1, most of the elements of an engineering enterprise's intellectual capital relate to the acquisition of information pertaining to either product or process and the tools for transforming this information. The infrastructure of PLM, as defined currently, centers on the integration of various software and associated hardware tools, ranging from CAD and analysis packages to PDM systems, *etc.*, used for capturing and processing product information. To some extent, these tools are also employed for capturing information relating to the underlying design processes.

In our opinion, PLM efforts thus far have been focused on integration and the improvement of interoperability. Although some of the relationships depicted by dashed and solid lines in Figure 10.1 have been implemented successfully, it is our belief that the effective management of a product's lifecycle extends beyond ensuring the seamless flow of information between tools and requires a system-based perspective of the entire engineering enterprise. Consequently we assert the importance of designing the design process alongside the product in PLM. Although design processes play a crucial role in PLM, integrating the design of "design processes" with the product has received little attention. Systematic methods for designing design processes have not been formalized. Additionally, while it is true that the potential of leveraging the components of existing products towards developing new products has been exploited, the possibility of leveraging PLM sub-processes in new product realization scenarios is substantial. Thus, as an engineering enterprise becomes increasingly concerned with meeting the dynamic requirements of a global marketplace, a closer attention must be paid to the mechanisms underlying the product development. Perhaps the most crucial of these mechanisms is the design process. In terms of the engineering enterprise, this translates to the need for a systematic means of development for original, adaptive, variant, and derivative products. Although much attention has been paid to addressing this issue from a product-centric perspective by exploiting the reusability and scalability of products through product platform and product family design, not much attention has been paid to an engineering enterprise's primary resource commitment – the design process and its design.

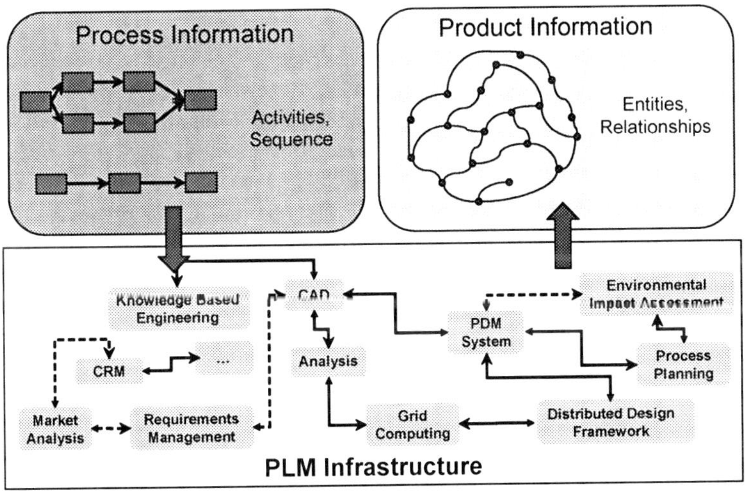

Figure 10.1. Integrating sources of intellectual capital in an engineering enterprise (product information, process information, and PLM tools)

Many emerging approaches to PLM are concerned solely with lifecycle considerations as they relate to a single product. Considering that most engineering enterprises strive to maximize product portfolio diversity, a perspective of PLM, focusing on the accommodation of the diverse and constantly changing needs of a global consumer base, may be appropriate. Taking a step back, the question becomes: "*How can a company ensure the effective use of resources across the entirety of its product portfolio, especially as markets evolve with time?*" To be successful in such continuously changing marketplaces, it is essential to not only address current customer requirements, but also accommodate impending changes. With this in mind, we emphasize that design processes should be viewed as constituting the strategy for developing a product, given a set of requirements. Satisfying changing customer requirements is thus subject to one's ability to adapt the underlying design processes. This is true whether referring to a single original design or an adaptive, variant, or derivative design, emanating there from. This is supported by the assertion of Herbert Simon that "... design process strategies can affect not only the efficiency with which resources for designing are used, but also the nature of final design as well" [53]. The design of design processes thus constitutes a fundamental prerequisite for the strategic deployment of products and the effective consideration of their respective lifecycle considerations. While the currently available methods and tools enable designers to effectively model, analyze and synthesize products, the means for applying the methods and tools to design the underlying *design processes* are non-existent. With this in mind, we believe that the following are important *requirements* in enabling the design of design processes:

1. Support for design information transformations
2. Support for design decision-making
3. Modeling and representation of design processes
4. Analysis of design processes
5. Synthesis of design processes

These considerations are addressed further in Section 10.2 through an illustrative example involving the design of LCAs, and mapped to the requirements for leveraging design process related intellectual capital and their design in Section 10.3. Research issues, emanating from these requirements and our strategy for addressing them are discussed in Section 10.4. Finally, we pose a number of questions regarding the future of lifecycle management as we expand our focus from products to processes and securing the intellectual capital associated with both. We assert that addressing these research issues would increase the adaptability of both products and design processes, thereby enhancing the enterprise agility with respect to changes in consumer demands.

10.2 Examples of Design Process Scenarios

In this section, we underscore the need for designing design processes by illustrating the effect of differing design processes on both the final design and the effectiveness with which the design goals are achieved. Specifically, design process related decisions are identified for each of the underlying scenarios. The chosen example is the multifunctional design of LCAs, where we identify different types of design process decisions, process goals, *etc.* involved in the achievement of the overarching objective pursued in the particular design scenario at hand.

10.2.1 Description of LCAs Design Problem

In this chapter, we rely on the design of LCAs [6, 20] in order to (1) emphasize the importance of leveraging existing intellectual capital for effectively designing design processes and (2) demonstrate the implementation of the design process models currently under development. LCAs are honeycomb materials (see Figure 10.2) that are processed through the extrusion of slurry through a multistage die. The slurry is composed of a binder, mixed with metal oxide powders. The structure resulting from extrusion is first dried and reduced into the metallic phase in a hydrogen-rich environment and then sintered to achieve nearly fully dense metal composites. A wide range of cell sizes and shapes, including functionally graded structures, can be achieved using this manufacturing process. These materials are suitable for multifunctional applications that require both strength and heat transfer capabilities [48]. Applications of these materials include heat sinks for microprocessors and combustor liners for aircraft engines. One of the main advantages of these LCAs is that desired structural and thermal properties can be obtained by designing shape, cell arrangement, cell wall thicknesses, and dimensions.

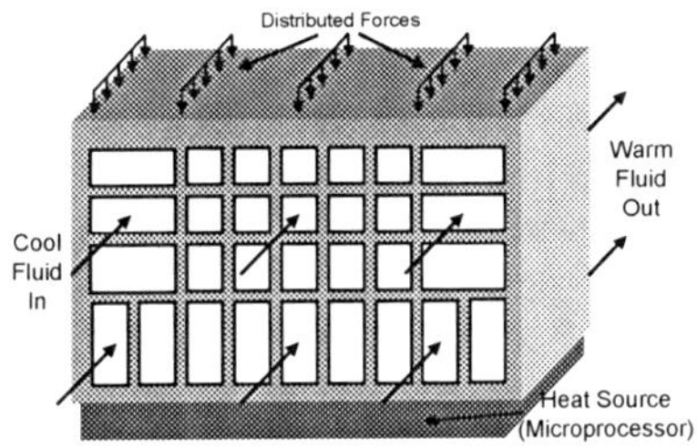

Figure 10.2. LCAs with rectangular cells

Although LCAs pertain to an emerging class of multifunctional structure-material systems, the underlying design process is clearly decomposable. It is because of this capability to clearly separate the different perspectives involved in the process that LCAs have been chosen as the illustrative medium. Each aspect of the multifunctional design is considered to be represented by different design experts. The problem is comprehensive enough to include various different activities involved in parametric design to illustrate the advantages of the proposed novel approach.

In this chapter, we model the design process of a LCA requiring high heat transfer rate and high stiffness. The design variables considered are: cell shape, total height of the LCA, thickness of the cell walls and fluid velocity. In order to further simplify the problem, we assume that the designers are restricted to use either triangular or rectangular cells. The design problem is summarized in Table 10.1 using the Compromise DSP word formulation [32].

The process of designing LCAs involves various steps such as cell shape selection, structural analysis, thermal analysis, design space exploration, geometry refinement, *etc*. The process can be structured quite differently depending on the designers' specific needs. Some example scenarios of LCAs design processes are discussed next.

Table 10.1. LCAs design problem

Given:
• FEM based thermal and structural models , boundary conditions, and design Requirements
Find:
• Design variables: cell shape, total height, thickness, fluid velocity, and deviation variables
Satisfy:
• Constraints and bounds on design variables, and goals
Minimize:
• Deviations from targets

10.2.2 LCAs Design Process Strategies

Many different strategies can be adopted for designing LCAs; four different design process approaches, each corresponding to differences in *a)* sequence of information transformations (see Strategy 1 and 2), *b)* information type (see Strategy 3), and *c)* model accuracy (see Strategy 4), are discussed repectively. It is the nature and configuration of the required information transformations that make up the design process. With this in mind, sequential design processes with 1) *thermal* considerations preceding *structural* considerations and 2) *structural* considerations preceding *thermal* considerations are illustrated in Section 10.2.2.1 and Section 10.2.2.2, respectively. A set-based design process is described in Section 10.2.2.3 and reliance on surrogate models in attaining required analysis results is discussed in Section 10.2.2.4.

10.2.2.1 Strategy 1: Sequential Design – Thermal First

In this scenario, the thermal goal assumes priority over the structural goal. Hence, the thermal designer fixes some variables in the design space and passes on the design to the structural designer. Given the choice between rectangular cell shape and triangular cell shape, the thermal designer chooses a rectangular cell shape because of superior forced conjugate (conduction and convection) heat transfer performance.

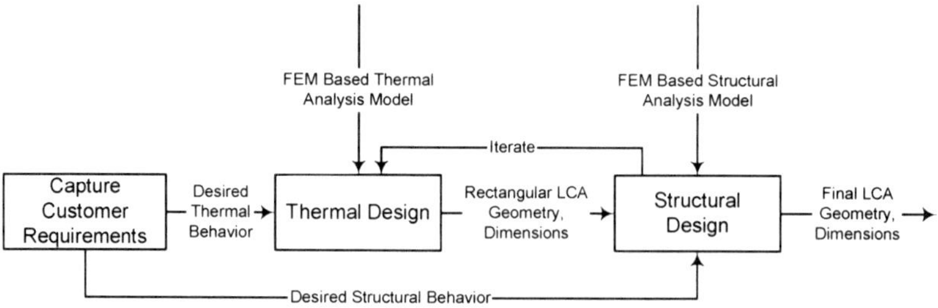

Figure 10.3. LCAs design scenario 1: thermal first sequential design

10.2.2.2 Strategy 2: Sequential Design – Structural First

In this scenario, the structural goal is more important than the thermal goal. The structural designer determines the structure and then passes the resulting geometry to the thermal designer for modification. In this case, the structural designer selects triangular cells in lieu of less stiff rectangular cells.

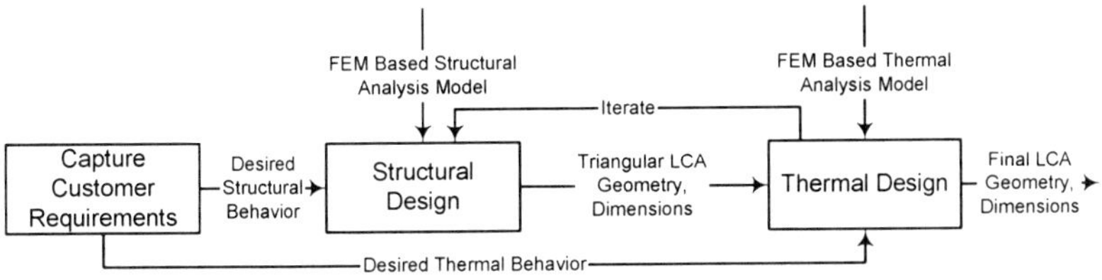

Figure 10.4. LCAs design scenario 2: structural first sequential design

10.2.2.3 Strategy 3: Set-based Design

In the set-based design scenario, designers consider sets of design alternatives rather than pursuing one alternative directly. The philosophy is to gradually narrow down the design space until a final solution is achieved.

In the LCAs design scenario, this may be implemented as one designer (thermal or structural) synthesizing a range of design parameters and then passing on this range to another designer to select the best value in that range. Since the designers do not pick a single alternative, the designers develop both cell topologies – triangular and rectangular. Although this approach is more likely to result in designs that show superior performance with regard to both thermal and structural considerations, the design effort involved in developing all alternatives is higher.

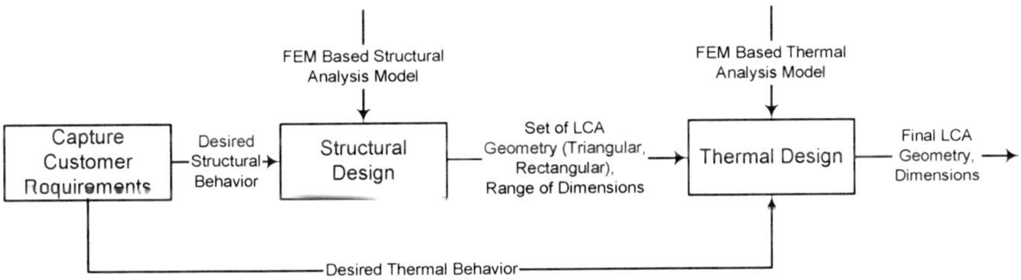

Figure 10.5. LCAs design scenario 3: set-based design

10.2.2.4 Strategy 4: Use of Surrogate Models

The computational intensity of analysis models associated with design is often substantial. In these instances, it becomes necessary to develop surrogate models to replace expensive computational runs. These surrogate models, however, are not exact and may introduce additional error. In the LCAs design example, simple response surface models can replace computationally intensive FEM analysis codes. The choice of appropriate models also depends upon progress made along the design process. In the earlier stages, it is not possible to use high fidelity analysis models because of limited knowledge regarding the design. However, in the latter stages of design, when the design specifications have been determined, high fidelity analysis models are usually more appropriate.

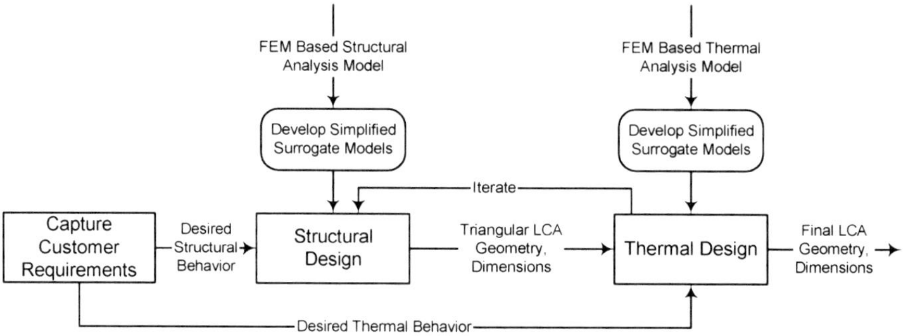

Figure 10.6. LCAs design scenario 4: use of surrogate models

10.2.2.5 Strategy 5: Parallel Iterative Design

Another design process option can involve performing design activities in parallel. In a parallel iterative design process for multifunctional applications, concurrent, point-based analysis is carried out for structural and thermal requirements. These analyses provide information about the simulated behavior for a given loading (both thermal and structural). This simulated behavior is compared to expected behavior. If these do not match, appropriate changes must be made to the geometric parameters to obtain the desired performance. The process is continued until the designers converge on a mutually acceptable solution.

In the design process scenarios discussed above, even though the product is the same, the design process is quite different. Needless to say, the results expected from these design processes are also different. Scenarios 1 and 2 highlight the fact that the design process has an effect on the final design (artifact). Scenario 3 highlights that the design process has an effect on the design effort. Finally, Scenario 4 underscores that the design of appropriate design processes is also dependent on the progress along a design process and affects both the efficiency and effectiveness of this design process.

In order to achieve the fast configuration of an appropriate process and to facilitate design process exploration, the elements of design processes should be modeled in a modular fashion with clearly defined interfaces. This idea is analogous to defining port-based models for design process elements. The input and output ports in design process elements are information- and knowledge-based. Reliance on modular design process building blocks will facilitate computer-based analysis of design processes in order to select the best design process option with regard to the context at hand. Modularity in design processes will also support synthesis of new processes from existing ones. The requirements for designing design processes are discussed in further detail in Section 10.3. Specifically, various considerations for a foundational framework for designing design processes are posed. The main factors are supporting synthesis and decision-making with respect to the product under consideration, as well as modeling representing, analyzing, and synthesizing the overarching design process, used to arrive at the final design.

10.3 Requirements and Critical Issues for Leveraging Design Process Related Intellectual Capital

Each of the five requirements established in Section 10.1 spurs a number of research issues that must be investigated in order to effectively manage and leverage an engineering enterprise's intellectual capital. The relationship among these requirements and the underlying research issues is summarized in Figure 10.7 and explored throughout this section. Specifically, *i)* support for design information transformations is discussed in Section 0, *ii)* support for design decision-making in Section 10.3.2, *iii)* modeling and representation of design processes in Section 10.3.3, *iv)* analyzing design processes in Section 10.3.4, and *v)* synthesizing design processes in Section 10.3.5.

10.3.1 Support for Design Information Transformations

1. Design processes should be decomposable into individual information transformations (design process building blocks) along with associated information flows such that the reuse of design processes (either in part or in their entirety) is facilitated.
2. Design process building blocks should guide designers by both the provision of structure and relevant information content.
3. The framework should facilitate capturing information flow constraints on design process elements.

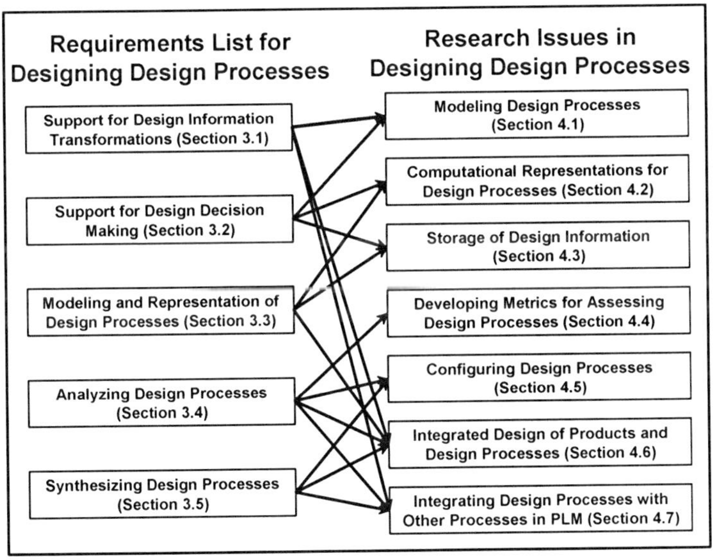

Figure 10.7. Mapping requirements list to research issues for designing design processes

The research issues involved in providing support for design synthesis include the development of computational models for information transformations in a manner that supports the integration of information about the product and the associated design processes. In the case of our LCAs example, this translates to supporting not only the activities associated with determining the cross sectional topology of the design, but also the sequence in which these considerations are taken into account.

10.3.2 Support for Design Decision-making

1. The framework should facilitate the identification of individual design decisions and any interactions.
2. The framework should support capturing stakeholders' perspectives in a consistent form and provide the structure for design processes to account for relevant information.
3. The framework should facilitate stakeholder interactions pertaining to the solution of independent, dependent, and interdependent decisions.

The research issues associated with providing support for decision-making include the development of generic and consistent computational models for engineering design decisions and modeling the interactions of different stakeholders involved in the product realization process. In terms of our LCAs example, it is vital to represent individual information transformations, such as the design decisions made by the structural and thermal experts, so that they can serve as self standing models that lend themselves to the integration in each of the myriad design process scenarios depicted in Figures 10.2 through 10.5.

10.3.3 Modeling and Representation of Design Processes

1. The design process elements must have clearly defined inputs, outputs and execution mechanisms. The framework should facilitate modeling information flows, dependencies and interactions. The framework should also support modeling design processes at various levels of abstractions.
2. The design process models should be computer interpretable, archivable, and reusable. The models should also support design process analysis.
3. The process elements should be domain independent and sufficiently generic to model complex design processes.
4. The design process modeling architecture must remain open to future extension and customization.

The research issues involved in modeling and representing design processes include the identification of key design process elements, formalization of the associated computational models and development of quantitative metrics for assessing the impact of these individual design process building blocks on the overarching design process. These models of design process elements should incorporate both product- and process-centric information. Recalling our LCAs design example, the key process elements are the various analyses and decisions required in each of the proposed design strategies (see Sections 10.2.2.1 through 10.2.2.4). Specifically, computational models are used to determine structural and thermal performance of the resulting design. These analyses are also used to support decisions regarding design variables and parameter values. The resulting sequence of design process elements can then be used to characterize the chosen design strategy comprehensively. Finally, appropriate metrics may be employed to gauge the adaptability of the resulting strategy to producing a range of products aimed at meeting myriad, differing functional requirements.

10.3.4 Analyzing Design Processes

1. In the framework of designing design processes, an important consideration is the development of metrics for analyzing the impact of constituent design process elements on the overarching design process.
2. The framework should support the composition of multiple evaluation criteria, as pertaining to constituent design process building blocks, to quantify the impact of the design process as a whole.

The research issues associated with analyzing design processes include the development of metrics for quantifying the impact of design processes on both the process goals and the final design. Returning to our LCAs design example, robust designs, for example, will accommodate a range of functional requirements (see [38]). Robust design processes should not be affected by these variations. The notion of design freedom [54, 55] is particularly important due to the quasi-uncorrelated relationships between functional requirements and topology. Design Freedom is defined as the extent to which a system can be adjusted while still meeting its design requirements [54]. Whether structural considerations precede thermal considerations or *vice versa*, will greatly affect the resulting structure. While triangular cells are more likely to be favored in the former sequence, rectangular cells favor the latter. The respective design problems (thermal and structural) must be designed so that their information flows are easily reversed. These considerations are addressed in Section 10.4.4.

10.3.5 Synthesizing Design Processes

1. The process elements must be modular in order to compose design processes from existing design process building blocks, as proposed in Requirements 1 and 2 of Section 0, so that *a)* existing design processes and *b)* design knowledge can be modeled and reused in a computer interpretable manner.
2. The configuration of the design processes from existing building blocks should be guided by product and design process related goals as defined by the engineering enterprise's strategy.

The research issues involved in synthesizing design processes include the identification of design process goals, development of metrics for evaluating the process performance against these goals, configuring the design processes appropriately to satisfy these goals and finally facilitate future adaptation as identified by the engineering enterprise's strategy. With regard to our LCAs design example, these concepts translate to determining relevant design drivers (*e.g.*, stiffness, compliance, and total heat transfer), reliance on measures of the resulting design freedom of the product (*e.g.*, robustness of the resulting topology to changes in boundary conditions), and determination of the more suitable sequence of required decisions to meet desired targets (*e.g.*, thermal design, followed by structural design, or *vice versa*).

Taking each of these needs into consideration, a mapping between the requirements for designing design processes and research issues is possible, as depicted in Figure 10.7. Having identified the underlying research issues involved in satisfying the requirements for designing design processes, we now turn our attention to each of these in detail.

10.4 Research Issues and Strategies for Designing Design Processes

Referring back to our motivating example in Section 10.2, the importance of PLM is evident when one considers the wide range of potential applications ranging from microprocessor heat sinks to aircraft engine combustor liners (see Figure 10.8). LCAs constitute an emerging family of complex material structure systems which are designed to uniquely satisfy a particular set of operational requirements. Hence, the traditional means of generating a product portfolio as suggested by Meyer [30], involving the development of product platforms and leveraging of modular components, are not suitable for generating a product variety. In other words, there are no product components that can be directly reused or scaled and product architectures cannot be taken advantage of. Hence, the following question arises:

How can designers leverage design knowledge so as to facilitate effective and efficient development of new products?

The answer lies in recognizing that the design process underlying the range of achievable products is common and thus, constitutes the primary resource for the extended design enterprise. Whether one focuses on product or process variety, the basic consideration remains the same – maximize external variety while minimizing internal variety [59]. *The means of achieving this goal for design process variety lies in designing design processes as open systems* [54]. According to Simpson and co-authors, the key to flexibility in open systems lies in modularity, mutability, and robustness. With this in mind, we focus on designing open design processes based on these assertions. The assumption here is that principles of open systems apply to both products and processes. In particular, we focus on developing baseline design processes that are easily adapted and reconfigured. The goals for ensuring the required flexibility include simple relationships, minimal interdependencies, clear and concise interactions, and generic process constructs that can be used to compose the processes. The expected result is the ability to model, capture, and analyze design processes both in part and in their entirety.

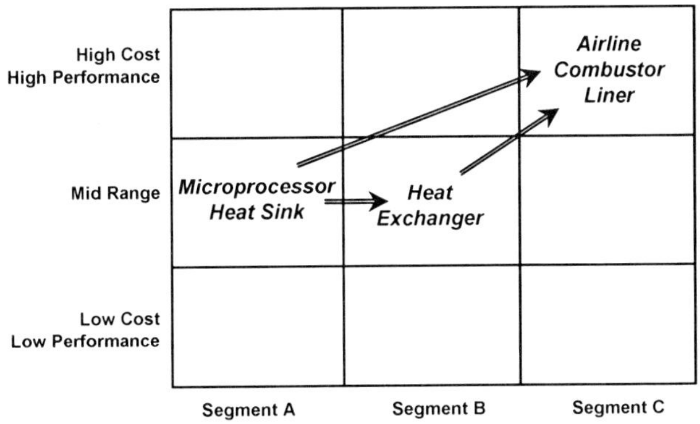

Figure 10.8. Market segmentation grid – leveraging design process knowledge across product variants

Considering the importance of design processes in determining a company's competitiveness, we discuss seven research areas in Sections 10.4.1 through 10.4.7 as outlined in Figure 10.7. These research areas and associated research objectives are summarized in Table 10.2. In each of the following sub-sections, we consider the underlying research issues, previous related work, open research questions and our strategy for addressing these.

Table 10.2. Research issues and objectives for leveraging design process related intellectual capital

	Research Issue	**Objective**
Section 10.4.1	*Modeling Design Processes*	Modeling the design processes in a modular fashion using generic process elements (transformations) that can be composed to form higher level design processes.
Section 10.4.2	*Computational Representations of Design Processes*	Representation of design processes in a manner that they can be reused either partially or in their entirety for either the same product or for different products
Section 10.4.3	*Storage of Design Information*	Capturing information by successfully separating product and process information at various levels of abstraction.
Section 10.4.4	*Developing Metrics for Assessing Design Processes*	Developing metrics for analyzing the effectiveness of design processes.
Section 10.4.5	*Configuring Design Processes*	Developing methodology for synthesizing new design processes from existing design process elements.
Section 10.4.6	*Integrated Design of Products and Design Processes*	Developing methodology for designing products in conjunction with the underlying design processes.
Section 10.4.7	*Integrating Design Processes with Other Processes in PLM*	Integrating process related intellectual capital pertaining to all aspects of engineering enterprise throughout a product's lifecycle.

10.4.1 Modeling Design Processes

10.4.1.1 Research Issue

In order to leverage knowledge about design processes, there is a need to devise a means of modeling these consistently and succinctly. Modularity of components is an additional requirement so that (1) component interfaces are clear, (2) component interactions are concise, and (3) both complete and partial reuse of design process components is facilitated. Design process components should also be computer interpretable, enabling analysis as well as execution. Finally, design process components should span all the hierarchical levels of a design process, ranging from planning at the organizational design process level to executing tasks at the computational level. In order to design the design process, the modeling technique should thus ensure that the *impact of design processes on products* is clearly determined, *modularity of design processes* is ensured, and *process elements are generic* (see Figure 10.7).

10.4.1.2 Previous Work

Design processes have been modeled from many different perspectives such as the activity based perspective [13, 14], the functional evolution perspective [52], the evolution of product states [58], the manipulation of knowledge [25, 26], and the decision based perspective [39]. Clearly, there is no single design process model that encompasses all required aspects of design. Some of the methods are focused on capturing processes to make organizational decisions (*e.g.*, [13, 14]), others towards understanding and capturing designers' intentions and rationale (*e.g.*, [39, 50, 51, 58]), while even others are focused towards artificial intelligence with the eventual intent of automating the process (*e.g.*, [25, 26]). These efforts thus far have been limited in terms of their reusability mainly because:

- There is a lack of consistency with regard to a single, domain independent set of design process building blocks that span all required levels of abstraction.
- Current design process modeling efforts do not support modularity either architectures or interfaces, both of which are essential for reuse at the sub-process level.

10.4.1.3 Research Questions

Considering the limitations of the methods listed in Section 10.4.1.2, a number of pertinent research questions related to modeling design processes arise:

- What are the key information transformations in design processes?
- How can design processes be modeled as hierarchical systems?
- How can interfaces between design process elements be defined?

10.4.1.4 Strategy: A Decision-centric Approach

Our approach to modeling design processes, aimed at addressing the research questions posed in Section 10.4.1.3, is rooted in the Decision Support Problem Technique, developed by Mistree and co-authors [31, 34-37, 39]. A fundamental assumption, from which many advantages with regard to addressing hierarchical interoperability are addressed, is that design processes are decision-centric. This is advantageous because decisions offer a consistent means of modeling processes regardless of domain, level of abstraction, perspective and discipline of the process considered. Bras and Mistree [4], in an extension of the DSP Technique, model design processes using a set of fundamental entities – *Phase*, *Event*, *Task*, *Decision*, and *System* Support Problems.

In our approach, *design processes are defined as networks of transformations of information from one state to another*. The state of information refers to the *amount* and *form* of that information that is available for design decision-making. For example, *analysis* is a transformation that maps the product form to behavior, whereas, *synthesis* is a mapping from expected behavior to the product form. An information model to support the design processes modeled in this fashion is shown in Figure 10.9. This information model is adapted from the CORE product model proposed by Fenves [15]. Analogous to the CORE product model, the process model shown is hierarchical and the entities are derived from a single *Core Design Process Entity*. The key entities in this process model represent the basic building blocks of design processes. There are different types of information *Transformations*, *Information*, and *Interfaces*. Transformations can be either *Decisions* or *Tasks*. It is important to realize that transformations themselves hierarchical and can be composed of other transformations. Design decisions are categorized as being either *selection* [17, 33] or *compromise* [32, 49] (see Figure 10.9). The former constitutes choosing among a set of feasible alternatives, the latter refining a given alternative. *Tasks* are categorized into abstraction, concretization, composition, decomposition, mapping, and evaluation (not shown in Figure 10.9). Transformations are defined by their function (intended role), structure (architecture), and behavior (actual performance). *Information* can be of two types – *Flow-Information* and *Meta-Information*. *Flow-Information* refers to the information processed by transformations (*i.e.*, Inputs and Outputs), whereas *Meta-Information* (Process Specification, Process Attributes) describes the characteristics of transformations. *Interfaces* connect transformations by processing the information shared among them.

With regard to the LCAs design example, described in Section 10.2, relevant decisions relate to the determination of topology based on structural and thermal considerations. To support these decisions, a number of tasks are required. Examples include the overall decomposition of the design problem, mapping of structure to behavior using analysis, and evaluation based on a comparison of simulated behavior with design requirements.

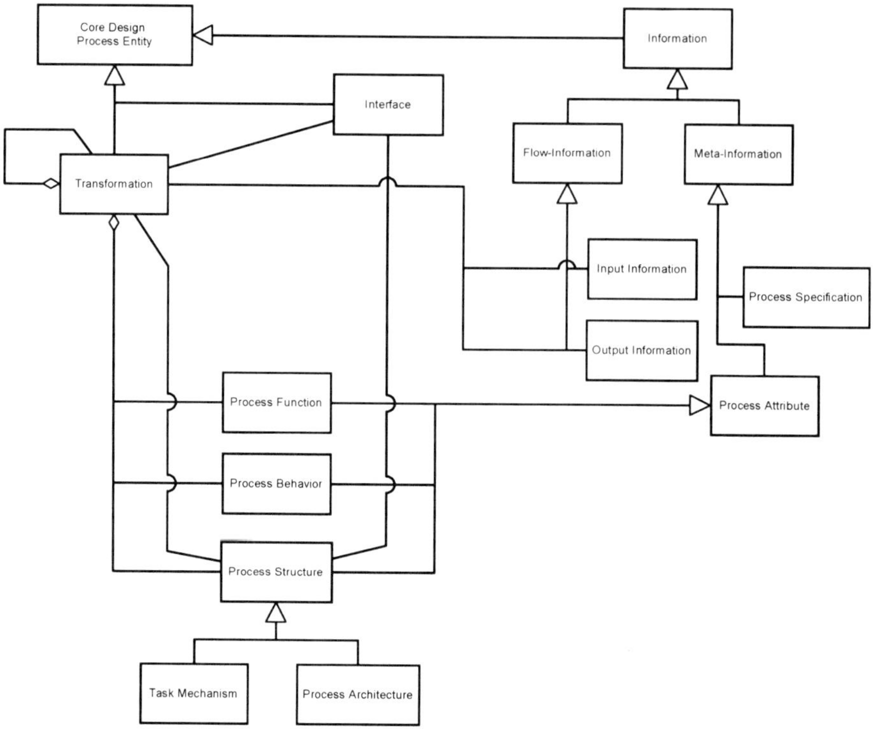

Figure 10.9. Design process information model

10.4.2 Computational Representations for Design Processes

10.4.2.1 Research Issue

In order to promote the reusability of design processes (see Figure 10.7), it is important to develop reusable computational constructs for design process building blocks that capture all relevant information content. These constructs for process building blocks can then be combined using appropriate interfaces to represent complete design processes and capture relationships among components as well as with the overall system. There is a lack of formal computational models for representing and reusing existing knowledge about design processes. The design process models currently used are either narrative or symbolic in nature [29].

10.4.2.2 Previous Work

Design processes are represented at a computational level in commercial software applications like ModelCenter® [42], FIPER [11], iSIGHT [12] and Hyperworks Process Manager™ [22]. The basic process element is a simulation code. The information captured using this process element in modeling processes with these applications is strictly related to the inputs, outputs, code to be executed, and the relationships between parameters. The design process is defined exclusively by the flow of parameter values between various software applications. This in effect links the *declarative* (*i.e.*, problem specific) information to the *procedural* (*i.e.*, design process specific) information. Consequently, reusability is limited to parametric design where the set of parameters and their relationships remain the same. A mere addition or deletion of a parameter requires reformulation of the underlying process. The design process descriptions cannot be reused even if the process remains the same and the parameters change.

10.4.2.3 Research Questions

Taking the drawbacks of current methods into consideration, the following pertinent research questions related to modeling design processes arise:

- How can information relevant to design process building blocks be captured as computational templates?
- How can these templates be combined to model complete design processes?
- How can the problem-specific (declarative) information be separated from the design process-specific (procedural) information?

104.2.4 Strategy: Separating Declarative Information from Procedural Information

Our strategy for modeling design processes is based on considering design processes as networks of information transformations (see Section 10.4.1.4). The architecture of our process modeling approach consists of three levels, namely a process specification layer, a declarative layer, and an execution layer. Their relationship is illustrated in Figure 10.10.

1. *Process Specification Layer:* In this layer, *a)* required information transformations are identified and *b)* required information flows are specified accordingly. In order to ensure that declarative information is separated from procedural information, information flows are clearly separated from information content. In other words, we capture only the mechanics of information transfer at this level, while problem specific information is defined separately at the declarative level. This results in a process map that remains the same irrespective of the application in which the process is used.
2. *Declarative Layer:* In this layer, problem formulation related information is captured. Consequently, the independence of information from process mechanics is guaranteed. Use of a standardized format ensures that the problem specific declarative information can be reused in different processes. Reliance on XML offers a convenient means of capturing information at this level.
3. *Execution Level:* In this layer, the details of code execution are captured. This level is specific to the design scenario and problem for which the process is used. Execution level codes interface only with the declarative problem formulation level. Thus, there is no direct link between the process

specification level and execution level. This preserves the modularity of processes.

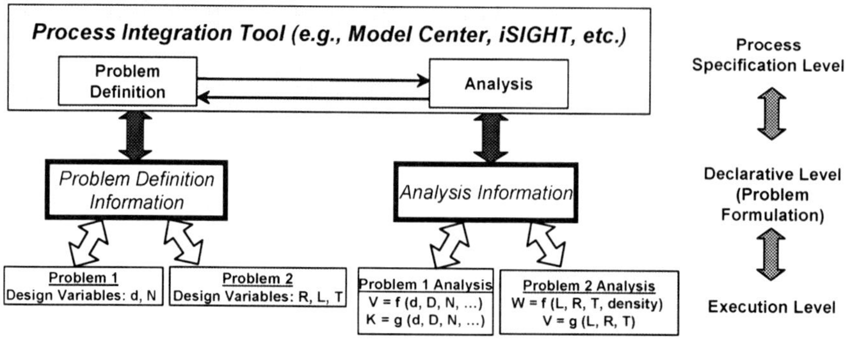

Figure 10.10. Architecture of process modeling framework

10.4.3 Storage of Design Information

10.4.3.1 Research Issue

It can thus be established that the design process is intimately linked to the product being designed. Many design process decisions can be identified only after some product related decisions have already been made. Hence, it is essential to consider the design of both the product and underlying design process simultaneously (see Figure 10.7). Neither aspect should be modeled in isolation. To enable the storage of product and process information separately, there is a clear need for a consistent information model that addresses both product- and process-centric concerns. There are numerous ways of evaluating the quality of a design process. While some focus on process dependent attributes such as individual activities and their sequence, as well as the underlying information flows, others focus on product specific attributes relating to the quality of the resulting design. Both perspectives are crucial. The challenge lies in *reconciling these concerns with respect to the overarching design drivers.*

10.4.3.2 Previous Work

There has been a significant amount of research with regard to modeling both products and processes. While some efforts have been concerned mainly with focusing on either product or process related aspects separately, others have pursued a more integrated perspective. On the product modeling side, notable efforts include the CORE product model by Fenves and co-authors [15]. On the process side, relevant endeavors include the proposition of a Process Specification Language (PSL) by Schlenoff [46, 47] and ISO 10303 STEP Standard AP 231 for Process-Engineering Data [40]. There are also a number of efforts that seek to reconcile product and process-centric perspectives. Examples include the Georgia Tech Process to Product Modeling Tool (GTPPM) developed by Lee, Eastman, and

Sacks [8, 45] and the Object-Oriented Modeling of Products and Processes, proposed by Gorti and co-authors [19].

As stated in Section 10.4.3.1, there is a definite need for an integrated product and process model in order to facilitate both partial and complete reuse of engineering design processes. Current instantiations of such models are limited in so far that they consider processes mainly with regard to components and their dependencies in terms of respective inputs and outputs. Design process execution, however, has not been a central concern thus far. Further, current design information models capture information in a manner that supports archiving *existing* processes and design rationale. Only information immediately relevant to the storage of a particular decision is captured. Since no additional information is available for future reference, the synthesis of new processes based on the existing sub-processes is not supported in current information models.

10.4.3.3 Research Questions

Considering the challenge posed in Section 10.4.3.1 and the limitations of current efforts, considered in Section 10.4.3.2, the following research issues arise:

- How can process information be separated successfully from product information?
- How can the structure and content of engineering design processes be captured effectively so that modular reuse of process components becomes feasible?
- How can design processes be stored so that their structure remains consistent throughout all levels of a process hierarchy?

10.4.3.4 Strategy: Process Templates

Our approach to storing design information is necessarily two pronged, focusing on a reconciliation of product with process-centric aspects of a design process. Necessarily, these must be separated clearly. Our current instantiation of this is an extension of the CORE product model [15] from solely modeling products to modeling design processes as well. In this extension, information transformations are modeled as hierarchical objects. A key feature of these information transformations is that they are derived based on a decision-centric perspective. Since decisions can be considered to constitute central elements of almost any process, regardless of domain and level of abstraction pursued within, a common means of modeling a process at all levels of a design effort results.

Constituent design process elements are synthesized to serve as reusable template to model the design process under consideration, either in part or in its entirety. Design Process Construct Templates[1] are computational objects that can be parsed, analyzed and/or executed on a computer [31]. Process Templates constitute compositions of interfaced design sub-process templates, as illustrated in Figure 10.11. It is important to note that the partial and complete templates shown

[1] Design Process Construct Templates are defined as computer based representations of design process elements, having well-defined inputs and outputs.

in Figure 10.11 share the common underlying architecture presented in Figure 10.10.

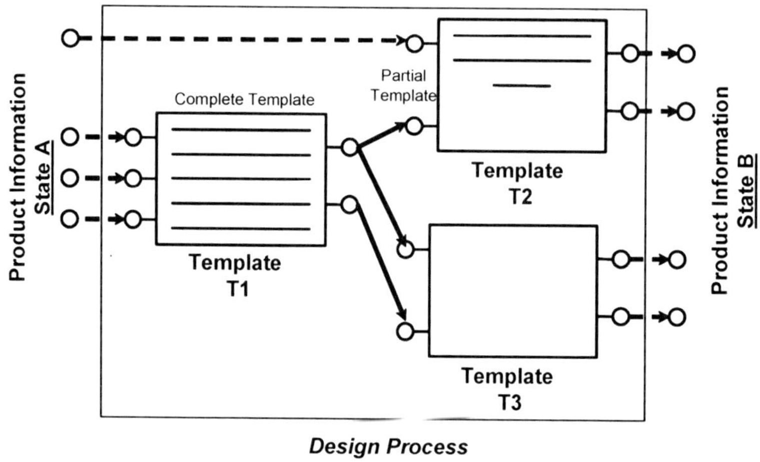

Figure 10.11. Modeling design processes using process templates

The design process in this figure involves three design process element templates: T1, T2 and T3. In the case of the LCAs example, this translates to series of decisions and supporting tasks. In progressing along the design process, these elements convert the information about a product from one state to another (*e.g.*, from State A to State B in Figure 10.11). In the case of analysis, for example, the first state (*i.e.*, State A) represents *geometry and loading conditions* considered in isolation and the second state (*i.e.*, State B) would represent their combination and the resultant *behavior*. The lines connecting these process elements represent flows of information. Process Templates can be *partial* or *complete*. *Complete* templates contain all the information required for carrying out a transformation and can be executed. *Partial* templates do not have sufficient information for executing a transformation. This point is illustrated in Figure 10.11. All the information required for performing T1 is available at information state A. Hence, the template associated with T1 is a complete template. Some of the information required for T2 is not available until T1 is executed, however. Hence, T2 is a partial template. Design processes using such a notation can be viewed as networks of Process Templates, connected by the information flowing between them.

10.4.4 Developing Metrics for Assessing Design Processes

10.4.4.1 Research Issue

In order to accurately evaluate alternative design processes, their analysis is required (see Figure 10.7). Design process alternatives result from changes in parameter values, information flows, and sequence. With regard to our LCAs example, these correspond to differences in design requirements, parameter form, and stakeholder succession, respectively, as shown in Section 0. This implies that design process options have to be explored and information required for making a design process decision has to be generated before the design process decision can be made.

In order to aid in the analysis, evaluation, and configuration of design processes with regard to differing design process goals, appropriate metrics are required. Additional considerations to those discussed in Section 10.4.1, may include reliability, reconfigurability, minimization of cost, minimization of risk, available resources, *etc*. As stated previously, we come from the perspective of designing products as well as design processes as *open engineering systems*. The key to designing open engineering systems is *adaptability to changes in the environment*. The environment for a product is the set of conditions under which it is being used. Hence, a product is open if it is adaptable to changes under the conditions in which it is used. The environment for a design process includes the product which is being designed, the considerations used to design the product (*e.g.*, robustness, reliability, *etc.*), and the environment in which the product is to be used. This implies that if a similar product is being designed or the same product is being designed with added considerations, the underlying process need not change. Hence, a process is open if it can be used to design both similar products and the same products with different design considerations. We conduct our analysis of design processes accordingly.

10.4.4.2 Previous Work

Cost, time and interdependencies are generally encountered metrics for design processes [5]. Braha and Maimon [2] have considered complexity for analyzing design process effectiveness. Rogers and Christine [44] have considered coupling strength as an indicator of design process effectiveness. The drawback of these process analysis metrics is that they do not quantify the effect of information transformations on the product.

10.4.4.3 Research Questions

Based on the points made in Sections 10.4.4.1 and 10.4.4.2, a number of research questions arise.

- How can design process components be characterized sufficiently to allow for their adaptation in the case of derivative and adaptive design processes?
- How can the effectiveness of design processes be quantified?
- How can the impact of design processes on products be measured?
- How can process effectiveness metrics be employed for synthesizing original, adaptive, variant, and derivative design processes?

10.4.4.4 Strategy: Open Systems Perspective

Our strategy is to model and analyze processes from an open systems perspective. Various techniques like robustness, modularity, maintaining design freedom, adaptability, *etc.* have been proposed for achieving openness in a system [54]. Hence, openness of systems can be measured by developing quantitative metrics for these. The quantitative measures related to openness of a product are: *design freedom* [23, 55, 60, 62, 63], *robustness*, *complexity* [2, 10, 43, 53] , *modularity* [41] (which is closely linked to complexity) and *coupling* [27, 28]. Previous research efforts are mainly focused on quantifying the openness of products but openness of design processes has not been addressed in the literature. It is here where we make our contribution towards metrics for design processes.

10.4.5 Configuring Design Processes

10.4.5.1 Research Issue

During the NSF Simulation Based Engineering Science (SBES) workshop [64], the need for designing complex products in a hierarchical fashion at multiple scales was surveyed, and the potential for the new field of SBES explored. It was noted that by integrating knowledge from different scales, we would be able to design products considered extremely difficult today. Most complex products are designed by their hierarchical decomposition into interacting components. This results in hierarchical products as well as processes. The possible ranges in scope and detail of the resulting design processes are illustrated in Figure 10.12. As the scope increases from involving a single designer, to teams and multiple organizations, the relevant detail of the design process changes from involving interactions among design variables to inter-organizational interactions. In cases such as the design of aircraft, processes can even extend to the multi-organizational level. In the design of systems of this complexity both top-down and bottom-up approaches are often combined. The research issue in configuring design processes from both top-down and bottom-up approaches is the ability to define processes as modular patterns that can be reused in different scenarios (see Figure 10.7).

10.4.5.2 Previous Work

The concept of modularity has been extensively studied in the product design domain. The role of product architectures has been established by Ulrich in [59]. Various methods have been developed to design families of products [7, 18]. The application of modularity, patterns and families to design processes has not been addressed to the best of our knowledge.

10.4.5.3 Research Questions

Relevant research questions emanating from the considerations brought forth in Sections 10.4.5.1 and 10.4.5.2 are:

- How can top-down design processes be captured for reuse?
- How can lower level design processes be leveraged to facilitate bottom up design of design processes?

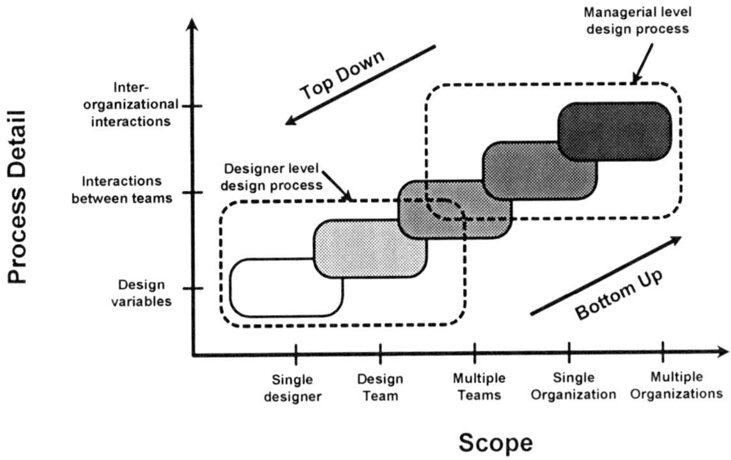

Figure 10.12. Processes represented at various levels of detail

10.4.5.4 Strategy: Process Families

In our approach we focus on developing a "baseline" process that can be extended to serve a number of different products, as identified in the engineering enterprise's strategy. We thus consider entire families of processes. Akin to the reliance on product families, there is an underlying emphasis on leveraging a common platform to the utmost extent possible – maximizing external variety, while minimizing internal variety. Often this is achieved through reliance on modularity and open architectures. This translates to a need for modeling and representing different products and the underlying process variants used to make them accordingly – in an openly extensible, standardized, modular, and adaptable manner.

Once such a baseline design process has been designed as envisioned here, arriving at process variants for designing new products is greatly facilitated, thereby significantly reducing the associated time and cost. This stands in marked contrast to designing processes to support the realization of products on a 1-to-1 basis. The engineering enterprise becomes significantly more agile in adapting to changes in consumer tastes and is better prepared to meet the challenges of compressed product lifecycles. Furthermore the management of Product Lifecycle considerations is greatly facilitated. Additional benefits are related to the fact that design chains can be designed in a manner reflecting the modularity of the underlying design processes. Consequently, interacting design chain stakeholders can be interfaced in a modular fashion, moving towards plug-and-play operation of processes across design chains.

10.4.6 Integrated Design of Products and Design Processes

10.4.6.1 Research Issue

Traditionally, design methodologies mainly focused on systematic design of products. The focus of the design community has expanded from the design of products to include considerations of manufacturing, maintenance, re-cyclability, *etc.* Integrated product and process design methods (IPPD) are mainly centered on designing both product and required manufacturing process simultaneously. In other words, design of products is carried out with consideration of additional factors that affect the product life cycle. Design of processes is limited to the design and simulation of manufacturing processes by industrial engineers. Integrating the design of *design processes* with the products has received little attention, however. A systematic method for designing design processes along with the product has yet to be formalized in the design literature.

10.4.6.2 Previous Work

Previous efforts aimed at addressing the need for a systematic concretization of design efforts are evident in the work of Bras and Mistree [3]. Specifically, they focus on devising a means of consistently modeling design processes from a decision based perspective. In doing so, they instantiate a set of primitives for modeling design process components. In their model, process components are connected by the information, energy and material flows between them. The technique involves two phases: Meta-design and design. Meta-design involves laying out the design process and the design phase involves executing the design process. The main drawback of this approach is that these two phases are interlinked and should ideally be carried out concurrently. The technique is also limited to mapping out the process of designing a product in terms of the required decisions. No considerations are given to the process related goals and the architecture of the design processes.

10.4.6.3 Research Questions

Relevant research questions emanating from the considerations brought forth in Sections 10.4.5.1 and 10.4.5.2 are:

- How can products and their associated design processes be designed in an integrated fashion?
- How can hierarchical considerations of products and processes be modeled consistently with regard to the required levels of abstraction?
- How can the evolution of products and processes be modeled along a design timeline?
- How can the various sources of uncertainty (*e.g.*, as pertaining to the environment, design variables, models, *etc.*) be accounted for as the quality and quantity of available information changes to reflect the current state of knowledge about (1) the product being designed and (2) the underlying process?

10.4.6.4 Strategy: Identifying Process Decisions

Decisions made during a design process are of two types: decisions about the product, discussed in Section 10.4.1.4, and decisions about the process through which the product is designed, as considered in this section. Such design process decisions can be divided into three main categories: a) concerning the architecture of the design process, b) the manner in which individual design activities are carried out, and c) the parameters of the design process. With regard to our LCAs design example, the sub-sections of Section 0 represent different architectures of the design process underlying LCAs development. The design process decisions form a three-tiered structure. More details about the nature of design process related decisions follow.

1. *Architecture of the Process:* The architecture of a design process refers to the network of design activities and associated information flows. The architecture of design processes can affect both the final product and process outcomes like time, cost, *etc*. Design processes can be partitioned in a variety of ways into individual activities and tasks. Partitioning a design process involves identifying activities and the information transfers between them. Some of the activities in a design process are coupled and others are uncoupled. It is preferable to have uncoupled tasks in a design process so as to reduce the number of costly iterations between tasks. Coupling can be categorized as being strong or weak depending on the amount of information dependencies between tasks. Decisions about design process architectures also include time sequencing of the tasks. Decisions such as which tasks need to be performed sequentially and which tasks can be performed concurrently are important when more than one design team is taking part in a given design process.
2. *Individual Design Activities:* Moving from process architecture decisions down to individual design tasks, decisions include the manner in which each activity is performed. One example of such design task level process decisions is the analysis task where the objective is to map form to behavior. Behavioral models can be developed at various levels of fidelity. Appropriate behavioral models needs to be selected depending on the information available, accuracy required and the progress along the design timeline. Mocko and co-authors [38] have shown that behavioral models can be organized hierarchically in a tree structure based on idealizations of the actual model. Another example of design task level design process decisions is synthesis.
3. *Design Process Parameters:* This is the lowest level of design process decision where designers are concerned with setting values of design process parameters like weights assigned to individual goals, optimization parameters, design of experiments parameters, *etc*.

10.4.7 Integrating Design Processes with Other Processes in PLM

10.4.7.1 Research Issue

A main aspect of Product Lifecycle Management (PLM) is the integration of processes and information throughout the value chain, including supply chains, design chains, demand chains, *etc*. In order to extend the value proposition of the effort of designing design processes (see Figure 10.7) presented in Sections 10.4.1 through 10.4.6, it is essential that design processes be integrated with other processes in the value chain. The underlying challenge lies in mapping the design processes with other processes.

10.4.7.2 Previous Work

A major thrust in PLM is the integration of the value chain throughout the extended enterprise. Design chains and supply chains form two essential components of the value chain. A significant amount of work is currently being undertaken by the Supply Chain Council[2] with regard to describing supply chains. For example, the SCOR model [57] is developed to represent and measure supply chains in a standardized manner to enable improvements in supply chain operations through analysis of current processes and best practice emulation. Along these lines, numerous case studies have been conducted. For example, SCOR model is currently being extended to the Enterprise Transaction Model by Streamline SCM [56].

10.4.7.3 Research Questions

Relevant research questions emanating from the considerations brought forth in Sections 10.4.7.1 and 10.4.7.2 are:

- How can decision-centric design process models be mapped into other processes within the value chain (*e.g.*, supply chain processes)?
- How can flexible interfaces be developed for enabling effective interactions among stakeholders in an enterprise?

10.4.7.4 Strategy: A Decision-centric Approach

Since the strategy outlined throughout this chapter is decision-centric, it is sufficiently generic to allow for adaptation to each of these contexts. Our research is focused on modeling design chains at various levels of scope and detail, ensuring domain independence and interoperability among the various stakeholders involved in a product realization process. Consequently, models and methods are being developed to address emerging design process needs on various levels of abstraction, so that the resulting hierarchy effectively supports the design activities of the enterprise.

Our approach to integrating design chains with the supply chains involve *(a)* identifying the decisions and information transformations involved in each of the level 3 elements in SCOR model, *(b)* modeling those decisions using the DSP

---

[2] www.supply-chain.org

Technique Palette [3], *(c)* using the decision elements, and *d)* developing object-oriented templates for elements in the SCOR models.

The research tasks involved in developing flexible interfaces between stakeholders are *(a)* modeling interactions between the design process elements and associated information flows, *(b)* modeling stakeholder relationships, commonly encountered throughout the value chain, *(c)* capturing design process interactions using object oriented templates that can serve as a springboard for knowledge capture, and *(d)* establishing communications protocols to represent the underlying interactions for enabling the required information transfers.

As a summary, throughout Section 10.4, we highlight seven key research areas (see Table 10.) that we believe are important in order to address the requirements (see Figure 10.7) posed by leveraging design processes as an important element of the intellectual capital. These research areas were explored in the light of existing literature to identify open research questions in Sections 10.4.1 to 10.4.7. We also outline some of our strategies for addressing these research questions. These strategies are focused on both decision-centric and systems-based view of design processes.

10.5 Conclusions

The future basis for competition is likely to rest on an enterprise's ability to anticipate and quickly respond to market shifts and changes. This requires the effective leveraging of resources. Considering that the bulk of the effort involved in product development lies in perfecting the underlying processes, these should be considered to constitute an enterprise's primary intellectual capital. Consequently, more attention must be paid to the manner in which these processes are designed.

Our starting premise, in this chapter, is that design processes are an integral part of its intellectual capital. Accordingly, we establish the design of design processes (together with product design) as a critical factor in addressing lifecycle considerations of an evolving product portfolio. Five key requirements for enabling the design of design processes are established and subsequently tied to underlying research issues – (1) identification of design process goals, (2) process related decisions, information transformations, and computational models thereof, (3) design process configuration, (4) quantification of design process impact, and (5) the integration of product and process-centric perspectives. We thus consider that attaining and retaining a competitive edge is likely to be a function of a company's agility in adapting existing design processes to the realization of adaptive, variant, derivative, and even original products. With this in mind, we provide a conceptual framework for addressing the underlying research issues involved in the development of a means to leverage design processes through composable, computer interpretable modeling techniques, facilitating their analysis, archival, and reuse. The proposed strategy is anchored in a decision-centric perspective of design processes; modular, computational template-based representation of design processes and their building blocks; utilization of existing standards for archiving of process information; metrics for assessment of design process performance; configuration-based techniques for design of design processes alongside products;

and integration with other processes in the value chain. Rather than proposing new technologies or standards under the PLM umbrella, it is our overarching objective to highlight design processes as key elements of an engineering enterprise's intellectual capital and to motivate fundamental research directions. We note that the vision articulated in this chapter is not meant to replace current efforts in the PLM arena. Instead, our aim is to augment these efforts via the inclusion of design process related intellectual capital, thereby enhancing the overall agility of the engineering enterprise.

We assert that managing the lifecycle of a design process will have much greater impact than merely considering the design of products in isolation. Hence, we believe that the vision and direction provided in this chapter are fundamental to the success of next generation agility in global enterprises. Considering the current scope of PLM, this thrust is extremely important. We thus envision extending the focus of PLM to include the lifecycle considerations of the design process, moving towards Design Process Lifecycle Management (DPLM). Considering the comprehensive nature of design processes, the underlying research problem is to manage and reuse *process knowledge* as a prime component of the intellectual capital. We must thus ask ourselves:

- To what extent can *families of processes* be modeled, captured, and reused?
- How can *top-down* design of engineering design processes be reconciled with *bottom-up* design of process components?
- How can all processes factoring into the value chain be *designed systematically* (*e.g.*, engineering design processes, supply chain processes, *etc.*)?
- How can product information be reconciled with processes at various levels of abstraction in an *entire global enterprise*?

In closing, we leave you with the following thought -

"Vision without action is merely a dream. Action without vision just passes the time. Vision with action can change the world." -- Joel A. Barker.

We have shared with you our vision – which is undoubtedly limited. We invite you, the members, visionaries, and practitioners of Collaborative Product Design and Manufacture, to enhance this vision and act so that we may collectively achieve our dream.

10.6 Acknowledgments

We acknowledge support from the National Science Foundation grants DMI-0085136 and DMI-0100123, as well as, Air Force Office of Scientific Research grants F49620-03-1-0348, and MURI 1606U81. Marco Gero Fernández was sponsored by a National Science Foundation IGERT Fellowship through the TI:GER Program (NSF IGERT-0221600) at the Georgia Tech College of Management and a President's Fellowship from the Georgia Institute of Technology.

10.7 References

[1] Berden, T. P. J., Brombacher, A. C. and Sander, P. C., 2000, "The building bricks of product quality: an overview of some basic concepts and principles," *International Journal of Production Economics,* 67, pp. 3–15.

[2] Braha, D. and Maimon, O., 1998, *A Mathematical Theory of Design: Foundations, Algorithms, and Applications*, Kluwer, Boston, pp. 241–278.

[3] Bras, B., Mistree, F., 1991, "Designing Design Processes in Decision-Based Concurrent Engineering," *SAE Transactions Journal of Materials & Manufacturing*, pp. 451–458.

[4] Bras, B. A. and Mistree, F., 1991, "Designing design processes in decision-based concurrent engineering," *SAE Transactions, Journal of Materials & Manufacturing (SAE Paper 912209)*, SAE International, Warrendale, Pennsylvania, 100, pp. 451–458.

[5] Browning, T. R. and Eppinger, S. D., 2002, "Modeling impacts of process architecture on cost and schedule risk in product development," *IEEE TRansactions on Engineering Management,* 49(4), pp. 428–442.

[6] Cochran, J. K., Lee, K. J., McDowell, D. L. and Sanders, T. H., 2002, "Multifunctional metallic honeycombs by thermal chemical processing," *Processing and Properties of Lightweight Cellular Metals and Structures*, pp. 127–136.

[7] Conner (Seepersad), C. G., DeKroon, J. P. and Mistree, F., 1999, "A product variety tradeoff evaluation method for a family of cordless drill transmissions," *ASME Advances in Design Automation Conference*, Las Vegas, NV. Paper Number: DETC99/DAC-8625.

[8] Eastman, C. M., Lee, G. and Sacks, R., 2002, "Deriving a product model from process models," *ISPE/CE2002 Conference*, Cranfield University, United Kingdom.

[9] Edwards, J. D., 2002, *Product Life Cycle Mangement: A White Paper.*

[10] Elmaghraby, S. E., 1995, "Activity nets: a guided tour through some recent developments," *European Journal of Operational Research,* 82(3), pp. 383–408.

[11] Engenious Inc., 2004, *FIPER,* http://www.engineous.com/product_FIPER.htm

[12] Engineous Inc., 2004, *iSIGHT, Version 8.0,* http://www.engineous.com/product_iSIGHT.htm

[13] Eppinger, S., Whitney, D. E., Smith, R. P. and Gebala, D. A., 1994, "A model-based method for organizing tasks in product development," *Research in Engineering Design,* 6(1), pp. 1–13.

[14] Federal Information Processing Standards Publication 183, 1993, *Integrated Definition for Functional Modeling (IDEF 0),* http://www.idef.com/Downloads/pdf/idef0.pdf

[15] Fenves, S. J., 2001, *A Core Product Model for Representing Design Information.* Gaithersburg, MD, National Institute of Standards and Technology.

[16] Fenves, S. J., Sriram, D., Sudarsan, R. and Wang, F., 2003, "A product information modeling framework for product lifecycle management,"

International Symposium on Product Lifecycle Management, Bangalore, India.

[17] Fernández, M. G., Seepersad, C. C., Rosen, D. W., Allen, J. K. and Mistree, F., 2001, "Utility-based decision, support for selection in engineering design," *13th Internation Conference on Design Theory and Methodology*, Pittsburgh, PA. Paper Number: DETC2001/DAC-21106.

[18] Gonzalez-Zugasti, J. P. and Otto, K. N., 2000, "Modular platform-based product family design," *ASME Advances in Design Automation Conference*, Baltimore, MD.

[19] Gorti, S. R., Gupta, A., Kim, G. J., Sriram, R. D. and Wong, A., 1998, "An object-oriented representation for product and design process," *Computer Aided Design,* 30(7), pp. 489–501.

[20] Hayes, A. M., Wang, A., Dempsey, B. M. and McDowell, D. L., 2004, "Mechanics of linear cellular alloys," *Mechanics of Materials,* 36(8), pp. 691–713.

[21] IBM, 2004, *IBM Solutions,* http://www-1.ibm.com/businesscenter/us/solutions/solutionarea.jsp?id=9351

[22] Inc., A., 2004, *HyperWorks Process Manager™ 6.0,* http://www.uk.altair.com/software/procman.htm

[23] Liker, J., Sobek, D., Ward, A. and Cristiano, J., 1996, "Involving suppliers in product development in the US and Japan: evidence for set-based concurrent engineering," *IEEE Transactions on Engineering Management,* 43(2), pp. 165–178.

[24] Lu, Y., 2002, *Analyzing Reliability Problems in Concurrent Fast Product Development Processes*, *PhD Dissertation,* Mechanical Engineering, Technische Unversiteit Eindhoven, Eindhoven, Netherlands.

[25] Maher, M. L., 1990, *Process Models for Design Synthesis*. AI Magazine. Winter, pp. 49–58.

[26] Maimon, O. and Braha, D., 1996, "On the complexity of the design synthesis problem," *IEEE Transactions on Systems, Man, and Cybernetics - Part A: Systems and Humans,* 26(1), pp. 142–151.

[27] Martin, M. V. and Ishii, K., 1997, "Design for variety: development of complexity indices and design charts," *ASME Design for Manufacturing Conference*, Sacramento, CA. Paper Number: DETC97/DFM-4359.

[28] Martin, M. V. and Ishii, K., 2000, "Design for variety: a methodology for developing product platform architectures," *ASME Design for Manufacturing Conference*, Baltimore, MD. Paper Number: DETC2000/DFM-14021.

[29] McGinnis, L. F., 1999, "BPR and logistics: the role of computational models," *1999 Winter Simulation Conference Proceedings (WSC), Dec 5-Dec 8 1999*, Phoenix, AZ, USA, pp. 1365–1370.

[30] Meyer, M. H., 1997, "Revitalize your product lines through continuous platform renewal," *Research Technology Management,* 40(2), pp. 17–28.

[31] Mistree, F., Bras, B. A., Smith, W. F. and Allen, J. K., 1996, "Modeling design processes: a conceptual, decision-based perspective," *International Journal of Engineering Design and Automation,* 1(4), pp. 209–221.

[32] Mistree, F., Hughes, O. F. and Bras, B. A., 1993, "The compromise decision support problem and the adaptive linear programming algorithm," *Structural Optimization: Status and Promise* (M. P. Kamat, Eds.), AIAA, Washington, D.C., pp. 247–286.

[33] Mistree, F., Lewis, K. andStonis, L., 1994, "Selection in the conceptual design of aircraft," *5th AIAA/USAF/NASA/ISSMO Symposium on Recent Advances in Multidisciplinary Analysis and Optimization*, Panama City, FL, pp. 1153–1166.

[34] Mistree, F., Muster, D., Shupe, J. A. and Allen, J. K., 1989, "A decision-based perspective for the design of methods for systems design," *Recent Experiences in Multidisciplinary Analysis and Optimization*, Hampton, Virginia. Paper Number: NASA CP 3031.

[35] Mistree, F., Smith, W. F., Bras, B., Allen, J. K. and Muster, D., 1990, "Decision-based design: a contemporary paradigm for ship design," *Transactions, Society of Naval Architects and Marine Engineers*, Jersey City, New Jersey, 98, pp. 565–597.

[36] Mistree, F., Smith, W. F. and Bras, B. A., 1993, "A decision-based approach to concurrent engineering," *Handbook of Concurrent Engineering* (H. R. Paresai and W. Sullivan, Eds.), Chapman & Hall, New York, pp. 127–158.

[37] Mistree, F., Smith, W. F., Kamal, S. Z. and Bras, B. A., 1991, "Designing decisions: axioms, models and marine applications," *Fourth International Marine Systems Design Conference*, Kobe, Japan, pp. 1–24.

[38] Mocko, G. M., Panchal, J. H., Fernández, M. G., Peak, R. and Mistree, F., 2004, "Towards reusable knowledge-based idealizations for rapid design and analysis," *45th AIAA/ASME/ASCE/AHS/ASC Structures, Structural Dynamics & Materials Conference*, Palm Springs, CA. Paper Number: AIAA-2004-2009.

[39] Muster, D. and Mistree, F., 1988, "The decision support problem technique in engineering design," *International Journal of Applied Engineering Education*, 4(1), pp. 23–33.

[40] Nell, J., 2003, *STEP on a Page (ISO 10303)*, http://www.nist.gov/sc5/soap/

[41] Newcomb, P. J., Bras, B. A. and Rosen, D. W., 1996, "Implications of modularity on product design for the life cycle," *1996 ASME Design Theory and Methodology Conference, ASME Design Engineering Technical Conferences and Computers in Engineering Conference*, Irvine, California. Paper Number: DETC-96/DTM-1516.

[42] Phoenix Integration Inc., 2004, *ModelCenter®, Version 5.0*, http://www.phoenix-int.com/products/ModelCenter.html

[43] Rechtin, E. and Maier, M. W., 1997, "The art of systems architecting," *Systems Engineering Series*, Boca Raton: CRC Press, pp. 119–136.

[44] Rogers, J. L. and Christine, B., 1994, "Ordering design tasks based on coupling strengths," *5th AIAA/USAF/NASA/ISSMO Symposium on Multidisciplinary Analysis and Optimization*, Panama City, Florida. Paper Number: AIAA-94-4326.

[45] Sacks, R., Eastman, C. M. and Lee, G., 2004, "Process model perspectives on management and engineering procedures in the north American precast /

prestressed concrete industry," *ASCE Journal of Construction Engineering and Management,* 130(2), pp. 206–215.

[46] Schlenoff, C., Knutilla, A. and Ray, S., 1996, *Unified Process Specification Language: Requirements for Modeling Process*. Gaithersburg, MD, National Institute of Standards and Technology.

[47] Schlenoff, C., Tissot, F., Valois, J., Lubell, J. and Lee, J., 2000, *The Process Specification Language (PSL): Overview and Version 1.0 Specification.* Gaithersburg, MD, National Institute of Standards and Technology.

[48] Seepersad, C. C., Dempsey, B. M., Allen, J. K., Mistree, F. and McDowell, D. L., 2002, "Design of multifunctional honeycomb materials," *9th AIAA/ISSMO Symposium on Multidisciplinary Analysis and Optimization*, Atlanta, GA. Paper Number: AIAA-2002-5626.

[49] Seepersad, C. C., Mistree, F. and Allen, J. K., 2002, "A quantitative approach for designing multiple product platforms for an evolving portfolio of products," *ASME Design Engineering Technical Conferences, Advances in Design Automation*, Montreal, Canada. Paper Number: DETC2002/DAC-34096.

[50] Shah, J. J., Rangaswamy, S., Qureshi, S. and Urban, S. D., 1998, "Formal ontologies and representation of design processes and rationale," *IFIP WG5.2 CAD Conference*, Tokyo, Japan, pp. 161–193.

[51] Shah, J. J., Rangaswamy, S., Qureshi, S. and Urban, S. D., 1999, "Design history system: data models and prototype implementation," *Knowledge Intensive CAD* (M. Tomiyama, Finger, Ed.), Kluwer, pp. 91–114.

[52] Shimomura, Y., Yoshioka, M., Takeda, H., Umeda, Y. and Tomiyama, T., 1998, "Representation of design object based on the functional evolution process model," *Journal of Mechanical Design,* 120(2), pp. 221–229.

[53] Simon, H. A., 1996, *The Sciences of the Artificial,* MIT Press, Cambridge, Mass.

[54] Simpson, T., Lautenschlager, U., Mistree, F., 1998, "Mass customization in the age of information: the case for open engineering systems," *The Information Revolution Current and Future Consequences* (Alan, W. H. R. and Porter, L. Eds.), Ablex Publishing Corporation, Greenwhich Connecticut, pp. 49–74.

[55] Simpson, T. W., Rosen, D., Allen, J. K. and Mistree, F., 1998, "Metrics for assessing design freedom and information certainty in the early stages of design," *Journal of Mechanical Design,* 120(4), pp. 628–635.

[56] StreamlineSCM, 2004, *Enterprise Transaction Model,* http://www.streamlinescm.com/

[57] Supply Chain Council, 2004, *Supply Chain Operations Reference (SCOR) Model,* http://www.supply-hain.org/slides/SCOR5.0OverviewBooklet.pdf

[58] Ullman, D. G., 1992, "A taxonomy for mechanical design," *Research in Engineering Design,* 3(3), pp. 179–189.

[59] Ulrich, K., 1995, "The role of product architecture in the manufacturing firm," *Research Policy,* 24(3), pp. 419–440.

[60] Ward, A., Liker, J., Cristiano, J. and Sobek, D., 1995, "The second Toyota paradox: how delaying decisions can make better cars faster," *Sloan Management Review,* 36(3), pp. 43–61.

[61] Wheelwright, S. C. and Clark, K. B., 1992, *Revolutionizing Product Development: Quantum Leaps in Speed, Efficiency, and Quality,* Free Press, New York.

[62] Wood, W. H., 2000, "Quantifying design freedom in decision based conceptual design," *ASME Design Engineering Technical Conferences*, Baltimore, Maryland. Paper Number: DETC2000/DTM-14577.

[63] Wood, W. H., 2001, "A view of design theory and methodology from the standpoint of design freedom," *ASME Design Engineering Technical Conference*, Pittsburgh, Pennsylvania. Paper Number: DETC2001/DTM-21717.

[64] Workshop Report, 2004, *Simulation-Based Engineering Science*. Arlington, VA, USA, National Science Foundation: pp. 1–18.

11

Manufacturing Information Organization in Product Lifecycle Management

R. I. M. Young and A. G. Gunendran

Wolfson School of Mechanical and Manufacturing Engineering
Loughborough University, U.K.

A. F. Cutting-Decelle

Ecole Centrale de Paris, Laboratoire de Genie Industriel, France

Progressive improvements in information systems offer the potential for radical improvements in manufacturing decision support systems as is evident by the uptake of modern Product Lifecycle Management approaches. However, drawing real value from these tools requires a clear understanding of how to organize information and configure systems to best advantage. This chapter discusses progress in the development of information frameworks, the importance of context awareness, the exploitation of manufacturing standards and future research requirements for the exploitation of product and process knowledge.

11.1 Introduction

Business competition means that the need for better, faster cheaper production is a never ending requirement. The identification of methods by which manufacturing improvements can be achieved is ongoing and has led to a range of approaches in recent years including Concurrent Engineering, Lean Manufacture and Agile Manufacture. In addition, the progressive improvements to information system capabilities continues to offer the belief that higher and higher levels of support for effective decision making can be achieved [1, 2].

Tools that can offer more effective breadth of information support are beginning to be developed in Enterprise Resource Planning (ERP) Product Lifecycle Management (PLM), and Customer Relationship Management (CRM). PLM, the focus of this chapter, offers the potential to provide sources of information that engineers can draw upon to offer rapid and effective support to their decision making [3, 4]. However to be fully effective these tools must be able to support the broad range of information needed to meet the needs of diverse teams of engineers working on problems in dynamically changing manufacturing environments.

While it is clear that the potential benefits of PLM are high, tapping these benefits is fraught with both short term and long term issues. Given the overhead involved in constructing an effective PLM environment it is particularly important for businesses to consider carefully the following challenges:

- What organization of information is needed to offer effective decision support?
- What range of views of the information is needed?
- What methods for integrated system design should be used?
- How can the level of information support be flexibly updated, maintained and extended?
- Can information be effectively shared across competitive software tools? For example, through the supply chain.

This chapter reflects on these challenges following a range of recent industry driven research projects at Loughborough University which have focused on the manufacturing information structures and methods needed to provide integrated life cycle support.

11.2 Information and Knowledge Infrastructures for Manufacture

It is clear that effective decision making in manufacturing businesses is influenced by a broad range of information and knowledge including knowledge of markets, existing products, design knowledge, manufacturing capability, product service and disposal. The majority of research in this area focuses on product information [5], while the work described in this chapter extends this to include a focus on manufacturing capability within an ICT environment and targets how infrastructures for manufacturing information and knowledge can be defined to support decision making. This is illustrated simply in Figure 11.1, which also shows the main target business areas of product development and manufacturing and procurement against which the research has been focused.

Manufacturing knowledge can be seen as being a significant part of a businesses capability. If we can identify computational methods by which key areas of manufacturing information and knowledge can be held within an infrastructure then we have the potential to speed up and improve the quality of decisions. It is important to note that manufacturing knowledge, however it is structured, is not static and has many facets. It is used during product introduction to aid design decisions, to plan manufacturing methods and to identify effective supply chain configurations. Once we start to manufacture actual parts and use products in service then we have the potential to learn and continuously improve our knowledge of what we have made and how well we can make things. It is this process which enables us to improve our knowledge and understanding of manufacture and therefore enhance our competitiveness as we continue through the product development lifecycle. An illustration of this lifecycle is provided in Figure 11.2.

Manufacturing knowledge is therefore not independent of products, although it can be applied to a broad range of products. It needs to be used in combination with the other key areas of business knowledge to support decision making. In considering the product lifecycle, rather than the development lifecycle, it can be seen that at each stage of the lifecycle it is important to have knowledge through the lifecycle. For example during design, design knowledge is of critical importance but manufacturing, service and disposal knowledge can and should have a significant influence on design decisions. It is therefore important that knowledge infrastructures should be able to support this type of interaction and not just provide isolated pockets of knowledge [6].

The concept which we have pursued in our research has been to look for a conceptual framework of information and knowledge which sits at the heart of the product lifecycle and therefore can support any area of decision with the lifecycle. This general view is illustrated in Figure 11.3.

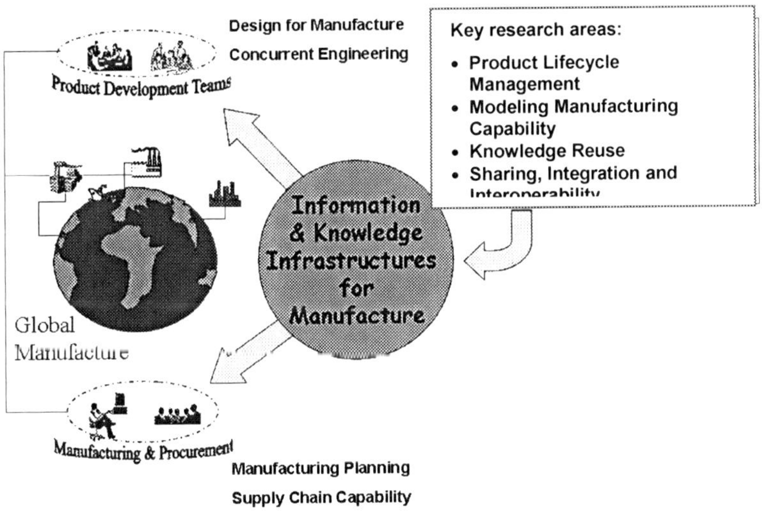

Figure 11.1. Target areas for manufacturing information & knowledge infrastructures

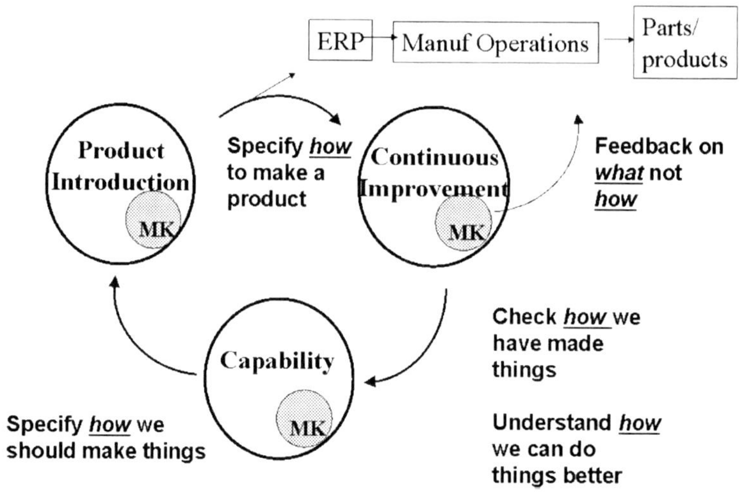

Figure 11.2. Manufacturing knowledge in the context of the product development lifecycle

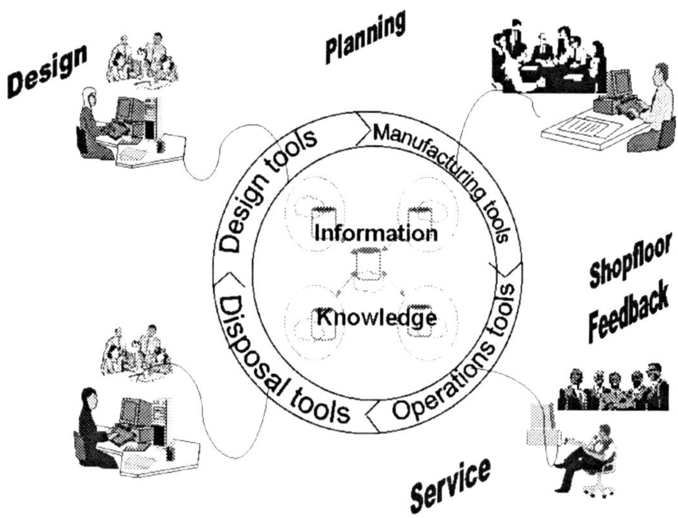

Figure 11.3. Information and knowledge at the heart of the product lifecycle

The effective structure of any information and knowledge infrastructure will be dependent on the use cases for the resulting system. We have taken the view that key two points provide the basis for the high level structure implied by the central part of Figure 11.3. These are firstly that product information is central to all decisions and secondly that each stage of the lifecycle has a core set of specific information and knowledge which reflect a business capability which is non product specific. For example businesses have design knowledge, manufacturing knowledge *etc.* which they use to improve their new product development and product support processes. This can be seen as a top level view of the lifecycle context within which an information and knowledge infrastructure should be defined. This, taken from a manufacturing perspective, along with more detailed views of product context is the subject of the next section.

Informally the use of the terms data, information and knowledge are often used ambiguously. It is worth noting that formally in our work we use definitions as follows: Data are simply symbols with no context and no relationships; information is data within a specific context; knowledge is information relationships within and across contexts [7].

11.3 Context Awareness: Its Significance for Information Organization

This section considers manufacturing information in relation to design for manufacture and in relation to process planning. It considers this information firstly from a product perspective and then from a life cycle perspective.

11.3.1 Product Context

Product development is typically a team-based exercise where members of the team all require similar but different sets of information in order to meet their specific tasks. The interpretation of product information in a form suitable for manufacturing decision making has typically been pursued through the use of features technology and part family variants. The results of manufacturing planning can then be captured in process plans for a product.

Features approaches are problematic in that they capture only a single context of information, *e.g.*, a machining feature is specific only to machining, it will not relate to assembly or to casting. Figure 4 illustrates a view of some simple features for machining, inspection and assembly which shows how, from each specific context the information of interest is different. The machining features identify shapes which have particular machining methods linked to them; the inspection features also relate to particular geometries, but in this case geometries which relate to inspection routines; the inspection features relate to methods of assembly and are not linked to a single component geometry but rather the geometric and tolerance relationships between components. It is possible in some cases where a dominant context can be defined that features can useful where they provide a focused and practical set of shapes which a design team can use. This is also the case where common features can be used to support multiple contexts. Hole features are probably the best examples of this as they can be used to represent multiple contexts such as function, casting, machining and assembly.

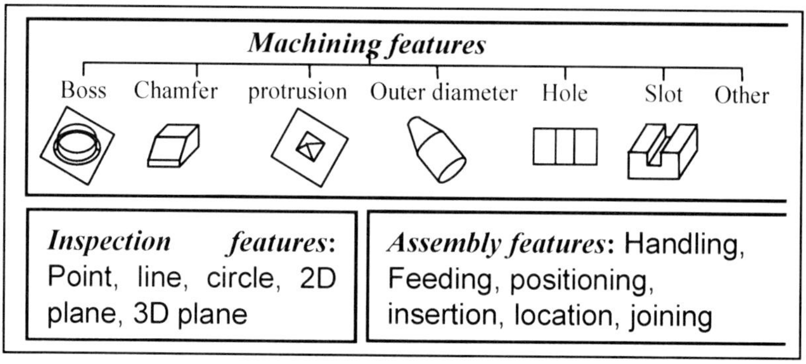

Figure 11.4. Examples of features in a process context

An alternative to dealing with shapes which link to an individual process is to consider complete parts which have an overall manufacturing method related to them *i.e.*, part families. Here the start point is to identify a part family where each product in the family has a similar set of functional requirements and each is made by a similar set of manufacturing processes. We illustrate this in Figure 11.5 with an example of a simple casing part which has two main functions; to withstand an internal pressure and to assemble with other components in the product. The flanges provide assembly relationships to other parts and the rings provide a wall strength capability to withstand the internal pressure. The overall manufacturing method for each complete part is similar but may change in detail, dependent on the allowable changes to the size of the specific rings and flanges. Changes to the flanges and rings are only allowed which stay within the context of the specified manufacturing method. If changes to the features are required which require a new method of manufacture then the product feature set are no longer relevant. Hence the product context has changed and the part family relationships are no longer valid.

In most cases there is a need for PLM systems to be able to support multiple contexts of information and the relationships between these contexts [9]. For example, a major step forward in potential functionality for manufacturing engineers would be achieved if part functional requirements could be linked to the relevant manufacturing views such as assembly, casting, forging, machining, heat treatment, grinding, *etc*. Our recent findings in this area as explained in Section 11.3.3. As well as developing an understanding of the relationship between views of a product it is also important to recognize that there should be clear relationships between manufacturing views of a product and the manufacturing capability of the business. This should provide links from features in a product model to manufacturing capability in a manufacturing model and also from resources and processes in a manufacturing model to the representation of process plans in a product model.

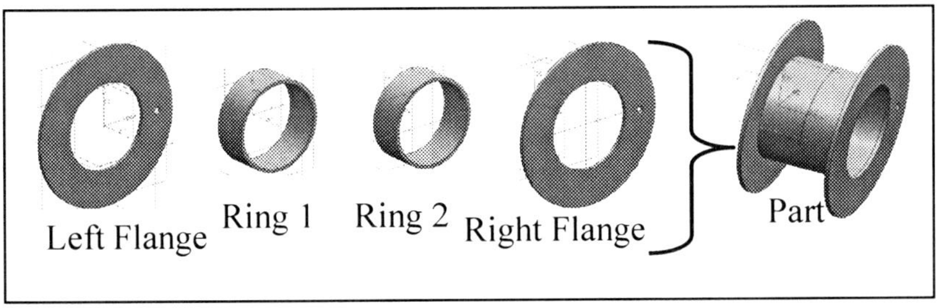

Figure 11.5. Example features in a product context

11.3.2 Life Cycle Context

Most PLM work appears, for historical reasons, to focus on a design perspective with at best the association of manufacturing documents for component parts. However, it is important to note that businesses have core information and knowledge on each aspect of the product lifecycle *i.e.*, the have information and knowledge on how to design products, information and knowledge on how to make products, information and knowledge on how to service products, and information and knowledge on how to dispose of products. Here we focus on the manufacturing context of the lifecycle and identify the need to represent manufacturing capability, independent of any specific product, as illustrated simply in Figure 11.6.

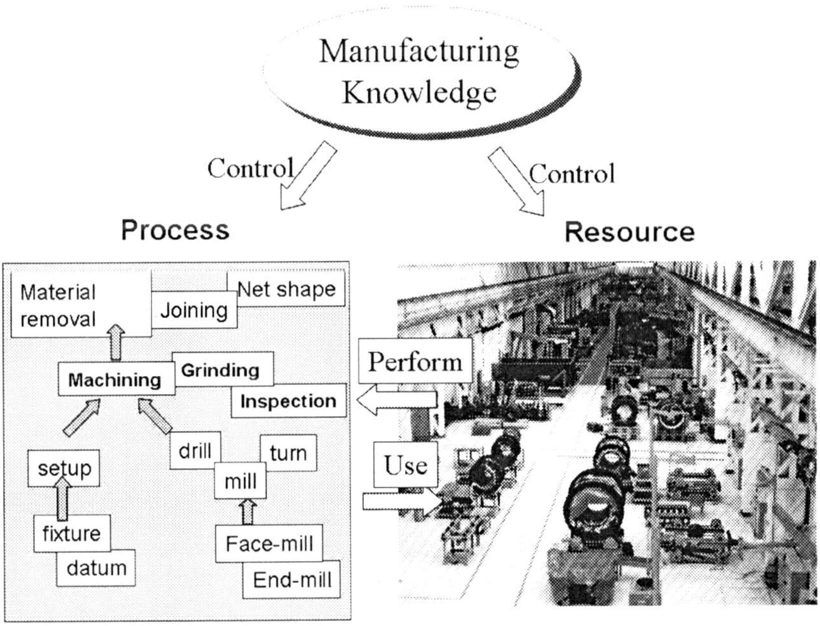

Figure 11.6. Manufacturing capability modeling

A Manufacturing Model, representing this core manufacturing capability is just as important as a product model from the manufacturing perspective of the life cycle. A manufacturing model should identify process and resource specifications, potential methods of manufacture and best practice for manufacturing [8]. These provide the manufacturing knowledge within the "capability" aspect of Figure 11.2, which can be used to support the new product introduction process. This manufacturing knowledge provides the basis by which the business controls the use of its processes and resources.

The importance of a manufacturing model is that it not only provides a common source of information to support design decisions, but it focuses the core competencies of the business so that as new understanding is generated during product manufacture, the model can be updated for future benefit. It therefore provides a clear integration link between PLM as a provider of manufacturing information and shop floor manufacturing systems in terms of data collection and feedback. This cycle and link between PLM and Manufacturing Systems is illustrated in Figure 11.7. Figure 11.7 also shows how information and knowledge models, through Product models, Manufacturing Models and Process Control Sets, support the product development lifecycle illustrated in Figure 11.2. The Product Model concept is well understood, the Manufacturing Model provides a manufacturing capability representation and the Process Control Sets provide a repository for shop floor data as it is collected and analyzed through the continuous improvement process. This last set of data is collected on a real time basis and is therefore considered to be the concern of manufacturing systems but outside the PLM scope. The relationship is driven through the understanding of manufacturing capability which is enhanced by analyzing the process control sets in order to update the knowledge contained in a manufacturing model.

11.3.3 Context Relationships

Sections 11.3.1 and 11.3.2 highlighted issues in relation to specific product and lifecycle contexts. This section uses an example product to explore some of the requirements for multiple contexts to be considered and goes on to propose an approach to integrate information across multiple contexts. Figure 11.8 illustrates a number of distinct, but related, contexts for a cylindrical part and a location rod.

The following list captures the main functions of the cylindrical part:

1) withstand high pressure
2) mount with other parts via the flanges
3) facilitate the attachment of other key parts through the attachment ring
4) facilitate the location of rods through the location ring

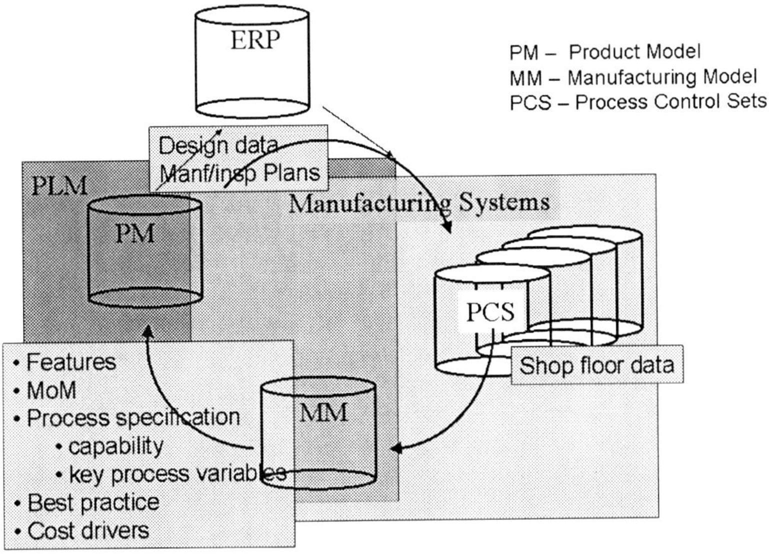

Figure 11.7. Information models in the context of PLM and manufacturing systems

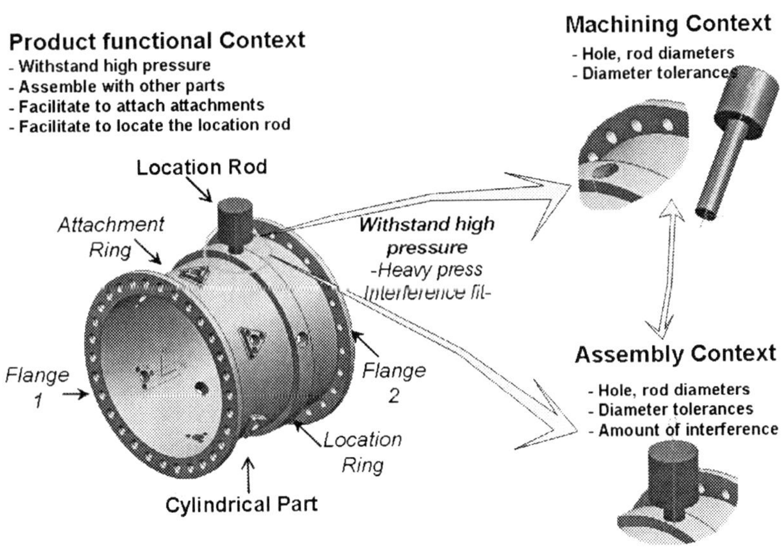

Figure 11.8. Multiple contexts of cylindrical part and location rod

The list of functions relate to a set of functional features on the component. However, the product needs to be considered from a number of different contexts to ensure an effective design. In this example we consider machining and assembly contexts alongside the functional context. Figure 11.8 also illustrates simple requirements for machining and assembly. The location ring facilitates the location of the location rod, which due to the high pressure requirement needs to be assembled with an interference fit. The interference fit between cylindrical part and location rod can be considered in the manufacturing process context as tolerances of the rod and the hole diameters of the cylinder. The parts then need to be machined to these required tolerances. While each of the function, assembly and machining contexts could be considered independently, it is important if effective design decisions are to be made that the relationships between them are maintained. Hence the transformation of information from one context to another is important for the support of all lifecycle activities of products.

It is important to have a core context to relate all others if relationships are to be maintained. In our work we have taken the functional context as the core context and explored the relationships from it to all other contexts of interest. This still requires knowledge of the relationships between all the contexts under consideration. The methodology followed to capture the product information and the transformation knowledge has been to use a two layer approach, where information contexts, like the normal features approaches, are held and a second knowledge layer, where knowledge of relationships between contexts as well as context specific knowledge is maintained [9, 10]. This is illustrated in Figure 11.9, again using the same example.

The required fit type can be identified from the functional requirements of the assembly of parts. In the particular example, the required fit is interference fit. The nominal diameter of the assembly of cylindrical part hole and the location rod can be used to identify the required tolerance set for the interference fit from the fits and tolerance table [11].The diameter and tolerance information can then be used to define the machining process requirements. The nominal diameter is a kind of common sharable information while the tolerance information is a kind of transformed information.

The transformation of information itself is not sufficient for the flexible integration of multiple contexts, because the transformed information may cause a problem for other context requirements. In the example, the required tolerance for the fit is derived from the fits and tolerance table, which is a kind of transformation knowledge, which transfers the fit function into a tolerance range. The tolerance range can then be used to define the manufacturing process. In the particular example, the machining process is considered as a manufacturing process. Hence, the tolerance information, which has been transformed from the functional information, can be considered as machining context information. However, the transformed information has not been verified for machineability. If the required tolerances cannot be manufactured with machining process, alternative manufacturing methods need to be identified. Therefore, the transformed information needs to be validated against the capabilities of each particular context. Hence, the transformed information is of limited value until the information is validated against the context requirements.

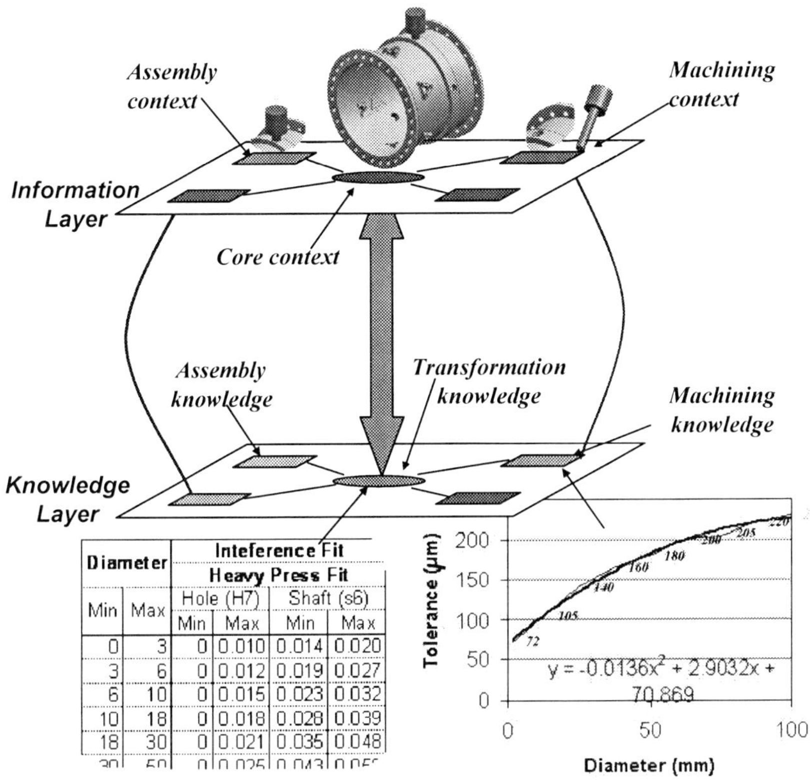

Figure 11.9. Information interaction of multiple contexts by utilizing transformation and context specific knowledge

Flexible integration between multiple contexts can be performed in two stages: firstly information transformation between the contexts and secondly the validation of the transformed information against each context. The validation of information against the context capabilities required the knowledge of the context capabilities. Therefore, two kinds of knowledge as transformation knowledge and context specific knowledge are required for the flexible information integration between multiple contexts.

In exploring multiple contexts there is a further issue which is critical to successful information sharing and that is the sharing of meaning. Within a common work environment it is reasonable to assume that the semantics are understood and shared. However, as we start to work across contexts and especially across businesses this assumption becomes less and less acceptable. Within software systems which are to support cross context and across business communication, there is therefore a fundamental need to provide a clear basis for sharing meaning [12]. This area of ontological engineering is now receiving substantial research attention [13].

11.4 Exploiting Manufacturing Standards

Where two similar systems have been configured it is unlikely that they will easily be able to interoperate and share information as the methods of information organization are likely to have been developed independently.

Problems in interoperation between software tools has led a number of large OEMs to insist that all their suppliers use the same tools in order to avoid this problem. However this simply moves the interoperability problem down the supply chain. The problem of interoperability is still a major problem as evident from a recent survey of the US automotive industry which suggests that such problems still cost in the order of $1 billion per annum [14].

International standards can be used to offer some flexibility as systems can interoperate, as long as they use standards to provide the basis for information sharing. As far as information sharing is concerned there are a number of ISO standards available which can aid this, especially in terms of resource information, but also in terms of process information. There are many standards that provide some level of information support for manufacture. Examples of these are ISO 10303, commonly known as STEP which focuses on product data; ISO 15531 (MANDATE) which offers data structures to capture views of manufacturing management data; ISO 13399 which provides a detailed representation of tooling systems; ISO 18629 (PSL) which offers a new approach to providing a semantically rich standard for process description.

11.4.1 STEP for Manufacturing

STEP product data representation

The biggest success of the ISO 10303 STEP standard started in 1996 with a data exchange standard for 3D product geometry, based on the Application Protocol AP 203. Today more then 2 million CAD stations contain STEP translators and for some industry segments it is fairly routine for an Original Equipment Manufacturer (OEM) to send a model to a machine shop and for that machine shop to process the geometry and make the part on its milling and turning machines [15]. The standard currently proposes more than 40 Application Protocols dealing with different aspects of product information or product manufacturing.

To increase the application of the standard to the domain of manufacturing, the STEP community set up a working group, called "STEP manufacturing", focused on the provision of standard information structures, related to products, with a manufacturing context *e.g.*, manufacturing features and process plan structures.

The manufacturing features aspect of their work has been captured in ISO 10303-224. This provides a useful and comprehensive set of over eighty feature definitions. However these are targeted at machining features only and therefore do not address any of the multiple manufacturing context issues raised in section 3.

The provision of standard data structures to represent process planning data is beginning to be developed, with ISO 10303-240 now available to capture machining process plans. Other standards for casting and inspection are under development as ISO 10303-223 and ISO 10303-119, respectively.

STEP-NC

STEP-NC supports the transfer of the required product and manufacturing information for NC code generation and is being developed within ISO 14649. This is planned to replace ISO 6983 "G-code" approach by specifying machining process rather than machine tool motion. The G-code approach is limited to specify the position and feed rate of axes and excluding valuable information such as part geometry and process plan. Therefore, the G-code approach is a low level programming approach and not portable from one NC controller to another. On the other hand, the STEP-NC is not only capable of representing full part description, but the manufacturing processes as well as the CAD design data with manufacturing information such as stock, cutting characteristics, and tool requirements [16].

11.4.2 Mandate – Resource, Time and Flow Models

ISO 15531 MANDATE is an International Standard for the computer-interpretable representation and exchange of industrial manufacturing management data. The objective is to provide a neutral mechanism capable of describing industrial manufacturing management data throughout the production process within the same industrial company and with its external environment, independent from any particular system. The nature of this description makes it suitable not only for neutral file exchange, but also as a basis for implementing and sharing manufacturing management databases and archiving.

This International Standard is organized as a series of parts, each published separately. The parts of ISO 15531 fall into the following series: resources usage management data, time model and manufacturing flow management data. All the parts of the MANDATE standard are written using the EXPRESS language to ensure better compatibility with ISO 10303.

ISO IS 15531-32 is the part which provides a representation for resources usage management data. Manufacturing resources form the basis and long-term potential of any company. The efficient use of these resources is one of the main goals in industrial management. Comprehensive information about available manufacturing resources is required in order to take the necessary decisions for efficient resource usage. Since many different enterprise functions and therefore also different IT-systems are dealing with manufacturing resources, a common, standardized model for resource description is necessary and provided by ISO 15531-32. This standardized model enables a company to communicate internally and externally about manufacturing resources and furthermore enables them to build up an industrial company's resource database.

A complete description of manufacturing resources is out of scope of the information model provided by the part 32 of the standard. Only data relevant for decisions concerning the usage of manufacturing resources (*e.g.*, within process planning or job scheduling) are considered. Therefore only data describing manufacturing resources in terms of their static and dynamic capabilities and capacities to perform manufacturing tasks are within the scope of this information model.

ISO IS 15531-42 is the part which provides a time model. Software applications related to factory or enterprise production, such as scheduling software, manufacturing management software, cost evaluation software, maintenance management, purchasing software, delivery software, *etc.* strongly require a reference to time related features such as point in time (date) and duration (interval of time). These references are needed to ensure the necessary time related relationships between the events dealt with by the applications. The availability of standardized time related references is particularly important for complex applications with multi-process environments, what is an environment commonly met in manufacturing.

In most of the standards, the time features are not independent from the events and the manufacturing management data they address. This leads to some difficulties in the way to handle time related relationships between events or data that include their own time relation and representation. In some of them the time related features may depend on events or objects addressed and their representation may change depending on the context, without any simple tool to identify the relation between them. This may be crucial in an environment where various processes are performed simultaneously or where many closely related software tools are used at the same time.

Developed in compliance with the system theory approach this part of MANDATE identifies the time as a constraint of the system environment and provides time related features included in a time model fully independent from the events handled by the manufacturing system. This time model is also fully independent from any manufacturing management data used by the manufacturing applications.

ISO IS 15531-43 is a data model for manufacturing flow management. A manufacturing management system manages the flow of information and materials through the whole production chain, from suppliers, through to manufacturers, assemblers, distributors, and sometimes customers. This part addresses the modeling of data for the management of manufacturing flows as well as flow control in a shop floor or factory. This manufacturing flow model is provided in the context of various processes that run simultaneously and/or sequentially, providing one or more products and/or components and involving numerous resources.

11.4.3 Process Specification Language

There are many standards available and these do not necessarily form a coherent set to support the needs of manufacturing. While these offer the best options available today, they also have limitations in the definitions of the concepts which they use. They provide very clear syntactic definitions but very basic semantic definitions. PSL is a standard which brings semantic rigor to manufacturing process modeling and has huge potential to offer improved information sharing across future manufacturing systems.

The Process Specification Language (PSL) project, whose development started at the National Institute of Standards and Technology (NIST, US), is a formal language aimed at creating a neutral, standard language for process specification to serve as a neutral representation to integrate multiple process-related applications throughout the manufacturing life cycle. ISO 18629 provides a generic language for process specifications applicable to a broad range of specific process representations in manufacturing applications.

ISO 18629 provides semantics to the computer-interpretable exchange of information related to manufacturing processes. Taken together, all the parts contained in the standard form a language for describing a manufacturing process throughout the entire production process within the same industrial company or across several industrial sectors or companies, independently from any particular representation model. The nature of this language should make it suitable for sharing process information related to manufacturing during all the stages of a production process.

The primary component of PSL is its terminology for classes of processes and relations for processes and resources, along with definitions of these classes and relations. Such a lexicon of terminology along with some specification of the meaning of terms in the lexicon constitutes what is known as an ontology. Within the ISO 18629 standard, the ontology is the PSL ontology for processes. The specification of models of PSL provides a rigorous mathematical characterization of the semantics of the terminology of PSL.

The current components of ISO 18629 are grouped into the following parts:

- Part 1: Overview and basic principles;
- Part 11: PSL-Core;
- Part 12: Outer Core;
- Part 13: Duration and ordering theories;
- Part 14: Resource theories;
- Part 41: Activity extensions;
- Part 42: Temporal and state extensions;
- Part 43: Activity ordering and duration extensions;
- Part 44: Resource extensions;

Additional extensions may be developed later according to industry needs by any standardization committee. All the parts of the standards listed have now reached the International Standard level.

11.5 Exploiting Product and Process Knowledge in Future

We propose that there are three major requirements to be set against information systems research in order for manufacturing businesses to exploit product and process knowledge in the future. These are:

1. Improved access to knowledge
2. Learning from manufacture

3. Tapping supply chain capability knowledge

Improved access to knowledge is a key requirement for manufacturing information systems in the future as continued improvement in support for people in their decision making will be necessary. The improvements from today are likely to be in terms of the quality of the information, the relevance of the information, the speed and ease of access to the information.

New systems are emerging in the area of shop floor data collection which not only support the immediate process control requirements of manufacture but also offer the potential to improve the understanding of how manufacturing processes can be optimized. Figure 11.10 illustrates how shop floor data collection can be used not only for product control but also process control. The combination of these and the ability to analyze both product and process data simultaneously offers an ideal route to building and enhancing manufacturing knowledge and feeding this back into a capability model as discussed in Section 11.2.

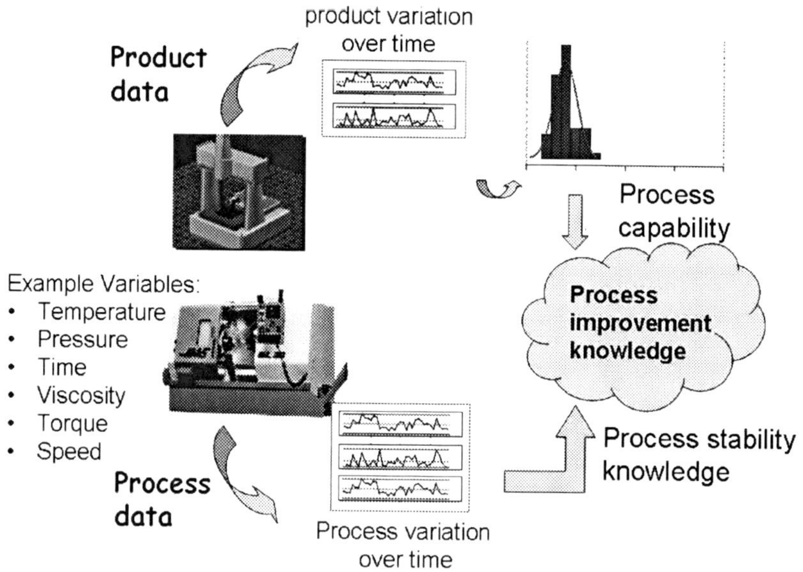

Figure 11.10. Learning from manufacture

As businesses strive to be more responsive there is a further need to tap and understand the capability of their supply chains and to be able to reconfigure their supply chains in a rapid, responsive and effective way. This leads to a requirement which is to understand the process capability of their suppliers as well as the product capability. In this way effective networks of collaboration can be introduced as illustrated in Figure 11.11.

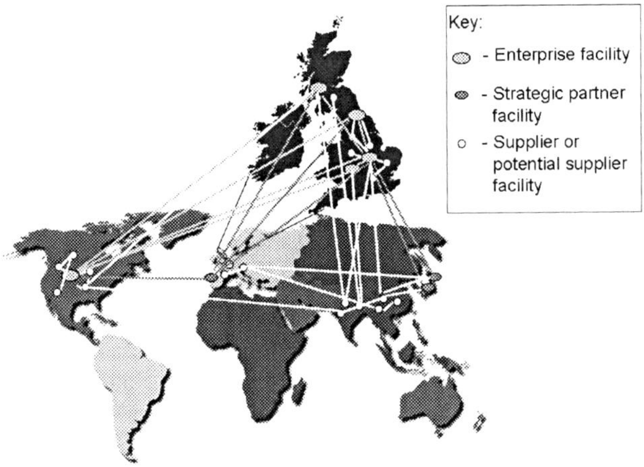

Figure 11.11. Collaborative capability networks

11.6 Conclusions

It is important that PLM configuration is undertaken with a clear view of the manufacturing aspect of the life cycle in mind. With current technology it is possible to construct manufacturing capability models, link these to product part families or features and to use these to support process plan generation and offer manufacturing support during design.

We have shown some early ideas on how to integrate multiple context approaches to support decision making. However there is a need for substantial research in this area to better support teams of engineers working together through a common PLM system. Similarly there will be a need to include higher levels of knowledge within PLM and provide mechanisms for the maintenance of that knowledge.

International standards play an important role in providing independent approaches to information sharing although the current standards have had limited uptake. In looking forward, there are substantial benefits to be gained once interoperation between PLM systems can be achieved. This needs to be based on flexible, rigorous methods which support shared meaning between systems.

Information organization is critical to decision support. Substantial progress has been made in terms of lifecycle models and product structure. However, for future advances to tap higher benefits for business there is a need for greater understanding of how to manage context relationships, knowledge maintenance and supply chain capability.

11.7 References

[1] Maropoulos, P. G., 2003, "Digital enterprise technology – defining perspectives and research priorities," *International Journal of Computer Integrated Manufacturing*, 16(7–8), pp. 467–478.

[2] Young R. I. M., 2003, "Informing decision makers in product design and manufacture," *International Journal of Computer Integrated Manufacturing*, 16(6), pp. 428–438.

[3] Abramovici, M. and Sieg, O., 2002, "Status and future trends of product lifecycle management (PLM) technology," *Proceedings of IPPD'2002 Conference*, Wroclaw, Poland, 21–22, Nov.

[4] Srinivasan V., 2005, "Open standards for product lifecycle management," In: *Product Lifecycle Management: Emerging Solutions and Challenges for Global Networked Enterprises*, Bouras, A., Gurumoorthy, B. and Sudarsan, R. (Eds.), pp. 475-484, (Inderscience Enterprises Ltd.).

[5] Sudarsan, R., Fenves, S. J., Sriram, R. D. and Wang F., 2005, "A product information modeling framewrok for product lifecycle management," *Computer Aided Design*, 37, pp 1399–1411.

[6] Young, R. I. M., Guerra, D., Gunendran, G., Das, B., Cochrane, S. and Cutting-Decelle A.F., 2005, "Sharing manufacturing information and knowledge in design decision support," In: *Advances in Integrated Design and Manufacturing in Mechanical Engineering*, Bramley, A., Brissaud, D., D. Coutellier, and C. McMahon (Eds.), pp. 173–188, (Springer).

[7] Mills, J. J. and Goossenaerts, J. 2001, "Towards information and knowledge in product realisation infrastructures," *Proceedings of Global Engineering, Manufacturing and Enterprise Networks*, Melbourne, Australia, pp. 245–254.

[8] Liu, S. and Young, R. I. M., 2004, "Utilising information and knowledge models to support global manufacturing co-ordination decisions," *International Journal of Computer Integrated Manufacturing*, 17(6), pp. 479–492.

[9] Gunendran, A. G., 2004, *An Information and Knowledge Framework to Support Multiple Viewpoints in the Design for Manufacture of Injection Moulded Products*, PhD Research Thesis, Loughborough University, U.K.

[10] Gunendran, A. G. and Young, R. I. M, 2006, "An information and knowledge framework for multi-perspective design and manufacture," *International Journal of Computer Integrated Manufacturing*, 19(4), pp. 326– 338.

[11] BS EN 20286-2, 1993, "ISO system of limits and fits, tables of standard tolerance grades and limit deviations for holes and shafts," *British Standard/ European Standard*, ISBN: 058018062 X.

[12] Young, R. I. M, Gunendran, A. G., Cutting-Decelle, A. F. and Gruninger, M., 2006, "Manufacturing knowledge sharing in PLM: a progression towards the use of heavy weight ontologies," accepted for publication in the *International Journal of Production Research*.

[13] Gomez-Perez, A., Fernandez-Lopez, M. and Corcho, O. (Eds.), 2004, *Ontological Engineering*, (Springer-Verlag).

[14] Ray, S. R. and Jones, A. T., 2003, "Manufacturing interoperability, concurrent engineering: enhanced interoperable system," *Proceedings of the 10th ISPE international Conference*, Madeira Island - Portugal, 26–30 July, pp. 535-540.

[15] Hardwick, M., 2004, "Experience with developing product, process and equipment models for manufacturing operations," *Proceedings of 2nd RTAS Workshop on Model-Driven Embedded Systems (MoDES '04)*, Toronto, Ontario, Canada, 25–28, May. http://www.cse.wustl.edu/~cdgill/RTAS04/mdes_program.html

[16] Sääski, J., Salonen, T. and Paro, J., 2005, "Integration of CAD, CAM and NC with Step-NC," Espoo, VTT. 23 p. VTT Working Papers; 28, ISBN 951-38-6580-0, http://virtual.vtt.fi/inf/pdf/workingpapers/2005/W28.pdf

12

Semantic Interoperability to Support Collaborative Product Development

Q. Z. Yang and Y. Zhang

Singapore Institute of Manufacturing Technology, Singapore

Semantic interoperability is a measure of the ability for heterogeneous processes and systems to understand, utilize, and transform meanings of product data in collaborative product development. Lack of formal and explicit semantics in digital product representations has imposed increasing difficulties in achieving semantic interoperability. This chapter presents a method to capture data semantics with product representations and to formalize data meanings with ontologies to support semantic interoperability. Both the object-based CAD modeling and STEP-compliant data modeling approaches are integrated with the OWL (Web Ontology Language) description framework. Common vocabularies, domain ontologies, and semantic schema mappings are defined to represent, interpret, map, and share the semantically interoperable product information across collaborating applications. Based on this method, a software prototype has been implemented and tested with collaboration scenarios in cross-disciplinary CAD, quality and reliability control, and product development process management.

12.1 Introduction

In Collaborative Product Development (CPD), a network of multi-functional disciplines (such as mechanical, electrical, optical and software engineering, quality assurance, and manufacturing) with their partners and suppliers works together to achieve common goals for competitive products. Numerous CPD processes and computer-supported applications are used in this collaboration environment, which is very heterogeneous because multi-disciplinary systems, proprietary product models, various data representation formats, and different domain terminologies and concepts are involved. A major challenge in achieving effective collaboration in such a heterogeneous environment is the lack of formal and explicit semantics in digital content to enable semantic interoperability.

According to Pollock [1], data semantics are the meaning of data. Meaning is subjective, and the interpretation of data semantics is constrained by context [1]. When contexts change, semantics also change. Product semantics issues have caused substantial difficulties in understanding and interpreting data meanings from one engineering application to the other. For example, ambiguous or implicit semantic content of geometric models (2D, 3D, solid, or shell) from a CAD application would cause different interpretations from a CAE application (*e.g.*, Ansys, Nastran, or Abaqus), which could lead to the failure of a design collaboration effort.

Over the years, a wide range of researches on product information exchange and data semantics integration has been conducted [2-5]. There have been widely accepted methodologies and frameworks that support neutral product/process information representation and exchange, such as the STEP [6] and ebXML [7] standards. The recent developments in grid technologies and middleware solutions have proposed new approaches and tools to semantically integrate the digital content and manipulate it over a network [4, 5]. The other advanced technologies, including Semantic Web and Web Services, have also been applied to engineering information management and knowledge sharing, innovative product design and engineering, design search and optimization, *etc*. [8, 9]. Among others, the product data model standardization and ontology engineering are widely recognized as an effective way to tackle the interoperability issues in multi-disciplinary applications.

In the standardization approach, all applications involved in a CPD project will adhere to standard product data specifications, such as STEP [6] Application Protocols or PSL XML [10] schemas. As the standards have specified meanings and structures for their data and terminologies, all collaborating parties and applications have to use the same sets of terms, labels, data schemas, and communication mechanisms. By agreeing on the use of the fixed interpretations, these applications could achieve disambiguity of data semantics in the context of the standards. However, the engineering meaning of standards is often implicitly encoded in the structures of their syntax, in the schemas of their data models, or in the priori agreements about interpretations of their terminologies. As lacking of formal and explicit semantic definitions in these standard specifications, they cannot ensure the consistent interpretation, understanding, and implementation of application semantics across disciplines. Furthermore, as the contexts of semantics in standards are fixed, the standardization approach is not flexible enough to resolve the semantic heterogeneity of multi-disciplinary applications. The approach may only be useful in restricted domains and relatively homogeneous environments [11].

Another approach to semantic interoperability is based on domain specific ontologies, in which semantics of terminology systems are specified in a well-defined and unambiguous manner [12]. Local vocabularies rather standardized global vocabularies are established to formally and explicitly capture, infer, and map the data semantics in particular product development contexts. Meanings of terminologies are encoded in formal languages such as OWL (Web Ontology Language) [13] and RDF (Resource Description Framework) [14]. Compared with the standard-based approach, the ontology-based approach is more suitable for use in non-restricted domains and heterogeneous environments [11]. Obviously this is an important development in semantic interoperability of heterogeneous systems, which has inspired our research reported in this chapter.

This chapter presents a method to capture data semantics with product representations and to formalize data meanings with ontologies to support semantic interoperability of CPD systems. The method integrates both the object-based CAD modeling and STEP-compliant data modeling approaches with OWL to represent, understand, interpret, and share the semantically interoperable product information across collaborating systems. It involves three areas: ontology engineering, object-based CAD, and product model standardization:

- Ontologies represent formal, explicit and shared understanding about application semantics, domain concepts and their relationships. They allow classification and precise description of the concepts/terminologies used in a domain and enable semantic mappings between them.
- Object-based CAD modeling captures both geometric and non-geometric data and their semantics as product properties, behaviors, inter-part relationships or constraints, and associates them with CAD objects to present more comprehensive multi-views to a wide range of CPD applications.
- Product data model standards serve as a common foundation for interoperating multi-disciplinary applications. In particular, the STEP and XML standards address the information sharability by classifying and defining the standardized information elements and their relationships, and facilitate the data communication between applications by the use of open and neutral file formats and databases.

The following sections focus on how these three areas can be integrated to resolve semantic interoperability issues in CPD processes and applications. The specific discussion includes: semantic interoperability concepts and enabling technologies in Section 12.2; product semantics capturing and STEP extension modeling in Section 12.3; vocabulary taxonomy and OWL ontology in Section 12.4; semantics-driven schema mapping in Section 12.5; software implementation of the approach in Section 12.6; collaboration scenarios in Section 12.7; and a conclusion remark in Section 12.8.

12.2 Semantic Interoperability Concepts and Technologies

Semantic interoperability is defined as a measure of the ability for heterogeneous processes and systems to understand, utilize, and transform meanings of product data in collaborative product development. Semantic interoperability mainly involves the issues of specifying, understanding/interpreting, and interoperating the heterogeneous product information in a semantically consistent manner. To address these issues, three key technologies are identified and used in this research: STEP standard for product information exchange, sharing, and interoperation; ontologies for explicitly specifying, understanding, and interpreting product semantics and their contextual constraints; and object-based product models for capturing product data semantics.

12.2.1 Data-driven Interoperability Standard

STEP is one of the most important product model standards for data-driven interoperation and integration of heterogeneous systems. STEP aims to provide an open, neutral product information representation and exchange mechanism for products throughout their lifecycles. Using the Express [15] information modeling language and the STEP integrated resources, the STEP community has developed more than 40 STEP Application Protocols (AP) to date for different lifecycle applications in mechanical, electronic, ship, and building design, engineering analysis, process planning, manufacturing, product lifecycle support, *etc.* Table 12.1 shows some of these APs.

STEP addresses the product information exchange, sharing, and interoperability issues by:

- Classifying and defining the standard information entities, rule relationships, and data inheritance models to capture the product information in individual application domains;
- Mapping the captured information into product lifecycle entities (defined in the STEP integrated resources) to be shared by all STEP APs, such as those shown in Table 12.1;
- Facilitating the exchange and sharing of the product lifecycle information across multi-disciplinary applications by standard data exchange formats (Part 21 and Part 28) and the Standard Data Access Interfaces (Part 22).

Some STEP APs, such as AP203 and AP214, have been implemented in commercial CAD systems to provide the capabilities for importing and exporting STEP Part 21 files. The downstream STEP-compliant applications will then use the neutral STEP files for CAD model data exchange and sharing. However, the downstream lifecycle applications are increasingly demanding more additional product information from CAD models, while advanced CAD systems can really generate models containing these additional information contents. The problem is that capturing the additional product data and their meanings in semantically-sound extensions of STEP models is very challenging. This has hampered the implementation and utilization of the STEP-based interoperability approaches in multi-disciplinary CPD systems.

12.2.2 Ontologies

Ontologies specify the semantics of terminology systems of product models and the meanings of product data formally and explicitly. In particular, OWL [13] provides rich ontological constructs including the OWL formal definitions and axioms to enable representation of and reasoning over the given concepts in an ontology for deriving their logical consequences.

Table 12.1. STEP application protocols

Part 201 Explicit drafting	Part 221 Functional data and schematic representation for process plans
Part 202 Associative drafting	Part 222 Design engineering to mfg. for composite structures
Part 203 Configuration controlled design	Part 223 Exchange of design and mfg. DPD for composites
Part 204 Mechanical design using boundary representation	Part 224 Mechanical product definition for process planning
Part 205 Mechanical design using surface representation	Part 225 Structural building elements using explicit shape representation
Part 206 Mechanical design using wireframe representation	Part 226 Shipbuilding mechanical systems
Part 207 Sheet metal dies and blocks	Part 227 Plant spatial configuration
Part 208 Life cycle product change process	Part 228 Building services
Part 209 Design through analysis of composite and metallic structures	Part 229 Design and manufacturing information for forged parts
Part 210 Electronic printed circuit assembly, design and mfg.	Part 231 Process engineering data
Part 211 Electronics test diagnostics and remanufacture	Part 232 Technical data packaging
Part 212 Electrotechnical plants	Part 233 Systems engineering data representation
Part 213 Numerical control process plans for machined parts	Part 234 Ship operational logs, records and messages
Part 214 Core data for automotive mechanical design process	Part 235 Material information for products
Part 215 Ship arrangement	Part 236 Furniture product and project
Part 216 Ship molded forms	Part 237 Computational fluid dynamics
Part 217 Ship piping	Part 238 Integrated CNC machining
Part 218 Ship structures	Part 239 Product life cycle support
Part 220 Printed circuit assembly manufacturing planning	Part 240 Process planning

Ontologies can be distinguished into two categories [16]: logic-based and non-logic-based, depending on whether logical axioms and definitions are used in ontologies or not. Typically a logic-based ontology explicitly specifies the semantics of terminologies through ontological definitions and axioms. The ontological definitions build a common understanding about terms, concepts and relations, while the axioms enable reasoning, mapping and matching of these definitions to admit or reject interpretations of terminologies and concepts. Non-logic-based ontologies, on the other hand, do not use axioms to specify the semantics of terminologies. Instead, they define the meanings of terminologies by reaching priori consensus and by fixing the interpretations with respect to pre-defined contextual constraints. Product data standards such as STEP specifications and XML schemas are often considered as non-logical ontologies.

There are two strategies in facilitating semantics interoperability of multi-disciplinary systems. The first one depends on a logic-based ontology to provide a reference of the transformation between terminology systems of heterogeneous applications. The semantics of the more specific terminologies from individual applications are then mapped to the meanings of the more generic terminologies in the reference ontology. Hence, the semantics of individual application model data can be understood and shared on the basis of the reference ontology (see Section 1.5 for an example of using a logic-based reference ontology in semantic schema mapping). In the second strategy, product data standards are used as shared non-logical ontologies. By using the same terminology system of a standard in an unambiguous way, all applications adhered to the standard will achieve interoperability.

In this research, a hybrid approach integrating both the standard-based and logic-based strategies is used to establish semantic interoperability among multiple application domains of CAD, quality and reliability control, and CPD process management. The STEP specifications are used as common non-logical ontologies for the representation of applications-specific information beyond standard STEP datasets, while a logic-based reference ontology and a set of domain ontologies in OWL are developed for the description, annotation, interpretation, and reuse of application semantics in these three application domains. To achieve this, the relevant application data semantics must be captured in CAD models, or more generally, captured in product models.

12.2.3 Product Models

Product models are the computer interpretable and processible digital representations of products. Product models formalize the syntax and semantics of product expressions, such as symbols, terminologies, concepts, or relationships. During the lifecycle of a product, different models are constructed to satisfy different information needs from different lifecycle applications. For example, product data models are often used for formalizing the product design information, bill of material, shape representation, product configuration, and so on; CAE/FEA/CFD models used for product functionality analysis or simulation; product development process models for design activity control, process information flow, task dependency description, process planning, *etc.*; and product performance models for quality, cost, time-to-market management and improvement. The present research has developed three types of product models (*i.e.*, product supplementary information models, product performance models, and

CPD process models) either in a STEP-compliant format or in an application-specific representation format for product semantics capturing and sharing in CAD, quality and process management applications.

(1) Supplementary Information Model

Supplementary information is defined as the additional product information described with explicit data semantics and embedded in CAD models. The supplementary information includes semantic properties (material, cost, quality criterion, process data, *etc.*), object behaviors (executable CAD manipulations, external programs, *etc.*), reference links to external sources (online product catalog, regulation standard, classification and specification system, *etc.*), interdisciplinary relationships (between design objects/components developed by different disciplines), and so on.

The STEP technology and object-based CAD modeling techniques are used for the development of supplementary design information models. The STEP AP203 [17] is used as the common data representation model for the mechanical domain, and AP210 [18] for the electronic domain in the present study. These two STEP APs are extended with supporting definitions for richer product semantics on entity types, object behaviors, relationships, constraints, *etc.* These supplementary definitions are modeled as STEP extensions and populated with product datasets extracted from the object-based CAD models or from user inputs [19]. Two STEP extension mechanisms are investigated to connect the supplementary definitions in the extension models to the relevant entities already existed in the standard STEP product models (details in Section 12.3.3). Semantic mappings are conducted to match the supplementary definitions with the extended STEP definitions (details in Section 12.5). In this way, the STEP-compliant product data models are developed to support semantic interoperation of CAD and other applications.

(2) CPD Process Model

A CPD process model describes the structures of collaborative product development activities, their dependencies, and interactions of information items among activities. Different process models in Express-G [15] and ARIS [20] are developed for modeling the new product introduction (NPI) processes of consumer products [21]. Through the use of ontologies for meta-data provisioning, the data meanings of the native NPI process models are annotated and extracted to populate the semantic definitions of the supplementary information models, so that they are capable of providing commonly understandable and semantically richer NPI process information for interoperating with other relevant applications.

(3) Product Performance Model

A performance model explicitly defines the relationships between a set of performance control variables and a performance measure. Three performance measures (*i.e.*, quality, time and cost) are identified to evaluate the performance of the NPI processes. Proprietary performance models, such as a quality matrix, a design cycle time model, and a relative cost index, are developed to measure, control and improve the performance of consumer product development [23]. The semantics of the performance data are classified and described by domain ontologies, according to which the native data pertaining to the performance models are related to the extended STEP models as semantic properties or as reference-linked performance criteria to be shared with other design applications.

(4) Relationships between Product Models

For these three application domains of CAD, quality/reliability control, and CPD process management, the information from each is organized in a native product model, which is either a CAD model, a process model, or a performance model. The model data from these models can be treated as some kinds of supplementary information to be organized in a supplementary information model, which will be mapped to a STEP extension model (details in Sections 12.3 and 12.5 respectively). The relationships of these product models with the supplementary information model and the extended STEP data model are illustrated in Figure 12.1.

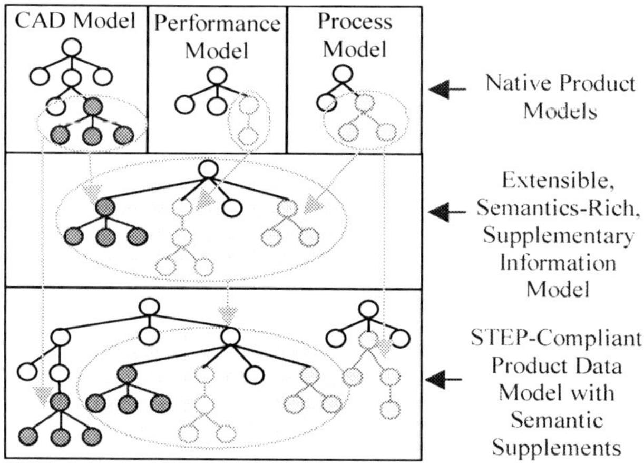

Figure 12.1. Product model relationships

The purpose of the supplementary information model in Figure 12.1 is to capture the data semantics from each product model, by specifying the data meanings and their contextual constraints in supplementary information definitions in order to provide interoperability. It is therefore not intended to cover all aspects of the information pertaining to individual product models, but to ensure that sufficient data semantics in well-defined collaborative contexts are captured and can be shared to any collaborating parties. The relationships in Figure 12.1 will be further described by domain ontologies and used for semantics-driven schema mapping in Section 12.5.

12.3 Product Semantics Capturing and STEP Extension Modeling

12.3.1 Representing Semantics in Supplementary Information Models

Semantic interoperation and integration requires that the product information generated from one application, such as a CAD system or a process management system, must be made understandable and transferable among other applications. By using a CAD system, 2D/3D design objects are created. On top of the CAD data, other product information such as the supplementary information (defined in Section 1.2.3) needs to be modeled, attached to and updated with the CAD objects, so that they would be able to support the data semantics needs from other downstream applications in the product lifecycle. For the supplementary information, it is essential to ensure the information carrying unambiguous semantics understandable and interpretable by other software applications. Toward this end, the following modeling method is developed.

In this method, the traditional CAD models generated from a CAD system are extended with the supplementary information. Implicit data semantics from CAD models are made explicit and captured in a set of entity type definitions, including the entity property and entity behavior definitions, so that the extended CAD objects can carry not only geometric representations and design characteristics from the CAD system, but also semantic instantiations from entity type definitions.

The entity types describe the object-based product supplementary data and precisely represent the intended meaning of these data. An entity type is defined by attributes and contextual constraints. An attribute has an identifier and a type indicator. A contextual constraint contains a set of explicit relations and methods to limit the validity of the attribute meanings, types and values that a hosting CAD system can support. All the attributes and contextual constraints of an entity type together describe the supplementary dataset semantics being defined. The entity types are configurable and may contain any number of attributes and constraints, depending on the needs for semantic description from particular viewpoints. For example, the following expression gives a definition for an entity property with five attributes and two contextual constraints. It is defined for a reliability testing application of consumer electronic components.

<PropertyName, Description, Value(>= 0), Unit(Enum(hour, min, g, ℃)), RefNo> (12.1)

Expression (1.2) below defines an entity behavior with six attributes without any context constraints. It is for the use in a new product development application.

<ObjectType, ObjectName, Description, Value, ExeType, RefNo> (12.2)

Using the entity property definition in Expression (12.1) as a template, the property objects can be instantiated. During instantiation, the contextual constraints in Expression (1.1) will not be presented in the property objects generated. Instead, they are implemented as constraint algorithms in the software add-on tools of CAD systems. Whenever a property object is instantiated, these constraint algorithms will be invoked to ensure that the auto-extracted or user-entered attribute values and types are compliant with the data scopes and types defined in Expression (12.1). Table 12.2 shows an example for modeling, according to Expression (12.1), a set of property objects to be used in the reliability testing of electronic components.

Table 12.2. Example of property objects for reliability testing

PropertyName	Description	Value	Unit	RefNo
Mass	Mass of sample component.	15	g	Test001
Temperature	Test temperature.	50	℃	Test010
ShelfTime	Time from a sample placed in a test chamber till a stabilized chamber temperature reached.	10	min	Test011

Similarly, the behavior objects can be modeled from the entity behavior definition in Expression (12.2). Table 12.3 lists some behavior objects for the supplementary information related to new product development in design and testing. It includes two CAD behavior objects for CAD manipulations of *Explode* and *Position*, two inter-part relationship objects for *LocatedOnTop* and *Enclosing* relations between CAD models, two constraint objects for the upper and lower time limits of *ShelfTime* and *TransitionTime* in the reliability testing, and one reference link object *PartDetail* pointing to a supplier's Website for product specifications.

Through the use of property objects and behavior objects in Tables 12.2 and 12.3, meanings of the supplementary information are captured in object-based representations. These semantics-rich objects will be encapsulated in CAD models for reuse across disciplines.

12.3.2 Embedding Supplementary Information in CAD Models

The supplementary information modeled by property and behavior objects in Tables 12.2 and 12.3 needs to be embedded into CAD models to make the information accessible and reusable for other downstream applications. Two methods are developed by which the property and behavior instances can be embedded in CAD models. The first method embeds these supplementary objects through CAD associations. Many commercial CAD packages such as the AutoCAD system used in this study, provide native facilities for constructing such associations. They provide facilities to store the CAD associations and their pointed property and behavior objects, together with the embedding CAD models, in data structures recognizable and processible by CAD systems. This method is simple, effective, and suitable for embedding CAD behaviors for use in CAD systems.

Table 12.3. Example of behavior objects

ObjectType	ObjectName	Description	Value	ExeType	RefNo
CadBehavior	Explode	Auto explosion of an imported CAD object into native objects.	Explode	dvb	B010
	Position	Smart positioning of an imported object in a CAD workspace.	Position	java	B011
Relationship	LocatedOnTop	Component on every layer is located one on top of the other to form a vertical tower.	(X_{ij},Y_{ij},Z_{ij}) [k]* = (X_{0j},Y_{0j},Z_{0j}) [k]* k = 1,2, …, MaxLayer	cc	R020
	Enclosing	Component 1 is enclosed in Component 2.	(X_1,Y_1,Z_1) [k]* ≦ (X_2,Y_2,Z_2) [k]* k = 1, 2, …, 8	cc	R021
Constraint	MaxShelfTime	Upper time limit for a sample placed in a test chamber.	30		C030
	MinTransitionTime	Lower time limit for temperature exposure of a sample.	5		C031
RefLink	PartDetail	Part specifications at a supplier's Website.	http://www.hrent.com/inv.htm	hyperlink	L041

* (X_{ij},Y_{ij},Z_{ij}) [ ] – a vertex coordinates array for a component at *j*th position on *i*th layer.
(X_{0j},Y_{0j},Z_{0j}) [ ] – a base vertex coordinates array for the *j*th vertical tower.
(X,Y,Z) [ ] – a vertex coordinates array of the enclosing or enclosed cube of a component.

The second method is used to embed complex behavior instances in CAD models, such as the *Explode* CAD manipulation and the *LocatedOnTop* relationship in Table 12.3. These instances are generated with complicated behavior information even external application programs. They are assigned to the relating CAD models by object links. A CAD add-on tool is needed to instantiate such object links on a hosting CAD platform. In the current research, a design Object Creation Wizard (details in Section 12.6.2) has been implemented as a CAD add-on tool for the AutoCAD system. When an object link between a behavior object and a CAD model is instantiated, this linkage relationship, rather the behavior object itself, will be embedded into the CAD model, retrievable and editable from the CAD modeling environment. The second method is suitable for embedding inter-part or inter-discipline relationships, constraints, and reference links needed for use in CAD and non-CAD systems, such as in CPD process management and quality assurance systems.

12.3.3 Modeling STEP extensions

To make the supplementary information STEP-compliant, the following STEP extension mechanisms, *i.e.*, property relationship extension and subtyping extension, are used.

(1) Property Relationship Extension Mechanism
Figure 12.2 illustrates the concept of the property relationship extension mechanism. In this method, the property and behavior objects defined in Section 12.3.1 are categorized into property sets (Pset), such as the reliability testing Pset or 3D representation maps Pset. The relationship entities are then specified to assign the relating Psets to the related entities in an existing STEP model. These relationship entities may be the subtypes extended from the standard STEP relationship entities by using the subtyping extension mechanism discussed in the next section.

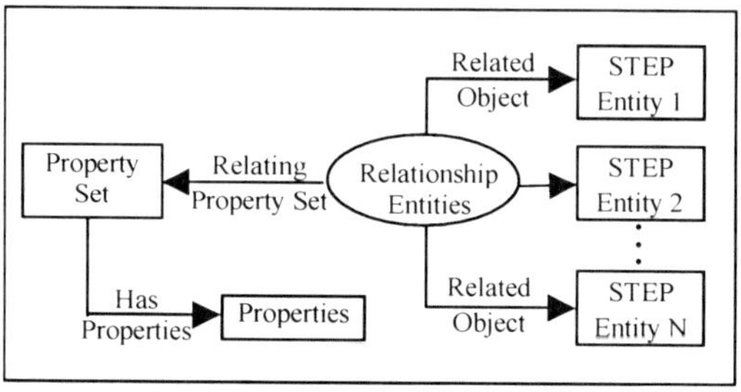

Figure 12.2. Property relationship extension mechanism

(2) Subtyping Extension Mechanism
In the subtyping extension, new entity definitions are created as subtypes of an existing STEP entity. Figure 12.3 describes this extension mechanism in Express-G [15] notation.

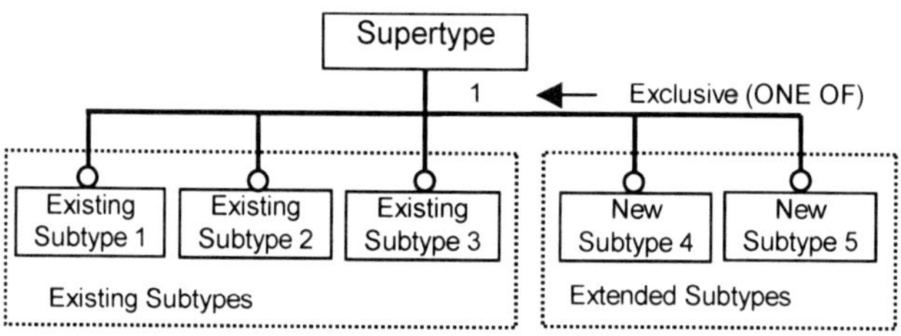

Figure 12.3. Subtyping extension mechanism

By using the exclusive (*ONE OF*) subtype/supertype constraint of the Express [15] language, new entities can be exclusively subtyped into the existing supertype entities. The extended entities are then used for example as relationship entities to connect the supplementary property set definitions to those entities already existed in the STEP product models.

12.3.4 Capturing Semantics in STEP-compliant Product Models

Using the information embedment approaches in Section 12.3.2, the product supplementary information is made available with CAD models in CAD native formats. The information is, however, only reusable on compatible CAD platforms.

This section discusses issues related to supplementary information capture in STEP-compliant product data models to make it applicable to STEP based CAD or non-CAD applications. To do so, the STEP Psets and inclusion properties are specified. For example, a single valued property, as an extended entity of STEP AP203, can be defined as follows.

<Name, Description, NominalValue, Unit> (12.3)

And a property set as:

<Name, Description, GlobalId, HasProperties> (12.4)

These extended entity definitions in Expressions (12.3) and (12.4), together with other STEP extension definitions will be instantiated with the supplementary information, through semantic mapping (details in Section 12.5), to form a STEP AP203 compliant product model that not only captures the semantics of the supplementary information, but also make the captured information interoperable to other STEP compliant applications.

12.4 Taxonomy and Ontology

12.4.1 Vocabulary Taxonomy

A vocabulary contains a collection of commonly agreed terminologies/concepts in a domain. Taxonomies characterize and organize complex domain vocabularies into hierarchical structures. A taxonomy is often used as a kind of semantic agreements to achieve an explicit naming approach to the shared use of data semantics, and to remove misunderstanding and misinterpretation of shard information. A CPD vocabulary taxonomy has been developed to classify and manage the shared terminologies together with their explicitly defined meanings used in the development of consumer products. The terminologies are mainly used for naming properties, CAD behaviors, object relationships and constraints of the supplementary information. All terminologies in the vocabulary taxonomy are inter-connected by "*is-a*" (super-sub) and "*part-of*" (part-whole) relations. Figure 12.4 depicts an excerpt of the vocabulary taxonomy.

The *is-a* relation of the vocabulary taxonomy in Figure 12.4 exists between domain concepts being classified, while the *part-of* relation between attributes within a concept. One of the logical properties of the *is-a* relation is transitive, which implies that if x is a subclass of y and y is a subclass of z, then x is also a subclass of z. This axiom can be illustrated by our example in Figure 12.4: *ShelfTime* is a subclass of *ReliabilityTestProperty* that in turn is a subclass of *SemanticProperty*. Due to the transitivity of the *is-a* relation, it can be inferred that *ShelfTime* is also a subclass of *SemanticProperty*, *i.e.*, every instance of *ShelfTime* shall be an instance of *SemanticProperty*. For example, Figure 12.4 characterizes that the *ShelfTime* has five attributes for:

<Property Name, Description, Value, Unit, RefNo.>

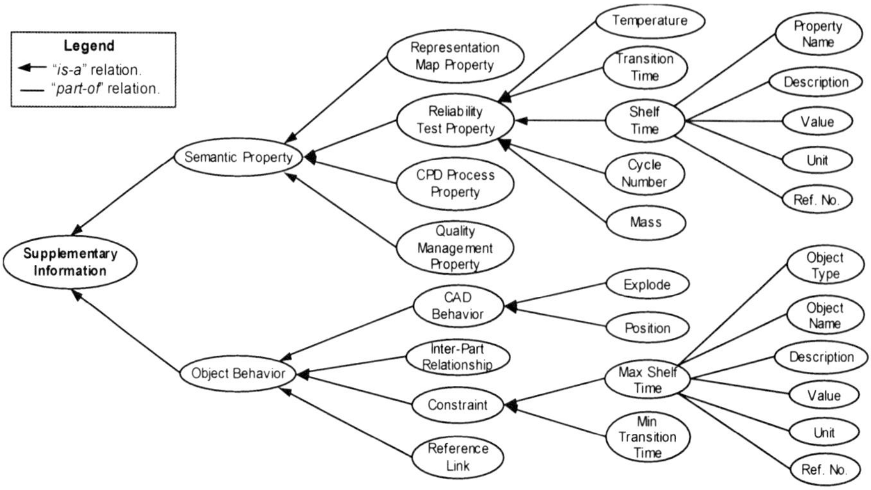

Figure 12.4. An excerpt of the vocabulary taxonomy

Suppose there exists a *ShelfTime* instance for an electronic component:

<ShelfTime_i1, Time from "Component_1" is placed in a test chamber till a stabilized chamber temperature reached, 10, min, Test011>

It can be inferred that *SemanticProperty* should also contain such an instance with exactly the same attribute values. When a reliability testing system searches for temperature cycling parameters, the query will be based on the semantics of the instance, rather on the particular terms used by a tester or used in a test parameter database. The vocabulary with its explicit hierarchical structure, terminology definitions, and inter-connection relations provides a semantic basis for OWL domain ontology modeling.

12.4.2 OWL Ontology

Domain ontologies in OWL are developed to make the intended meaning of CPD domain concepts (terminologies) explicit, through attaching information about the properties of relations to the terminologies. The domain ontologies are then organized in a reference ontology used for mapping and sharing the product supplementary information and semantics across CAD, quality assurance, and CPD process management applications. The *is-a* and *part-of* relations in the vocabulary taxonomy in Section 12.4.1 are formalized and expanded with the OWL constructs, which provide rich semantics for the development of domain specific ontologies, including the logical relations (such as *owl:differentFrom*, *owl:disjointWith*, *owl:inverseOf*); properties of relations (*owl:TransitiveProperty*, *owl:SymmetricProperty*, *owl:FunctionalProperty*); restriction types (*owl:hasValue*, *owl:cardinality*); and so on.

Each domain concept in the vocabulary taxonomy in Figure 12.4 is defined by an OWL class. Attributes of a concept class are formalized with the OWL datatype properties (*owl:DatetypeProperty*), while relations between classes are modeled with the OWL object properties (*owl:ObjectProperty*) and other relation constructs. Figure 12.5 shows an OWL class for the definition of the *ShelfTime* concept.

```
<owl:Class rdf:ID="ShelfTime">
  <rdfs:subClassOf rdf:resource="#ReliabilityTestProperty"/>
  <owl:equivalentClass rdf:resource="&onto2;ChamberingTime"/>
  <owl:disjointWith rdf:resource="#Temperature"/>
  <owl:disjointWith rdf:resource="#TransitionTime"/>
  <owl:disjointWith rdf:resource="#CycleNumer"/>
  <owl:disjointWith rdf:resource="#Mass"/>
</owl:Class>
```

Figure 12.5. OWL class definition for *ShelfTime*

The *is-a* relation between concepts of *ShelfTime* and *ReliabilityTestProperty* in Figure 12.4 is formalized by *subClassOf* axiom (Class1 $\subseteq$ Class2) in Figure 12.5. The other OWL axioms used in Figure 1.5 are *equivalentClass* (Class1 $\equiv$ Class2) and *disjointWith* (Class1 $\subseteq$ $\neg$Class2). The logical axioms provide mechanisms for query, matching and reasoning about concepts, their attributes and relations. For example, the specification of the disjointedness of *ShelfTime* with other four classes in Figure 12.5 ensures that an individual cannot be an instance of more than one class of the five. And the equivalent class, specifying that two classes have exactly the same instances, is often used to indicate synonyms for the use of domain concepts across disciplines.

An OWL property, such as *ObjectProperty* or *DatatypeProperty* is a binary relation to relate instances of two OWL classes, or relate an OWL instance to a RDF [14] literal or an XML Schema datatype [22]. Figure 12.6 shows that the *ShelfTime* instances are associated with the *DesignComponent* instances by an *ObjectProperty*, and with the string datatype of XML Schema by a *DatatypeProperty*.

```
<owl:ObjectProperty rdf:ID="associatedWith">
  <rdfs:domain rdf:resource="#ShelfTime"/>
  <rdfs:range rdf:resource="#DesignComponent"/>
</owl:ObjectProperty>
<owl:DatatypeProperty rdf:ID="propertyName">
  <rdfs:domain rdf:resource="#ShelfTime"/>
  <rdfs:range rdf:resource="&xsd;string"/>
</owl:DatatypeProperty>
```

Figure 12.6. OWL properties for ShelfTime

Both the object and datatype properties in Figure 12.6 use the *domain-range* pairs to restrict the binary relations specified. Besides the *domain-range* pair, there are a variety of other OWL restrictions such as cardinality restrictions (*maxCardinality*, *minCardinality* and *cardinality*) that can be used to constrain the range of an OWL property in specific contexts. For example, the class definition for *ShelfTime* in Figure 12.5 can be further extended with a *minCardinality* restriction to constrain that the *ShelfTime* class has at least one *associatedWith* property, as shown in Figure 12.7.

```
<owl:Class rdf:about="#ShelfTime">
  <rdfs:subClassOf>
    <owl:Restriction>
      <owl:onProperty rdf:resource="#associatedWith"/>
      <owl:minCardinality rdf:datatype="&xsd;nonNegativeInteger">
        1
      </owl:minCardinality>
    </owl:Restriction>
  </rdfs:subClassOf>
</owl:Class>
```

Figure 12.7. OWL restriction on a property of ShelfTime

Described by these formal, explicit and rich semantics shown in Figures 12.5, 12.6 and 12.7, the domain concept of *ShelfTime*, its properties and relationships with other concepts can be incorporated into the shared reference ontology to be queried, reasoned or mapped across disciplines to support the interoperability of heterogeneous systems.

12.5 Semantics-driven Schema Mapping

Schema mappings from the supplementary information definitions to the extended STEP Pset definitions are established based on the common vocabulary and the OWL ontologies developed in the previous section. The difference between conventional schema mapping and semantics-driven schema mapping is that the latter uses ontologies to capture the relationships and contexts of maps. The scope of the semantic assertions in the semantics-driven approach is also constrained by ontologies. Figure 12.8 shows our semantics-driven schema mapping approach.

In the semantics-driven approach, ontology can be used either as a global schema or as a semantic transformation reference in schema mapping. In the former case, the ontology, either a standard-based or a logic-based, should be more abstract than any of the application schemas being mapped to it. During the mapping process, maps between the vocabulary of each application and the vocabulary of the global ontology are defined in order to transform the data semantics. By interpreting the underlying ontology and applying the semantic definitions, relationships, and axioms described in the ontology to the translation process, the semantic maps transform the data semantics from application to application, or from local schemas to the global one.

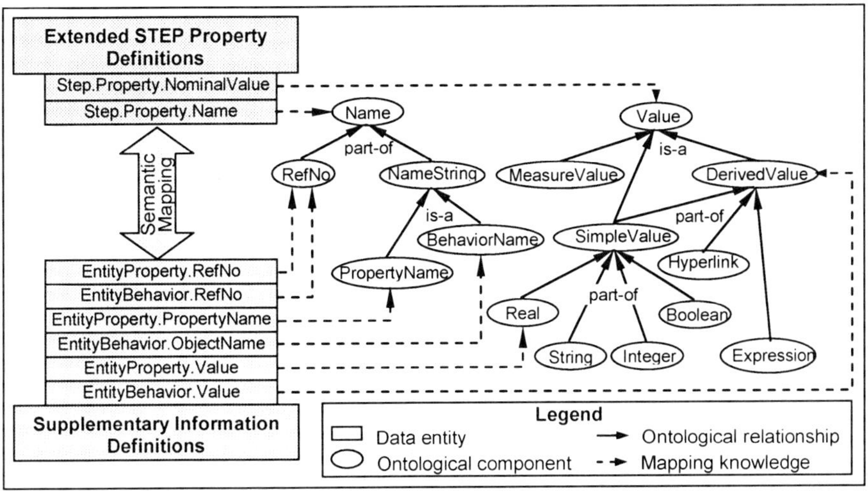

Figure 12.8. Semantics-driven schema mapping

In our approach in Figure 12.8, the product supplementary information definitions are mapped to the extended STEP property definitions via domain ontologies. However, the ontologies themselves are not used as the global schema. Instead, they are used as a reference to provide a pivot point for the reinterpretation of the data meanings in different schemas from different applications. Our approach uses the STEP model as a global schema in semantic mapping. The maps from the supplementary information definitions in Expressions (12.1) and (12.2) to the extended STEP property definitions in Expressions (12.3) and (12.4) are established based on the ontological relations in domain ontologies. These ontologies are defined by drawing terminologies from the common vocabulary. In the mapping process, semantically equivalent components in the expressions above are identified and compared based on the semantic descriptive definitions of terms/concepts, and the semantic relationships and rules in the domain ontologies. The semantic mapping knowledge is then established at both the entity and attribute levels.

Based on the mapping knowledge, semantically equivalent concepts in both the supplementary and the STEP extension definitions are mapped to each other. A few semantic maps between the STEP extension in Expression (12.3) and the entity property definition in Expression (12.1) and the entity behavior definition in Expression (12.2) are shown in Figure 12.8.

Figure 12.9 below presents an excerpt of the reference ontology used in Figure 12.8 to define the semantic maps. It specifies the relations between the domain concepts of *Name*, *RefNo*, *NameString*, *PropertyName* and *BehaviorName* in Figure 12.8. There are *is-a* and *part-of* relations among these concepts. The *is-a* relation is specified by *subClassOf*. The *part-of* relation is defined using one of the OWL axioms for transitivity (*owl:TransitiveProperty*). The properties of the *part-of* relation are restricted by *owl:onProperty* and *owl:allValueFrom*. These ontological definitions and relations in Figure 12.9 are used to detect whether semantic conflicts exist during the mapping process.

```
<owl:Class rdf:ID="Name">
  <!-- Other definitions for the class -->
</owl:Class>
<owl:Class rdf:ID="NameString">
  <rdfs:subClassOf>
        <owl:Restriction>
          <owl:onProperty rdf:resource="#part-of"/>
          <owl:allValueFrom rdf:resource ="#Name"/>
        </owl:Restriction>
  </rdfs:subClassOf>
</owl:Class>
<owl:Class rdf:ID="RefNo">
  <rdfs:subClassOf>
        <owl:Restriction>
          <owl:onProperty rdf:resource="#part-of"/>
          <owl:allValueFrom rdf:resource ="#Name"/>
        </owl:Restriction>
  </rdfs:subClassOf>
</owl:Class>
<owl:Class rdf:ID="PropertyName">
  <rdfs:subClassOf rdf:resource="#NameString"/>
  <owl:disjointWith rdf:resource="#BehaviorName"/>
</owl:Class>
<owl:Class rdf:ID="BehaviorName">
  <rdfs:subClassOf rdf:resource="#NameString"/>
  <owl:disjointWith rdf:resource="#PropertyName"/>
</owl:Class>
<owl:ObjectProperty rdf:ID="part-of">
  <rdf:type rdf:resource="&owl;TransitiveProperty"/>
</owl:ObjectProperty>
```

Figure 12.9. An excerpt of the reference ontology

12.6 Software Prototype Development

12.6.1 Software System Architecture

The semantic interoperability approach described in the previous sections has been implemented in a Web-enabled prototype system. It is mainly used to support the product semantics definition, capturing in product models, and sharing cross CPD applications of multiple platform CAD systems, quality and reliability control, and CPD process management. The prototype uses a multi-tier architecture consisting of a collaboration server and a set of client-side CAD add-ons and interfacing software tools. Figure 12.10 shows the architecture of the prototype.

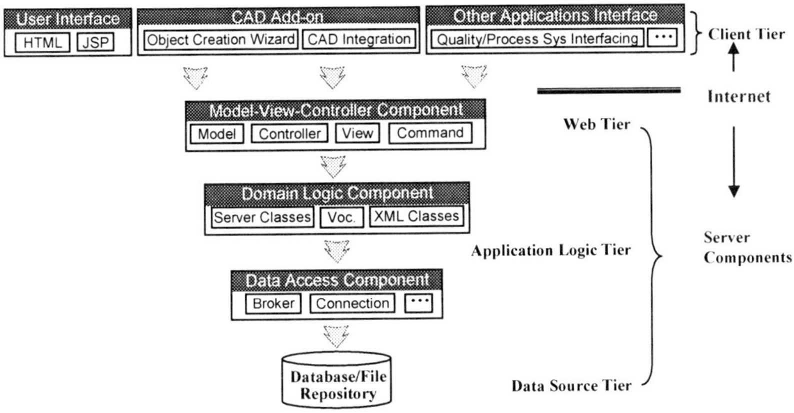

Figure 12.10. Prototype architecture

The client applications in Figure 12.10 include three toolkits: a design Object Creation Wizard and a CAD Integration Toolkit, both being used with CAD systems as add-on tools; and a Quality and Process Systems Interfacing Tool. The collaboration server includes several functional modules for Web-based presentations and communications, domain application services, and data access control. Both the client toolkits and server components work together to provide online services in supplementary information definition and embedment in CAD objects, semantics capturing in product models, common vocabulary and ontology library maintenance, management of design objects with semantic supplements, and semantics reusing in CAD and non-CAD applications through sharing of the semantic supplements and other product data in neutral formats (STEP and XML). The software development of the prototype is detailed in the following two sections for the client and the server respectively.

12.6.2 Client Toolkits

(1) Object Creation Wizard

An Object Creation Wizard is designed as a CAD add-on tool for use with the AutoCAD system. It mainly provides functions for: defining the semantics of entity properties for CAD models; instantiating the entity property and entity behavior definitions with design parameters extracted automatically from AutoCAD models or supplemented by users; validating the supplementary information elements against the implemented contextual constraints in the wizard; linking property objects with CAD models or styles; and semantically mapping the CAD and STEP entity definitions. Functional modules are developed with the wizard to implement these capabilities. Figure 12.11 shows one of the UML [24] sequence diagrams to depict how the Pset Definition Module works. Figure 12.12 shows the implemented Pset Definition Module with an illustration on how the vocabulary library is used to support the unified naming of properties.

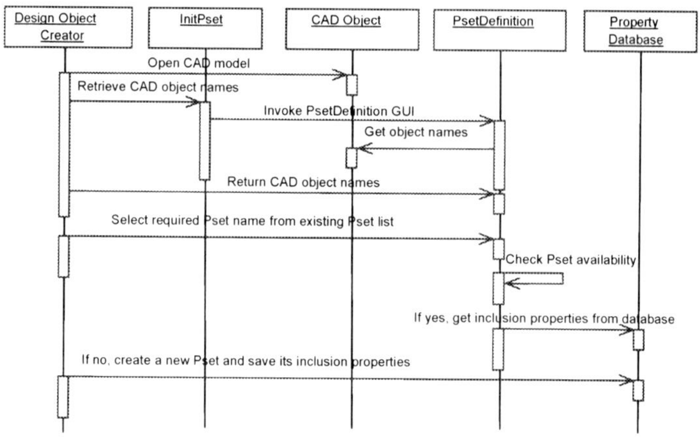

Figure 12.11. UML sequence diagram for Pset definition

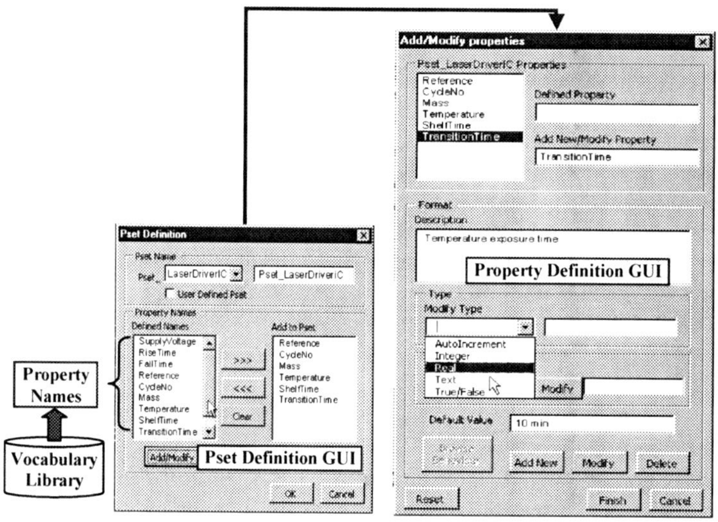

Figure 12.12. Defining Pset and inclusion properties

(2) CAD Integration Toolkit

The CAD Integration Toolkit is another CAD add-on to provide more value-added functions to the CAD users who are using semantics of supplementary information in their native CAD systems. The main functionality of the toolkit include: identifying and retrieving the supplementary information semantics from the CAD representation maps, inter-part relationships, CAD behaviors, *etc.*, in the downloaded CAD models for their reuse in a CAD system; interpreting the retrieved information semantics according to the common vocabulary and reference ontology and linking them back to the corresponding CAD objects/symbols when they reside within the native CAD system; and providing interfaces for CAD users to access external digital information sources and for them to invoke the built-in object behaviors directly from within a CAD design environment. The accessing and invoking capabilities of the toolkit are depicted in Figure 12.13.

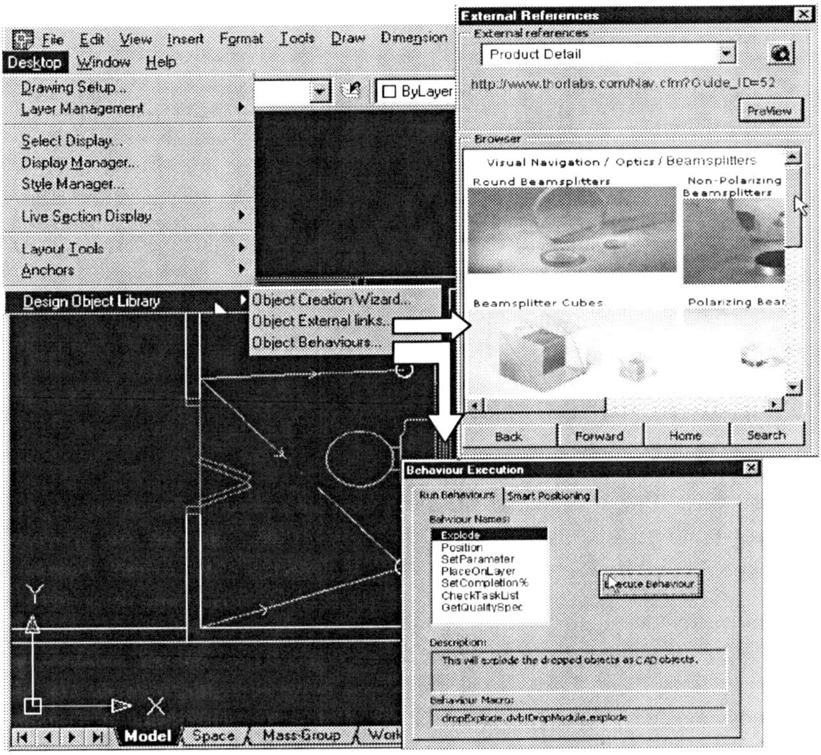

Figure 12.13. Invoking object behavior and external link in CAD environment

(3) Quality and Process Systems Interfacing Tool

This tool uses a Java3D-enabled Web browser to render geometries of design objects. Quality Assurance (QA) engineers and project managers can use it to view the 3D CAD models without using a CAD system. The tool provides facilities for editing and populating the Pset definitions with the data extracted from a set of quality tools and from a CPD process management system. It also supports the product development activity coordination and design negotiation through CPD process configuration and XML messaging between designers, QA professionals, and product development managers. Figure 12.14 shows one of the GUIs of the tool for extracting/adding CPD process attribute values from a process configuration to populate a CPD process Pset definition.

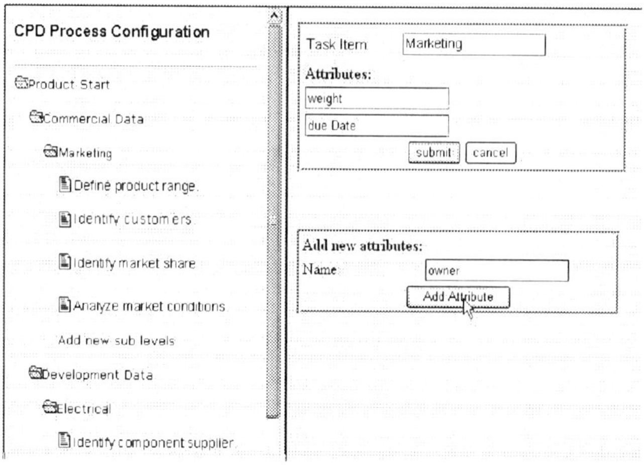

Figure 12.14. Extracting/adding CPD process data for a process Pset definition

12.6.3 Collaboration Server Components and Services

(1) Model-View-Controller component

This component is responsible for the client and server communication, including to route incoming client requests to the Controller Servlets for generating Command objects required for the requests; to instantiate the Model objects for execution of domain logics (functions) of the prototype system; to present the processing results to client browsers through View objects; and to map the client next action from View to Model for new responses of the server. A UML class diagram in Figure 12.15 shows the relationships between a *ControllerServlet* class, a *CommandManager* bean, and other related classes for generation of *Command* objects to process client requests.

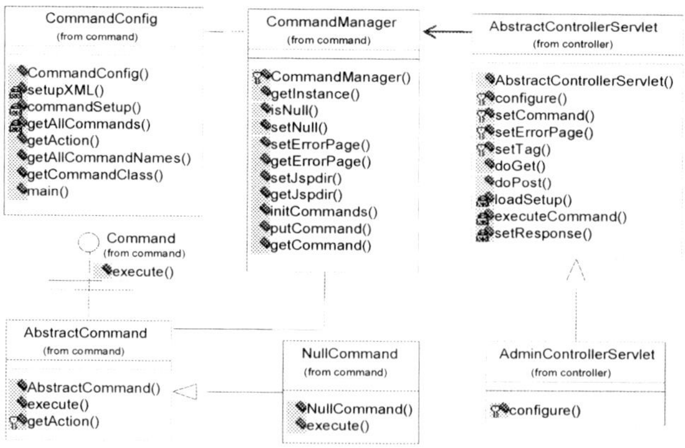

Figure 12.15. UML class diagram for *Controller* and *Command*

(2) Domain logic component

The domain logic component establishes the basic structure of domain objects (*i.e.*, the extended design objects with semantic supplements, vocabulary objects, property and behavior objects, *etc.*). For example, each domain object must have at least one broker to process client requests for accessing the semantic interoperability services and the server database or file repositories. Use cases are designed to analyze the manipulation scenarios of domain objects for realization of various domain application logics. Figure 12.16 is a UML use case diagram to illustrate the scenarios of defining, editing and maintaining the vocabulary objects with this component.

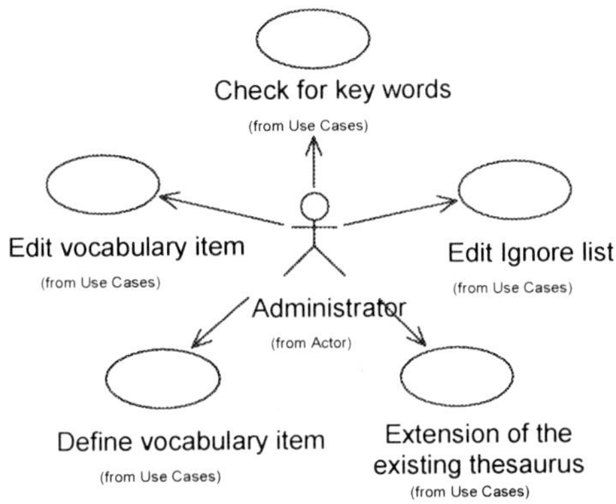

Figure 12.16. UML use case for vocabulary management

(3) Data access component

The component is used for database connection and brokers management, such as to help a domain object to get a database connection; to release the database connection; to retrieve the appropriate broker; and to manage the number of brokers to be instantiated. Any domain objects can only manipulate the server database objects through brokers. For example, in order to edit a vocabulary item in the server vocabulary database, a database connection object, a *VocabularyObjectBroker* and a *VocabularyObjectAttributeBroker* need to be instantiated for interacting with the database and conducting the *find()* and *update()* operations to the existing vocabulary item.

(4) Portal services

The implementation of the server portal provides the following services:

- Uploading, downloading, searching, and updating design objects with the semantic supplementary information, based on users' roles and access rights assigned;
- Definition, editing, search and validation of common vocabulary;
- Design object viewing and dragging-dropping in multiple CAD formats and using neutral data structures (STEP and XML);

- Object profile management, such as browsing the design objects in the server repositories, online visualizing CAD models, retrieving vocabulary definitions, *etc.*;
- Message broadcasting and conflict logs to facilitate conflict detection, negotiation, and collaboration in product development; and
- Design team setting and activity coordination through member profile management, role and task assignment, access privilege control to the shared information, online collaboration information query.

Figure 12.17 is a screenshot of a portal page for one of the above services: object profile management.

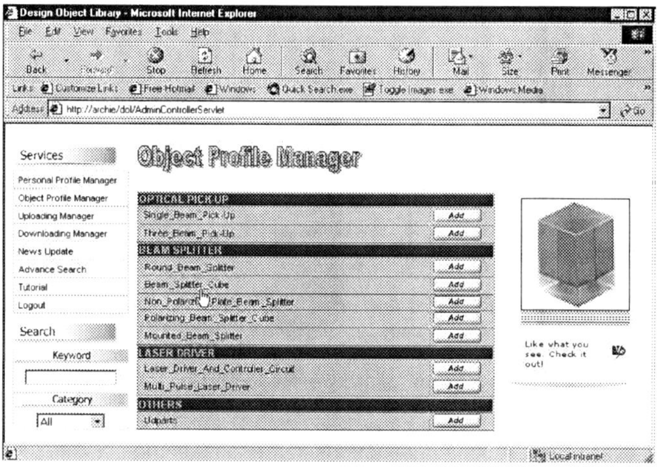

Figure 12.17. Portal service for object profile management

12.7 Collaboration Scenarios

Supporting and enhancing product development collaboration is the major objective of this research. This is illustrated by the following collaboration scenarios through the use of the semantic interoperability method and the prototype system developed in the previous sections. These scenarios concern the collaboration in multi-disciplinary design of consumer products, such as CD and DVD players. Mechanical, electrical, optical design disciplines, quality assurance, and product development process management are involved in the scenarios for collaborative design of a DVD Optical Pick-up System (OPS) in a networked heterogeneous environment.

12.7.1 Support of Collaborative Design Process

This scenario focuses on the collaborative OPS design processes and interactions between collaborating participants. Table 12.4 lists the involved participants, their responsibilities and individual software systems used in this collaboration.

Table 12.4. A collaborative design environment

Collaboration Participant	**Responsibility**	**Native Client-Side Software System Used**	**Access Right to the Collaboration Server**
Project Manager A	Project coordinator & design process planner.	A design process management system.	Use design objects; "Read" access only.
Mech. Designer B	Mechanical design & OPS model integrator.	AutoCAD	Create/use design objects; "Full" access to all objects.
Ele. Designer C	Electrical & electronic component design.	Mentor	Create/use design objects; "Full" access to electrical components.
Optical Designer D	Optical system design and analysis.	OptiCAD	Create/use design objects; "Full" access to optical components.
Quality Engineer E	Reliability testing & quality assurance.	A set of testing & quality tools.	Use design objects; "Read" access to all objects.

Project Manager *A*, as the CPD coordinator, defines the collaborative design processes, the owner of tasks, the responsibility of each team member, and the participants' access privileges to the collaboration server (as shown in Table 1.4). The native design process information from a process management system is annotated according to a domain ontology for the CPD process. The process ontology instances are kept in an XML instance file and uploaded to the collaboration server. The server manages updates of the XML instance files for any design process changes.

Each participant is working on his/her design issues in a distributed private workspace at the client side with preferred domain-specific tools as indicated in Table 12.4. However they can access and share the same resources for design task assignments, design constraints, design objects with the embedded semantic supplements, and design support services from the collaboration server, if they join in a collaborative design session. Once logging in a project workspace of the server, online notification presents the updated design process progress and new design requirements. The mechanical, electrical, and optical design objects with rich semantics can be shared via the server, but each design discipline has to follow the common vocabulary to name their supplementary information elements in order to make the semantics understandable to other disciplines. The vocabulary library of the server provides facilities to support this requirement. The design disciplines interact with each other through the server, which provides the messaging and online viewing mechanisms to facilitate design communication, negotiation and visualization. Hence, the individual's design work can be evaluated/commented and their activities coordinated.

12.7.2 Design Objects Modeling and Semantics Capturing

This scenario involves the use of the design Object Creation Wizard (details in Section 12.6.2) for product semantics capture in design models of the OPS.

Mechanical Designer *B* performs mechanical part design and acts as the OPS model integrator. According to OPS studies on its functional requirements, design constraints and previous product structures, *B* creates a preliminary OPS assembly drawing in which the OPS configuration is defined. The OPS assembly includes mechanical, electrical and optical components, such as the base of OPS, laser diode, photo diode, beam splitter, and grating lens, *etc*.

The CAD representations of the OPS assembly are modeled by the AutoCAD system. The mechanical parts of the assembly will be further modeled by AutoCAD during the detailed design stage. However, the electrical and optical components only have symbolic representations in the OPS assembly drawing. As such, the additional information required for the electrical and optical detailed design, and for the component reliability testing, *etc*., needs to be embedded into the preliminary OPS assembly drawing and continually updated with the progress in design and analysis of its components.

By right-mouse click on the assembly drawing in the AutoCAD modeling space, the Object Creation Wizard is invoked. Figure 12.18 illustrates how the wizard specifies and captures the native geometric information from the AutoCAD system to define the supplementary data semantics for *Representation Maps* of the *base* component of the OPS. The selected multi-view blocks from the base CAD model are aggregated into a representation map called *Rmap_3D_Base-A*. It has two inclusion objects for *Rmap_3D_Base-A_Default* and *Rmap_3D_Base-A_Top* respectively.

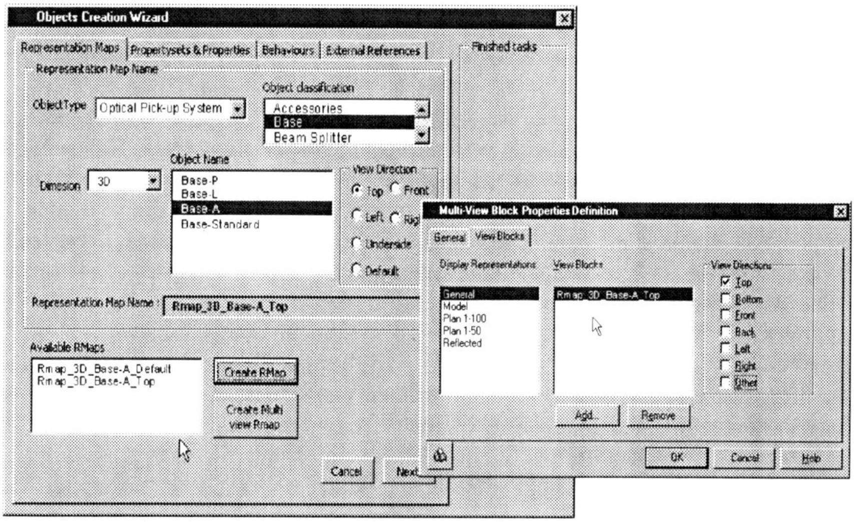

Figure 12.18. Specifying and capturing supplementary data semantics for a *base* CAD model

Besides specifying and capturing supplementary data semantics with design objects, the wizard also provides standard procedures and facilities to guide the XML instance file generation for these CAD models. In the current example, the supplementary information and the XML content of the OPS assembly includes geometry representation maps, semantic properties, object behaviors, and external links for its constituent components. The supplementary information is used such as for:

- Specifying and embedding OPS integration requirements and quality criteria for the OPS multi-disciplinary design components;
- Assigning representation maps for the base multi-view blocks using the terminologies in the common vocabulary library;
- Defining semantic properties for the beam splitter and an inter-part relationship, AdjacentWith, between the beam splitter and the base;
- Defining the semantic definitions for the temperature cycling test constraints of the laser driver IC at the qualification phase of the OPS; and
- Embedding external reference links pointing to online sources for suppliers and products information of the OPS components.

Once the OPS assembly drawing is created with the above supplementary information embedded, it can be uploaded (together with its XML instance file as the neutral representation of the attached supplements) to the server for sharing the information and semantic definitions with other participants in the collaboration.

12.7.3 Semantics Sharing with Heterogeneous Systems

This scenario describes semantics sharing across Electrical CAD (E-CAD), optical CAD, and reliability testing systems. The semantic content defined with the supplementary information definitions, captured in the product models, and delivered through the neutral XML instance files are shared with these heterogeneous systems. This is done through querying, inferring and matching the ontological definitions and relations of the domain concepts involved, in order to allow these computer systems to understand and utilize the meaning of semantic content embedded in the design objects.

The Optical Designer *D* visualizes the OPS assembly drawing to understand the OPS configuration and downloads its XML instance file. *D* selects the symbolic representation of the beam splitter to develop it into a detailed design using the OptiCAD system. The beam splitter has been attached with an inter-part relationship for *AdjacentWith* in Section 12.7.2. The term *AdjacentWith* is retrieved from the XML instance file. The semantics of the term are interpreted through querying the ontological relations defined in the shared reference ontology, which the OptiCAD add-on tool follows. For example, given the following excerpt of an ontology shown in Figure 12.19, the add-on tool can reason about the instance of the OWL relation *AdjacentWith* being an inter-part relationship because the domain of the OWL object property *hasExeType* is *InterPartRelationship* according to the ontology in Figure 12.19.

```
<owl:ObjectProperty rdf:ID="hasExeType">
  <rdfs:domain rdf:resource="#InterPartRelationship"/>
  <rdfs:range rdf:resource="#exeType"/>
</owl:ObjectProperty>
<owl:Class rdf:ID="exeType">
  <owl:oneOf rdf:parseType="Collection">
    <owl:Thing rdf:about="#dvb"/>
    <owl:Thing rdf:about="#cc"/>
    <owl:Thing rdf:about="#java"/>
    <owl:Thing rdf:about="#hyperlink"/>
  </owl:oneOf>
</owl:Class>
<owl:Thing  rdf:ID="AdjacentWith">
  <hasDescription rdf:datatype="&xsd;string">
     Two multi-disciplinary design objects are adjacent with each other
  </hasDescription>
  <hasValue rdf:datatype="&xsd;string">
     Relatioship_adjacentWith
  </hasValue>
  <hasExeType rdf:resource="#dvb"/>
</owl:Thing>
```

Figure 12.19. Ontological definition of *AdjacentWith*

The tool can further identify and retrieve the *hasDescription* and *hasValue* (both are defined elsewhere in the shared reference ontology) information through other properties of *AdjacentWith* in Figure 12.19. The information will then be executed by the add-on tool to control the multi-disciplinary design relationship during the modeling of the beam splitter.

Electrical Designer *C* retrieves the XML instance file of the OPS assembly together with the relevant integration and quality requirements for the electrical/electronic component design. *C* also extracts the semantic definitions of the testing constraints of the *laser driver IC* sub-assembly specified in Section 12.7.2. With a "full" access right to the electrical/electronic design objects, *C* is able to create or modify these objects. The completed design objects are uploaded to the server. Based on the design parameters of the *laser driver IC* from the Mentor E-CAD system, *C* instantiates the semantic definitions of the testing constraints (as shown in Table 12.5) and updates the reliability test property values in the server database.

Table 12.5. Reliability testing constraints

Mass (M)	Shelf Time (T1)	Transition Time (T2)	CycleNo
M <= 15 g	T1 >= 10 min.	T2 <= 5 min.	10
15 g < M <= 150 g	T1 >= 30 min.	T2 <= 15 min.	
150 g < M <= 1500 g	T1 >= 60 min.	T2 <= 30 min.	

Quality Engineer *E* drags and drops the detailed design of the *laser driver IC* sub-assembly (submitted by Designer *C*) from the server to his testing workspace, together with this model's XML instance file. Figure 12.20 below shows an excerpt of the XML file instantiated from the ontological definitions of the concept *ReliabilityTestProperty* in Figure 12.4.

```
<ReliabilityTestProperty rdf:ID="LaserDriverIC_Mass">
  <propertyName rdf:datatype="&xsd;string">Mass</propertyName>
  <description rdf:datatype="&xsd;string">Mass of laser driver IC</description>
  <value rdf:datatype="&xsd;positiveInteger">80</value>
  <refNo rdf:datatype="&xsd;string">Test001</refNo>
  <hasUnit rds:resource="#g"/>
</ReliabilityTestProperty>
```

Figure 12.20. XML instance file for *Mass*

The testing application system understands the meaning of terminologies used in the XML file by referring to the terminology definitions. The logical relations of the terminologies are queried based on the semantic descriptions of the definitions. For example, by querying the XML instance in Figure 12.20, the testing system interprets the semantics of the concept for *LaserDriverIC_Mass* as: "mass of the laser driver IC is 80g" set by Electrical Designer *C* during the design stage. Based on this interpretation and the constraints in Table 12.5, the testing system conducts proper settings for the temperature cycling test of the *laser driver IC* sub-assembly. In this way, the data semantics from different product models are shared across heterogeneous systems explicitly and flexibly.

12.8 Conclusions

Collaborative product development needs semantically interoperable product models and design objects supplemented with formal and explicit engineering meanings to support semantic interoperability. A new ontology-driven, STEP-based solution has been developed for specifying, capturing, understanding, and sharing the product semantics to facilitate heterogeneous information integration and interoperation among CPD applications in multi-disciplinary CAD, quality and reliability control, and product development process management. The solution can enhance the collaborative product development by providing a semantic interoperability method and a software prototyping system to the cross-functional CPD participants and systems as illustrated in the collaboration scenarios. Our next efforts will be focused on the further development of the method and the prototype to provide the semantic rule modeling and ontology composition services for more effective collaboration in product development.

12.9 Acknowledgements

The authors wish to thank Mr. Rajesh Babu, Mr. Xingjian Xu and Mr. Jason Cheng for their contributions to the software development in this research.

12.10 Acronyms

CPD	Collaborative Poduct Dvelopment
STEP	STandard for the Exchange of Product model data
PSL	Process Specification Language
OWL	Web Ontology Language
XML	eXtensible Markup Language
RDF	Resource Description Framework
NPI	New Product Introduction
CAE	Computer-Aided Engineering
FEA	Finite Element Analysis
CFP	Computational Fluid Dynamics
Pset	Property Set
OPS	Optical Pick-up System

12.11 References

[1] Pollock, J. T., Hodgson, R., 2004, *Adaptive Information: Improving Business through Semantic Interoperability, Grid Computing, and Enterprise Integration*, Wiley-Interscience.

[2] Bruijn, J., Ding, Y., Arroyo, S., Fensel, D., 2004, "Semantic information integration in the COG project," *Corporate Ontology Grid Project White Paper*, available at http://www.cogproject.org/publications/sii-wp.pdf

[3] Yoshioka, M., Umeda, Y., Takeda, H., Shimomura, Y., Nomaguchi, Y. and Tomiyama, T., 2004, "Physical concept ontology for the knowledge intensive engineering framework," *Advanced Engineering Informatics*, 18, pp. 95–113.

[4] Alexiev, V., Breu, M., Bruijn, J., Fensel, D., Lara, R. and Lausen, H, 2005, *Information Integration with Ontologies: Experiences from an Industrial Showcase*, Wiley, West Sussex, U.K.

[5] Tao, F., Chen, L., Cox, S., Shadbolt, N., Puleston, C. and Goble, C., 2003, "Semantic support for grid-enabled design search in engineering," *Proceedings of the First GGF Semantic Grid Workshop, the Ninth Global Grid Forum (GGF9)*, Chicago IL, USA.

[6] *ISO TC184/SC4, ISO 10303*, 1994, "Industrial automation systems and integration – product data representation and exchange," Geneva.

[7] Gibb, B. and Damodaran, S., 2002, *ebXML: Concepts and Application*, Wiley.

[8] Stork, A., Smithers, T. and Koch, B., 2002, "WIDE – semantic Web-based information management and knowledge sharing for innovative product design and engineering," *CG Topics*, 4, pp 23–24.

[9] Tao, F., Chen, L., Shadbolt, N. R., Pound, G. and Cox, S. J., 2003, "Towards the semantic grid: putting knowledge to work in design optimisation", *Proceedings of the 3rd International Conference on Knowledge Management I-KNOW '03*, Austria.

[10] *ISO TC184/SC4 and TC184/SC5*. "Process Specification Language (PSL)," available at http://www.nist.gov/psl/

[11] Ciocoiu, M., Gruninger, M. and Nau, D. S., 2001, "Ontologies for integrating engineering applications," *Journal of Computing and Information Science in Engineering*, 1(1), pp 12-22.

[12] Guarino, N., 1998, "Formal ontology and information systems", in Guarino, N., (Eds.), *Formal Ontology in Information Systems*, IOS Press.

[13] *W3C*, "OWL Web ontology language overview", available at http://www.w3.org/TR/2003/PR-owl-features-20031215/

[14] *W3C*, "RDF/XML syntax specification", available at http://www.w3.org/TR/2004/REC-rdf-syntax-grammar-20040210/.

[15] *ISO TC184/SC4, ISO 10303 Part 11*: "Description methods: The EXPRESS language reference manual," Geneva, 1994.

[16] Bittner, T., Donnelly, M. and Winter, S., 2005, "Ontology and semantic interoperability", In Prosperi, D. and Zlatanova, S. (Eds.), *Large-scale 3D data integration: Problems and challenges*, CRCpress (Tailor & Francis).

[17] *ISO TC184/SC4, ISO 10303 Part 203*, 1994, "Configuration controlled design," Geneva.

[18] *STEP AP210 Website*, available at http://www.ap210.org/

[19] Yang, Q. Z. and Lu, W. F., 2005, "A Web-enabled engineering object modeling environment to support interoperability and intelligent services in collaborative design", *Proceedings of ASME 2005 Design Engineering Technical Conferences*, Paper No: DETC2005-84240.

[20] Scheer A-W, *ARIS - Business Process Modeling*, 3rd Edition, Springer, 2000.

[21] Yang, Q. Z., Zhu, C. F., Ming, X. G. and Lu, W. F., 2005, "Configurable product design processes with integrated knowledge maps", *Proceedings of the 12th ISPE International Conference on Concurrent Engineering: Research and Applications*, pp.161–166.

[22] *W3C*, "XML schema part 2: datatypes second edition," available at http://www.w3.org/TR/xmlschema-2/.

[23] Yang, Q. Z., Li, X., Lu, W. F., Ganesh, N. and Brombacher, A. C., 2005, "Design quality process management for improvement of engineering design performance," *Proceedings of the 15th International Conference on Engineering Design*, Melbourne, Australia.

[24] Rational Software Corporation, *Rational Rose 2000e*, Cupertino, CA, USA, 2000.

13

A Proposal of Distributed Virtual Factory for Collaborative Production Management

Toshiya Kaihara, Susumu Fujii and Kentaro Sashio

Department of Computer and Systems Engineering, Kobe University, Japan

In this chapter, a Distributed Virtual Factory (DVF) concept has been introduced. DVF consists of distributed precise manufacturing simulation systems connected by Time Bucket synchronization mechanisms. Different from the conventional manufacturing system simulations that deal only with material and information flow, the DVF concept focuses also on detailed product cost analysis to facilitate profitable factory management. Meanwhile, the distributed manufacturing simulation is integrated with Activity-Based Cost (ABC) estimation method to estimates the precise manufacturing activity cost. It has been confirmed that DVF integrated with ABC successfully provides effective means to monitor product costs throughout the entire factory. The experimental results show that the DVF concept is effective enough to provide the strategic benefits with respect to both accurate shipping date and detailed product cost in VE environment.

13.1 Introduction

There is a growing recognition that the current manufacturing enterprises must be agile, that is, capable of operating profitably in a competitive environment of continuously changing customer demands [1]. Supply Chain Management (SCM) or Virtual Enterprise (VE) has increasingly become a common idea for enterprises to survive in the agile environment. However, in the construction of effective SCM or VE, there exists a lack of methods, tools, and environment to support the integration of process models from multiple organizations.

Manufacturing system is one of the core business units to form an effective SCM or VE coalition, and it is crucial to present attractive and collaborative opportunities to other business units. There are several attractive benefits, such as cycle time, fulfilment of due date or accurate and quick shipping date, cost, and quality assurance for volatile ordered products. It is well known that manufacturing system simulation is a powerful technique which can provide the above-mentioned benefits [2, 3]. Distributed simulation model concepts provide practical solutions to facilitate a globally precise simulation model in a SCM or VE environment, because it is constructed as the integration of several manufacturing simulation models of production modules scattered worldwide. The total behavior of the whole manufacturing system is only attainable using the distributed simulation concept.

In this chapter, we first introduce a Distributed Virtual Factory (DVF) concept [4, 5], which consists of several distributed precise simulation models connected by several synchronization mechanisms called Time Bucket algorithms [6, 7]. DVF is applied to the precise evaluations of the whole manufacturing system under two major types of manufacturing operational logics, the PULL and PUSH methods. In this study, we newly apply Activity Based Costing (ABC) method [8] to the DVF architecture to estimate the detailed cost analysis of the products. The methodology facilitates strategic enterprise management to prepare the request for the bids in the VE environment. The effectiveness of the proposed concept in agile manufacturing is finally examined using several simulation experiments.

13.2 Distributed Virtual Factory

13.2.1 Concept

As manufacturing systems have become automated, the so-called automation islands appear in factories. Nowadays, most manufacturing systems consist of several automation islands, such as Direct Numerical Control systems (DNC), Flexible Manufacturing Systems (FMS), automated cell systems, Automated Guided Vehicle (AGV) systems and so on. In this course of development, many simulation studies have been performed to obtain the effective design of the automated systems. Some simulation systems developed for those studies have been extended for use in operational decision making at the shop floor in which the new system is installed. This results in the advent of simulation islands, each corresponds to the real automation island in a factory.

Attempts have been made to integrate the automation islands by connecting the computers in the islands using information network systems for the information flow and by connecting the storages in the islands with some transportation systems for the material flow. Presently, however, most of the simulation islands remain as they were. We consider the integration of simulation islands as a virtual manufacturing system or a Distributed Virtual Factory (DVF) to utilize the potential effectiveness of simulation islands for overall improvement of the design and management of a factory [9, 10]. In this study, we propose the DVF concept which consists of several distributed precise simulation models implemented in different CPUs and linked altogether with a computer network.

To be used as an effective tool for evaluation in the circumstances described above, a DVF needs to satisfy the following features as well as the basic ones for the conventional simulation languages:

(1) The system at the factory level is a large-scale system in reality. The simulation system needs to model the total system with the same details as of the subsystems.
(2) The scope covered by the system includes most of the factory-wide or even world-wide activities, consisting of machining and assembly shops, warehouses, and transportation systems which are referred to as subsystems in the above and are located at the area level in ISO/TC184/WG1 CIM reference model of the manufacturing system with six layers.
(3) The level of automation differs amongst the target factories, the target shops, and the target machines. The system needs to cope with all such situations.
(4) To ease the model building of a factory wide simulation system, it is essential to provide the capability to extend the model gradually in terms of size and detail. Upgradeability of an old subsystem to a new one is also essential.

To develop a DVF, we can take two approaches, *i.e.*, to develop a completely new simulation system from scratch [11] or to construct a system utilizing the existing simulation systems [4]. The latter can integrate the simulation islands in a factory and will effectively save the effort of developing a totally new simulation, because it can utilize the existing standalone simulation systems with minor modifications.

13.2.2 Structure

A distributed simulation system should mostly satisfy the structural features of a factory and the requirements for a factory-wide simulation system when it is developed on a distributed computer system installed as the infrastructure of DVF.

A DVF structure is shown in Figure 13.1. Each subsystem at the area level can be modeled with a processor and a transportation system. T-Process, connecting areas are modeled with a processor transferring works. Functions for information exchange amongst areas are also necessary to model the collection and the dispatch of information occurring from time to time. One processor can be allocated to model a global decision making system collecting ordinary status reports of areas for decision making as the factory level management system in Figure 13.1.

13.2.3 Time Bucket Mechanism

A basic function, termed the synchronization mechanism, is required to synchronize the timing of event execution amongst simulation processes in a distributed simulator on multiple processors, since simulation processes of each area and T-Process have different simulation clocks in their computational contents at a specific time in the real world.

The authors have proposed a Time Bucket Mechanism for the distributed manufacturing system simulation by integrating simulators. The time bucket method has various extensions to fit the DVF environment as follows:

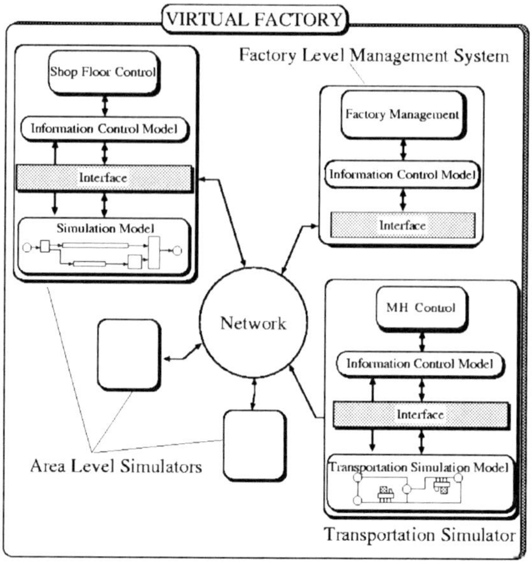

Figure 13.1. A DVF structure

(1) Simple Time Bucket mechanism (TB): In this method, all the simulators execute simulation processing independently and concurrently at the interval called Time Bucket, which is not necessarily equal to the time bucket in MRP (Manufacturing Resource Planning). Each simulator has its own simulation clock and stops its execution at the interval of one Time Bucket. When all simulators come to the end of one Time Bucket, messages on the material flow are exchanged among them, and they restart the simulation execution. The larger the size of Time Bucket is, the more efficient the execution will be, since all the simulators execute their simulation processing independently in a Time Bucket period. This indicates that the size of Time Bucket should be larger considering the independency of the different areas.
(2) Single-Phased Bucket mechanism (SPB): In the SPB algorithm, a simulation processing for one Time Bucket is divided into two phases. One is an Area simulation processing phase, and the other is a simulation phase of transportation. They are executed alternately. Since the T-Process starts its processing of one Time Bucket after all the Area simulators come to the end of the Time Bucket, it can receive transportation request messages appropriately.
(3) Double-Phased Bucket mechanism (DPB): DPB requires the status saving function and loading function in T-Process for roll back operations. The difference between SPB and DPB is the transportation simulation phase, *i.e.*, the performance of T-Process. In the DPB algorithm, T-Process that receives messages from the Area simulators executes its processing for two Time Bucket periods, and stops. The T-Process then rolls back to its

status at the end of the first Time bucket, i.e., before the Area simulators start simulation processing of the second Time Bucket, T-Process predicts the time when any work arrivals to Area simulators so that the Area simulators can register the time of arrival events.

Readers can refer to [6, 7] for the details of the mechanisms.

13.3 Cost Analysis

13.3.1 Cost Analysis in Manufacturing Systems

Conventional costing methods are related to cost factors to determine what the total manufacturing cost is going to be. To make a reasonable estimation of the manufacturing cost, there are some factors that need to be considered. Manufacturing costs include costs for material, processing, inventory, labor, R&D, rent for use of third-party testing facilities, and logistics. A cost model has a very mathematical nature and the size of this model can grow to tremendous proportions, if the number of the operations is high.

Some costing methods rely on integer programming methods. Here, a mathematical model is set up where there is an objective function of the cost. The aim is to minimize this objective function subject to a set of constraints. These constraints differ from company to company, but generally consist of the cost factors stated above.

13.3.2 Activity-based Costing (ABC)

ABC is a method of cost management that identifies business activities performed, tracks costs associated with these activities, and uses various cost drivers to trace the costs of those activities to products [8]. The cost drivers reflect the consumption of activities by the products. ABC provides a far more accurate portrayal of cost than conventional methods. Given a better understanding of cost, management can make far better decisions in terms of competitive advantages. Furthermore, the improved understanding and localization of costs can be used to eliminate low value but high cost activities and hence reduce cost. It is an aid to Business Process Reengineering (BPR).

ABC systems focus on activities required to produce each product or provide each service based on each product or service consumption of the activities. The fundamental difference between ABC and conventional costing is that conventional costing assumes that products cause costs, whereas ABC assumes that activities cause cost and the cost objects create the demand for activities. The ABC systems encourage significant breakdown of work activities and the proper allocation of costs, automatically making a number of potentially hidden costs more visible.

In [9], it is shown that how ABC permits the very important distinction between resource usage and resource spending. The difference is unused capacity. Elimination of this unused capacity permits costs to be reduced. ABC deals much better with large-scaled and integrated manufacturing systems than basic costing methods. With ABC, the activities are determined and associated with their specific costs. The eventual costs depend on the number of activities (each with their specific costs) which are taken to compete the product.

13.3.3 DVF and ABC

ABC provides an accurate portrayal of costs under current factory conditions. The integration between the DVF concept and the ABC method realizes an additional attractive merit in factory management. The integration obviously enables to estimate detailed product costs in all over the factory, because DVF evaluates all the manufacturing data in a distributed manner and provides the detailed manufacturing data throughout the factory into the ABC analysis.

It is sometimes required to estimate the cost analysis in the near future by observing current data, especially in an agile manufacturing environment. A strategic operation is executable only by the future estimation including cost analysis. DVF and ABC are effectively integrated, because simulation can provide detailed manufacturing activity data, which are required but normally difficult to estimate, for the ABC analysis. Several issues to discuss on product cost through the entire factory are provided by the proposed approach. Precise cost analysis about the whole factory is promptly realised in a VE environment for bidding an attractive offer in VE.

13.3.4 Manufacturing Model

For easier explanation, we prepare a large-scaled factory, which consists of 7 sub-modules (Figure 13.2(a)), such as Factory Management, Cooperative Factory, Processing A, Processing B (Figure 13.2(b)), Material Storage & Parts Storage, Assembly Line (Figure 13.2(c)), and Distribution Center.

Each sub-module is implemented into different processors and they organize DVF with the synchronization mechanism as a whole. The factory handles 5 types of products and 50 types of materials, which consist of 20 types in k-part (k1-k20) and 30 types in m-part (m1-m30). The bill of materials is described in Table 13.1. The process flow of each material is shown in Figure 13.3. In this chapter, ABC is applied to the area covered from the Material Storage to the Parts Storage via two types of the Processing (A, B) as a basic study.

13.3.5 Formulations for Cost

The following given constants are defined:

Direct cost

Cm : Material cost (m: material)

DC_L : Direct labor cost (L: labor)

DC_E : Direct energy cost (E: energy)

Indirect cost

CL	: Indirect labor cost
C_R	: Repair cost (R: repair)
C_E	: Indirect energy cost
$C^{MA}{}_D$	: Facility cost (D: Depreciation, MA: machine ID)
C_S	: Stock cost

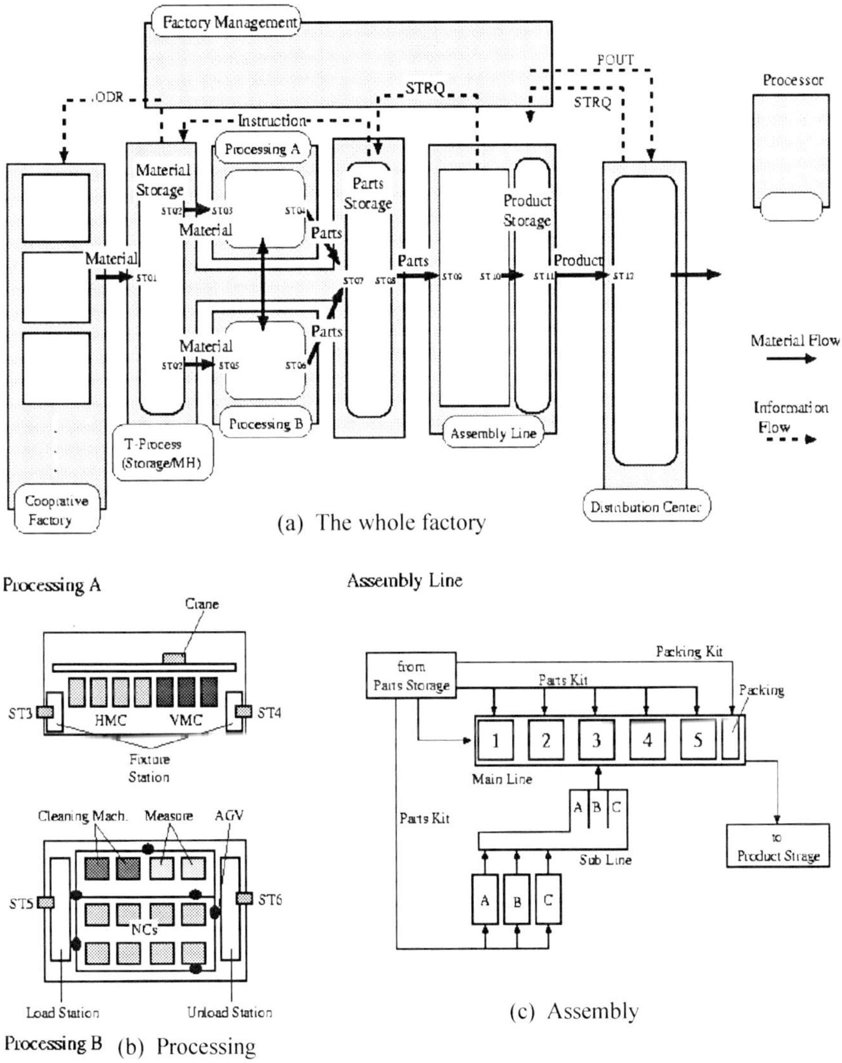

Figure 13.2. Target factory

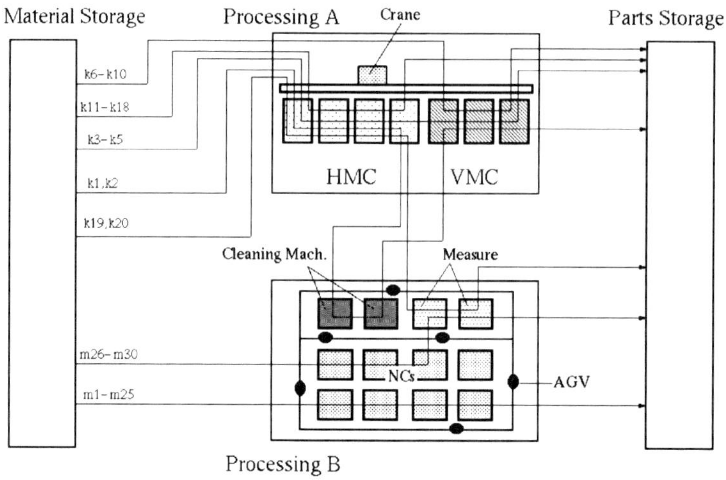

Figure 13.3. Material flow in the process modules

Table 13.1. Bill of materials

A	B	C	D	E
a10 a20 a30	b10 b20	c10 c20	a10 b20	a10 c20
k01 k19 k06 k20 k11 k16	k02 k19 k07 k20 k12 k16	k03 k17 k08 k18 k13 k19 k16 k20	k04 k18 k09 k19 k14 k20 k17	k05 k19 k10 k20 k15 k18
m01 m26 m06 m29 m11 m30 m16 m19 m20 m21	m02 m22 m07 m26 m12 m27 m16 m29 m17 m30 m19 m20	m03 m20 m08 m23 m13 m26 m16 m27 m17 m28 m18 m29 m19 m30	m04 m24 m09 m27 m14 m28 m17 m29 m18 m30 m19 m20	m05 m28 m10 m29 m15 m30 m18 m19 m20 m25
d32 d08 d33 d09 d01 d10 d02 d11 d03 d12 d04 d05 d07	d30 d19 d31 d13 d14 d15 d16 d17 d18	d28 d22 d29 d23 d16 d24 d17 d25 d18 d26 d19 d27 d20 d21	d32 d33 d01 d02 d03 d04 d05 d06	d30 d13 d31 d14 d07 d15 d08 d16 d09 d17 d10 d18 d11 d19 d12
p01 p04 p02 p05 p03	p01 p04 p02 p05 p03	p01 p04 p02 p05 p03	p01 p04 p02 p05 p03	p01 p04 p02 p05 p03

The data which are estimated and acquired by simulation are as follows:

$MP^{MA}{}_{P}$	: Total processing time for product type *P* in machine *MA*
$LP^{MA}{}_{P}$	: Total set-up time for product type *P* in machine *MA*
$V^{AR}{}_{P}$	: Total transporting time for product type *P* in area *AR*
$S^{ST}{}_{P}$	: Total queuing time for product type *P* in stock *ST*
MF^{MA}	: Total breakdown time in machine *MA*
LF^{MA}	: Total maintenance time in machine *MA*
$M^{MAN}{}_{P}$	: Total amount of product *P* processed in machine *MA*
$V^{ARN}{}_{P}$	: Total amount of product *P* transported in area *AR*
$S^{STN}{}_{P}$	: Total amount of product *P* stored in stock *ST*
MF^{MAN}	: Total number of breakdowns in machine *MA*

Where we have 5 types of machines as follows:

MA: *H*(HMC), *V*(VMC), *N*(NC), *C*(CLEAN), *M*(MEASURE)

2 types of parts are as follows:

P: *k*, *m*

3 types of areas at the processing units in our factory model are as follows:

AR: *a*(Processing A), *b*(Processing b), *out*(amongst area)

Generally, ABC consists of two major steps and each step in our approach is formulated as follows:

STEP 1: Indirect cost allocation to the activities

i) Indirect labor cost (cost driver: operational time)

Total operational time of all the operators is

$$L=\Sigma\ LP^{H}{}_{P} + LF^{H} + \Sigma\ LP^{V}{}_{P} + LF^{V} + \Sigma\ LP^{N}{}_{P} + LF^{N} + LF^{C} + LF^{M} \qquad (13.1)$$

Then, indirect labor cost I for the set-up operation in machine *MA* is

$$CL^{MA}{}_{L1} = (\Sigma\ LP^{MA}{}_{P} / L\ ) * C_{L} \qquad (13.2)$$

Indirect labor cost II for machine repair operations in machine *MA* is

$$CL^{MA}{}_{L2} = (LF^{MA} / L\ ) * C_{L} \qquad (13.3)$$

ii) Repair cost (cost driver: breakdown time)

Total operational time of all the operators is

$$MF= MF^{H} + MF^{V} + MF^{N} + MF^{C} + MF^{M} \qquad (13.4)$$

Then, repair cost in machine *MA* is

$$C^{MA}{}_{R} = (MF^{MA} / MF) * C_R \qquad (13.5)$$

iii) Energy cost (cost driver: processing time)

Total processing time in machine MA and transporting time in area AR are:

$$MP=\Sigma\, MP^{H}{}_{P} + \Sigma\, MP^{V}{}_{P} + \Sigma\, MP^{N}{}_{P} + \Sigma\, MP^{C}{}_{P} + \Sigma\, MP^{M}{}_{P} + V^{a}{}_{P} + V^{b}{}_{P} + V^{out}{}_{P} \qquad (13.6)$$

Then, energy cost in machine *MA* is

$$C^{MA}{}_{E} = (\Sigma\, MP^{MA}{}_{P} / MP) * C_E \qquad (13.7)$$

And, energy cost in area *AR* is

$$C^{AR}{}_{E} = (\Sigma\, VP^{AR}{}_{P} / MP) * C_E \qquad (13.8)$$

STEP 2: Activity cost allocation to the products

In this step, the cost price of each product is calculated using activity cost allocation. The final cost price in product *P* is attained as follows:

$$P_P=C_m + IC^{L1}{}_{P} + IC^{L2}{}_{P} + IC^{E}{}_{P} + IC^{D}{}_{P} + IC^{S}{}_{P} + IC^{R}{}_{P} \qquad (13.9)$$

where $IC^{*}{}_{P}$: cost related to activity * in product *P*

As an example we describe the formulation to attain $IC^{L1}{}_{P}$, which is indirect labor cost I for set-up operation in product *P*. At first, from (2) indirect labor cost I ratio is:

$$R^{MA}{}_{L1}= CL^{MA}{}_{L1} / \Sigma\, LP^{MA}{}_{P} \qquad (13.10)$$

And, set-up time per a part in machine *MA* is

$$D^{LPMA}{}_{P}= LP^{MA}{}_{P} / M^{MAN}{}_{P} \qquad (13.11)$$

Then, indirect labour cost I per a part in machine *MA* is obtained as

$$IC^{L1MA}{}_{P}= R^{MA}{}_{L1} * D^{LPMA}{}_{P} \qquad (13.12)$$

Finally

$$IC^{L1}{}_{P}= IC^{L1H}{}_{P}+ IC^{L1V}{}_{P} +IC^{L1N}{}_{P} \qquad (13.13)$$

Cost drivers of the cost items at each step are shown in Table 13.2.

Table 13.2. Cost Drivers

Indirect Cost	Cost Driver STEP 1	Cost Driver STEP 2
Labour Cost	Working Time	Working Time Operating Time
Energy Cost	Operating Time Transporting Time	Operation Time Transporting Time
Facility Cost	Operating Time Transporting Time	Operating Time Transporting Time
Stock Cost	Time in Stock	Time in Stock
Repair Cost	Repairing Time	Operating Time

13.4 Experimental Results

13.4.1 Simulation Model

A simulation model based on the concept of DVF has been developed in order to evaluate its effectiveness in a VE environment. The model is precise enough to provide the offer for VE contracts, and includes a manufacturing management system for the whole factory. There are two types of major management policies that are implemented in the simulation model to investigate a practical DVF in a VE environment. One is the PUSH logic, which is operated by the top-down strategic plan, such as the MRP system (Figure 13.4), the other is the PULL logic, in which down stream business processes send their requests to the preceding up stream ones (Figure 13.5). JIT (Just In Time) system is one of the most popular systems categorized in the PULL logic.

As we described in the previous Section 13.3.4, the target factory has 7 areas, including Factory Management, Cooperative Factory, Processing A, Processing B, Material Storage & Parts Storage, Assembly Line and Distribution Center. These area models are developed separately with different software and implemented on different CPUs independently shown in Table 13.3. The precise simulation model calculates the possible shipping date of the required products in VE from the current inventory conditions including WIP (Work In Process) under various management policies. 5 types of products (A, B, C, D, E) are considered to be manufactured in the simulation model shown in Table 13.1. We assume the part types of k and m are produced in the factory and the others are purchased. In this experimental model, we selected TB as the synchronization mechanism.

A product order pattern in the experiment is shown in Figure 13.6. We assume the order amount ratio in each product type is (A:B:C:D:E)=(6:4:2:1.5:1) . The daily amount of orders varies drastically between 10 and 50 in the simulation conditions followed by uniform random distribution.

13.4.2 Total Factory Management in DVF

The distributed simulation of the precise factory model is executed to confirm the validity of the model in a VE environment. The estimation of the management policy influence in the inventory conditions in each manufacturing area is quite difficult but important to propose strategic offers in VE. Precise estimation of the dynamical inventory changes facilitates the accurate control of the product shipping date.

The simulation results of part *k*01 inventory transition in the part storage under the PUSH and PULL management policies are shown in Figures 13.7 and 13.8, respectively.

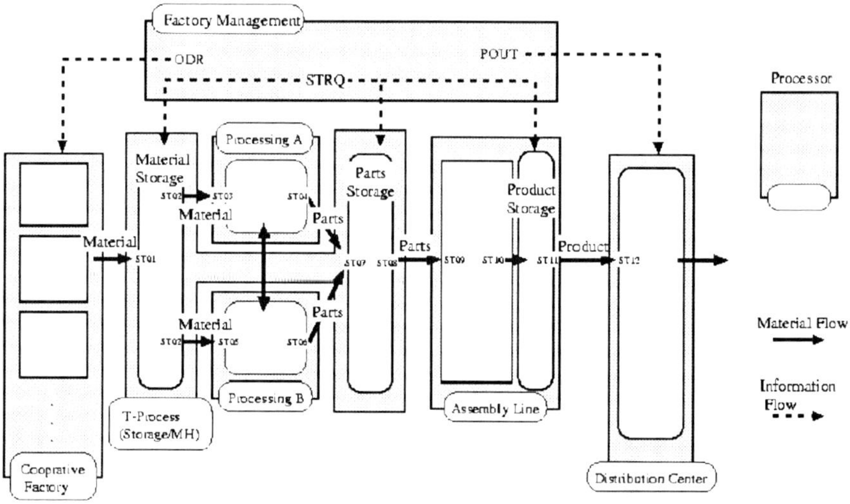

Figure 13.4. Simulation model in PUSH logic

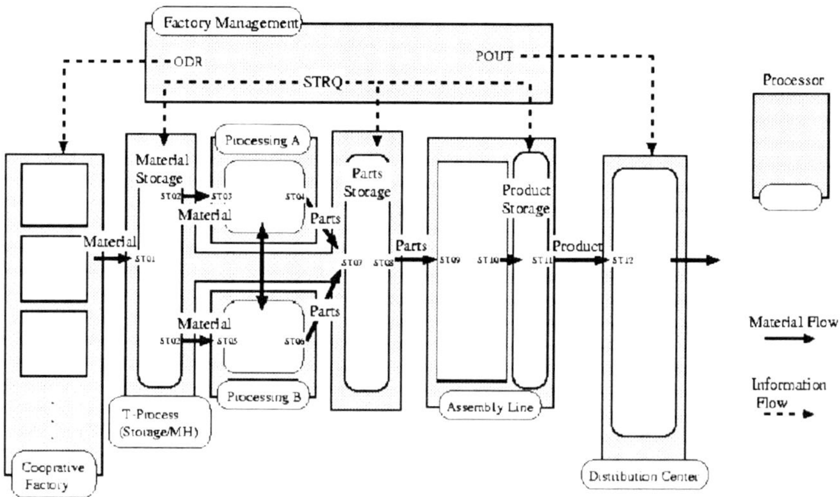

Figure 13.5. Simulation model in PULL logic

Table 13.3. Configuration

Area	EWS	Simulation software
Factory Management	Axil: Axil Server 400	C, Microsoft ACCESS 97
Cooperative Factory	SUN: Spark Station 4	Smpl, C
Processing A	SUN: Spark Station 4	SLAM II, Fortran, C
Processing B	SUN: Spark Station 4	SLAM II, Fortran, C
Material & Parts storage	SUN: Spark Station 4	Smpl
Assembly Line	SUN: Spark Station 4	SLAM II, Fortran, C
Distribution Center	SUN: Spark Station 4	Smpl, C

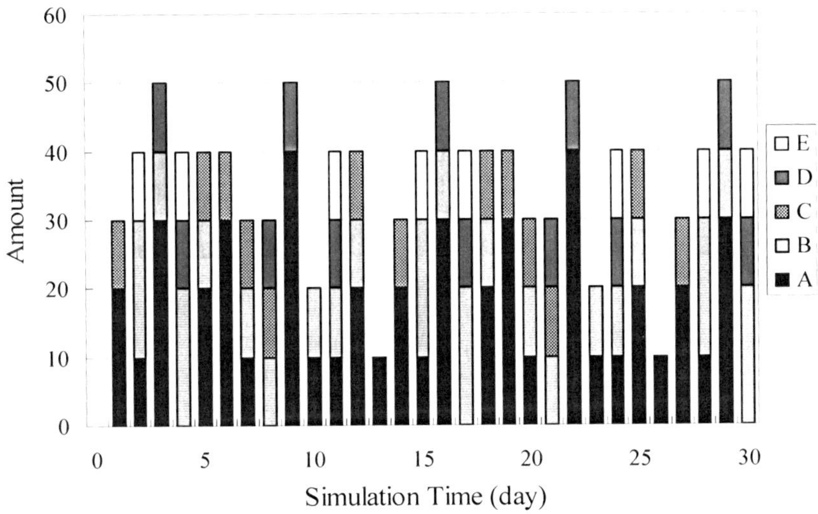

Figure 13.6. An example of product order

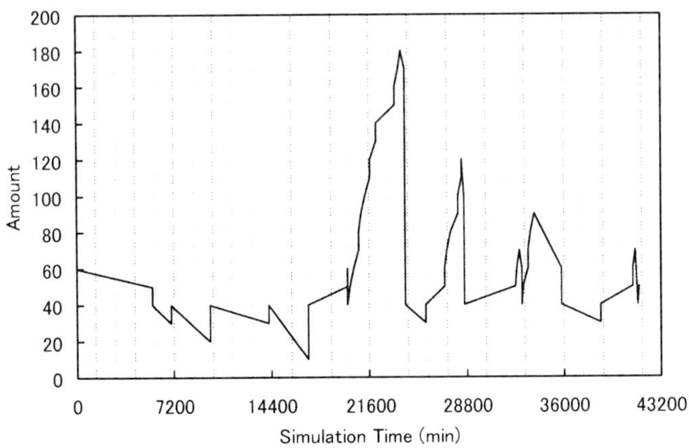

Figure 13.7. Part *k*01 inventory in PUSH management

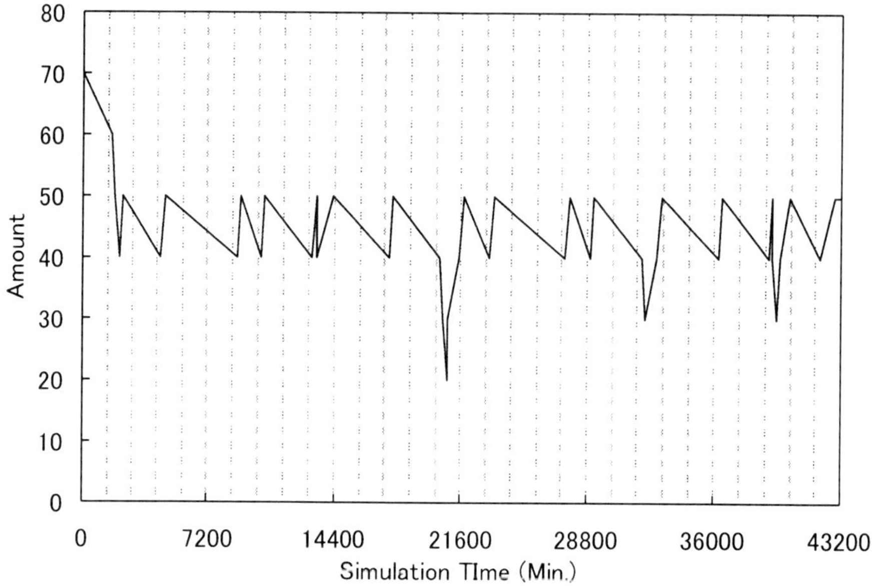

Figure 13.8. Part *k*01 inventory in PULL management

It is obvious that the inventory change in the PUSH logic fluctuates more than the change in the PULL logic, because PUSH logic easily suffers from unexpected dynamical demand changes. On the other hand, the inventory in PULL logic is stable even in a volatile situation. However, in the other simulation experiments, the PUSH logic has been proven to be effective, if we can accurately grasp and estimate the product demand, *i.e.*, the dynamical product demand change is relatively small.

Simulation experiments confirmed that the DVF provides rational materials to discuss the managerial strategy of the entire factory as a whole. The accurate possible shipping date of all the products is acquirable by the precise inventory estimation so as to make an appropriate bid in the VE environment.

13.4.3 Cost Analysis

For bidding an attractive offer in VE, we introduced the ABC approach into the DVF architecture. As ABC is a procedure that enables the estimation of product costs more accurately, we think the procedure is essential to realize an effective enterprise management in agile manufacturing, which is characterized by small batch sizes and high customer satisfaction.

In this chapter, ABC is applied to the area covered from the Material Storage to the Parts Storage via two types of the Processing (A, B) shown in Figures 13.2(a) and 13.2 (b) as a basic study. The distributed simulation of the target factory was executed to confirm the validity of the ABC approach with the DVF model. It is essential to simulate the dynamical changes of the inventory conditions in each manufacturing area in order to analyze the accurate and precise cost analysis in the lot level.

Figures 13.9 and 13.10 show the results of conventional cost analysis and ABC cost analysis of Part *k*01 ~ *k*05 (currency is Yen) in the PULL management, respectively. In the conventional approach, a single cost driver, operating time, is used to evaluate each product cost. In the experiment, these results are completely different, and the difference is mainly caused by the stock cost evaluation. The stock cost is calculated accurately in ABC because it is directly proportional to the actual stock time. By observing the ABC result, shop floor managers realise that it is essential to decrease the inventory level of Part *k5* to reduce the product cost in this case.

In the DVF model, it is also possible to estimate and collect all the lot level data from each discrete-event transaction, such as stock-in, stock-out, process-start, process-finish, *etc*. This enables the ABC approach in the lot level as well as in the part type level as shown in Figure 13.11. This figure shows the lot level cost of each produced part in type *k*01. Lot level ABC clarifies a transitional cost analysis and facilitates subtle management of the entire factory.

Precise lot level cost estimation enables us to operate appropriate price bids for a product demand in the VE environment.

It has been confirmed that detailed cost analysis of each product through the whole factory is attainable by the proposed approach. DVF integrated with ABC successfully provided effective materials to discuss on product cost in the entire factory. Since the strategic operation is executable only by the future estimation including cost analysis in agile manufacturing environment, our approach is quite promising as the factory management in the next VE age.

13.5 Conclusions

In this chapter, we introduce a Distributed Virtual Factory (DVF) concept, which consists of distributed precise manufacturing simulation systems connected by Time Bucket synchronization mechanisms. Although conventional manufacturing system simulations normally deal only with material and information flow, the proposed concept focuses also on detailed product cost analysis to facilitate profitable factory management. The integration of distributed manufacturing simulation and ABC method is quite rational, because the simulation productively estimates the precise manufacturing activity data, which is required but normally difficult to estimate for the ABC analysis. It has been confirmed that DVF integrated with ABC successfully provides effective means to monitor product costs throughout the entire factory. The experimental results show that the DVF concept is effective enough to provide the strategic benefits with respect to both accurate shipping date and detailed product cost in VE environment. The DVF concept is concluded to play an important role in factory management in the coming collaborative manufacturing era.

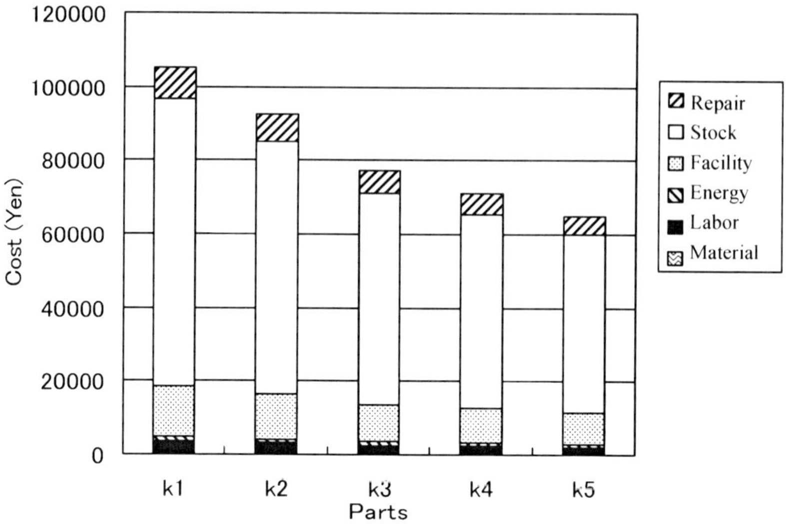

Figure 13.9. Conventional cost analysis

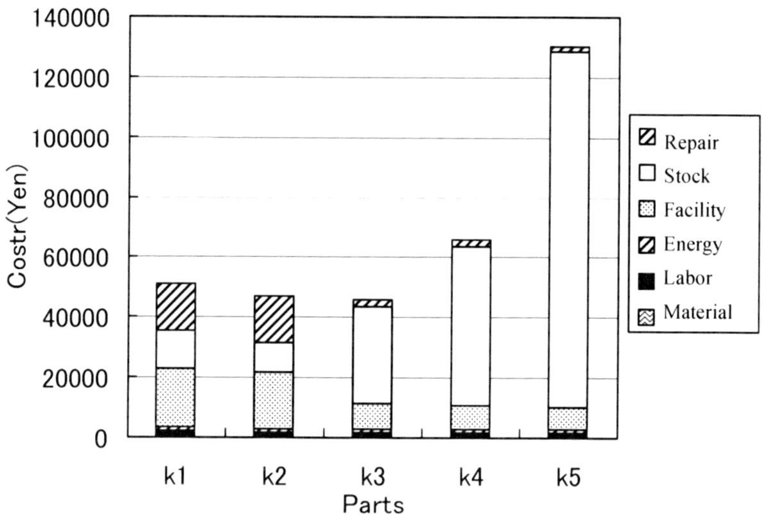

Figure 13.10. ABC-based cost analysis

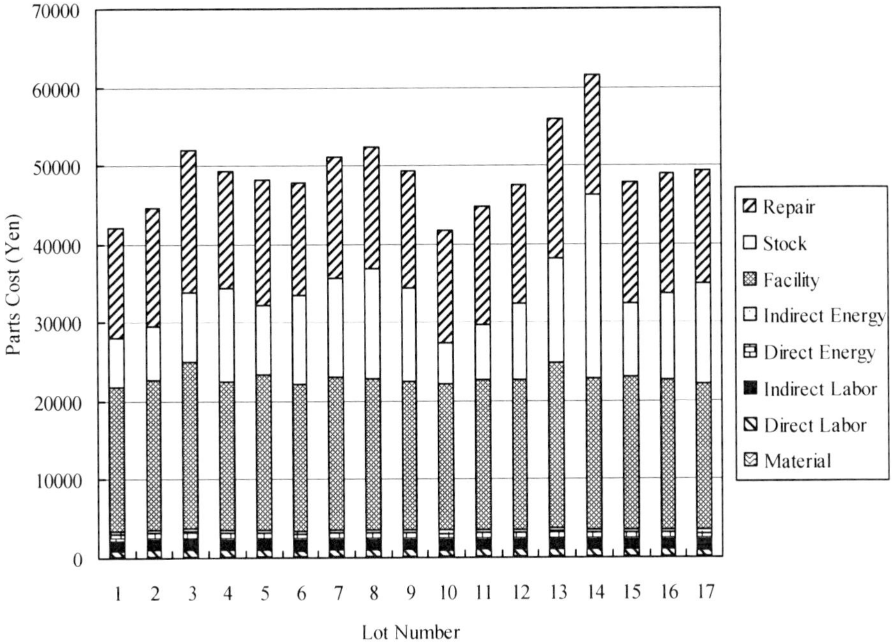

Figure 13.11. Lot level ABC analysis in Part *k*01

13.6 References

[1] Iwata, K. and Fujii, S., 1997, "New manufacturing era – adaptation to environment, culture, intelligence and complexity," *Summary of the Questionnaire of the 29th CIRP International Seminar on Manufacturing Systems*.

[2] Law, A. M. and Kelton W. D., 1990, *Simulation Modeling and Analysis*, McGraw-Hill.

[3] Grant, H. F., 1987, "Simulation and factory control – an overview," *Proceedings of the IX ICPR*, pp. 576–583.

[4] Fujii, S., Kaihara, T. and Tanaka, M., 1998, "Manufacturing system simulation in virtual enterprise," *Simulators International XV, Simulation Series*, 30(3), The Society for Computer Simulation International, pp. 179–184.

[5] Sashio, K., Fujii, S. and Kaihara, T., 2005, "A study on data handling mechanism of a distributed virtual factory," Knowledge *And Skill Chains in Engineering and Manufacturing*, Springer, New York, pp. 293–300.

[6] Fujii S., et al., 1994, "Synchronization Mechanisms for Integration of Distributed Manufacturing Simulation Systems under CIM Environment," *Proceedings of Advances in Intelligent Computer Integrated Manufacturing*

System, pp. 268–274.

[7] Fujii, S., Kidani, Y., Ogita, A. and Kaihara, T., 1999, "Synchronization mechanisms for integration of distributed manufacturing simulation systems," *International Journal of Simulation, The Society for Computer Simulation*, 71(3), pp. 187–197.

[8] Cooper, R. and Kaplan, R.S., 1992, "Activity-based systems: measuring the costs of resource usage," *Accounting Horizons*, September, pp. 1–13.

[9] Fuii, S., Kaihara, T., Morita, H. and Tanaka, M., 1999, "A distributed virtual factory in agile manufacturing environment," *Proceedings of the 15th conference of the IFPR (ICPR-15)*, 2, pp. 1551–1554.

[10] Sashio, K., Fujii, S. and Kaihara, T., 2004, "Distributed virtual factory under e-business environment," *Proceedings of 11th International Conference Simulation in Production and Logistics*, pp. 451–460.

[11] Kaihara, T. and Besant, C. B., 1993, "Manufacturing system modeling structure based on object oriented paradigm," *Proceedings of the 1993 Summer Simulation Conference*, pp. 831–836.

Index

Printed in the United States
73641LV00003B/85-138

9 781846 288012